| $f(t)u(t)$ | $F(s)$ |
|---|---|
| 3. $\delta'(t)$ | $s$ |
| 4. $t^n u(t), \quad n = 0, 1, 2, \ldots$ | $\dfrac{n!}{s^{n+1}}$ |
| 5. $e^{-bt} u(t)$ | $\dfrac{1}{s+b}$ |
| 6. $t^n e^{-bt} u(t)$ | $\dfrac{n!}{(s+b)^{n+1}}$ |
| 7. $\sin \beta t \, u(t)$ | $\dfrac{\beta}{s^2 + \beta^2}$ |
| 8. $\cos \beta t \, u(t)$ | $\dfrac{s}{s^2 + \beta^2}$ |
| 9. $t \cos \beta t \, u(t)$ | $\dfrac{s^2 - \beta^2}{(s^2 + \beta^2)^2}$ |
| 10. $t \sin \beta t \, u(t)$ | $\dfrac{2\beta s}{(s^2 + \beta^2)^2}$ |
| 11. $e^{-bt} \sin \beta t \, u(t)$ | $\dfrac{\beta}{(s+b)^2 + \beta^2}$ |
| 12. $e^{-bt} \cos \beta t \, u(t)$ | $\dfrac{s+b}{(s+b)^2 + \beta^2}$ |
| 13. $te^{-bt} \cos \beta t \, u(t)$ | $\dfrac{(s+b)^2 + \beta^2}{[(s+b)^2 + \beta^2]^2}$ |
| 14. $te^{-bt} \sin \beta t \, u(t)$ | $\dfrac{2\beta(s+b)}{[(s+b)^2 + \beta^2]^2}$ |
| 15. $\sinh \beta t \, u(t)$ | $\dfrac{\beta}{s^2 - \beta^2}$ |
| 16. $\cosh \beta t \, u(t)$ | $\dfrac{s}{s^2 - \beta^2}$ |

# ANALYSIS OF ELECTRICAL NETWORKS

## SHLOMO KARNI
University of New Mexico

**JOHN WILEY & SONS**
New York   Chichester   Brisbane   Toronto   Singapore

*Library of Congress Cataloging in Publication Data:*

Karni, Shlomo, 1932–
    Analysis of electrical networks.

    Includes indexes.
    1. Electric network analysis.     I. Title.
TK454.2.K27     1986     621.319′2     85-12206
ISBN 0-471-80268-9

Printed in the United States of America

10  9  8  7  6  5  4  3  2  1

# ABOUT THE AUTHOR

**Shlomo Karni** was born in Poland in 1932. He received his BSEE degree (cum laude) from the Technion, Israel Institute of Technology, in 1956; his M.E. from Yale University in 1957, and a Ph.D. in electrical engineering from the University of Illinois, Urbana, in 1960.

Since 1961 he has been with the Department of Electrical and Computer Engineering at the University of New Mexico in Albuquerque, New Mexico, where he is a Professor and the Director of Graduate Studies. During 1954–1956, Professor Karni was employed by the Palestine Power Company, Haifa, Israel; 1956–1957 was spent as a teaching assistant at Yale University. From 1957 to 1961, he was Instructor, then Assistant Professor, at the University of Illinois. He was Visiting Professor, Department of Electrical Engineering, University of Hawaii, during 1968–1969; a Visiting Professor and the first Acting Dean at the School of Engineering, Tel Aviv University, 1970–1971; and a Visiting Professor at the Technion, Israel Institute of Technology, during the summer of 1977.

Professor Karni has served as a consultant to Westinghouse Learning Corporation, Gulton Electronics Industries, the Air Force Weapons Labs, and the Los Alamos ERDA (now DOE) Labs.

Professor Karni was on the editorial board of the IEEE *Spectrum* (1969–1972) as Associate Editor and is currently Editor of the IEEE *Magazine on Circuits and Systems*. He has also served as a consultant to major publishing houses (John Wiley & Sons, McGraw-Hill, Prentice-Hall, Houghton Mifflin, Addison-Wesley, and others).

He is the author of three books: *Network Theory: Analysis and Synthesis*, (Boston: Allyn & Bacon, 1966; Russian translation: Moscow: Svyaz Publishing Co., 1973); *Intermediate Network Analysis*, (Boston: Allyn & Bacon, 1971); and *Analysis of Continuous and Discrete Systems*, with W. Byatt (Holt, Rinehart & Winston, 1981). In addition, he has written over 60 technical articles.

Dr. Karni is a Fellow of the IEEE, and a member of Tau Beta Pi (board of directors, 1972–1976); the New York Academy of Science (1970); Eta Kappa Nu; and The Academy of Hebrew Languages, Jerusalem, (correspondent, 1970).

# PREFACE

This book is intended for a one-semester (or two-quarter) second course in circuit analysis. The student is expected to have had the following topics as prerequisites: $R$, $L$, $M$, and $C$ as circuit elements, analysis of resistive networks via node and loop equations, dc Thévenin and Norton equivalent circuits, power and energy, superposition, classical solutions of first-order ($RL$, $RC$) circuits and of second-order ($RLC$) circuits, simple operational amplifier circuits, and sinusoidal steady-state (phasor) analysis.*

To help the student and the instructor in the review of some of this material, Chapter 1 provides a summary leading to the general time-domain formulation of dynamic equations for $RLC$ networks. Appendix A contains the essentials of linear algebraic equations, matrices, and determinants. In Appendix B, a brief survey is given of the op-amp, together with some of its common circuits. Students and instructors should use their discretion in studying this review material and referring to it as needed. Appendix C, on scaling, may be conveniently studied or reviewed at this early stage.

The rest of the book is devoted to more advanced topics in circuit analysis. Interconnections of networks, topology, and signal flow graphs (Appendix D) stress the uniformity and the organized formulation of network equations. Convolution reiterates the principle of superposition. The Laplace transform and the Fourier transform serve as powerful tools in the solution of network equations; in addition, they provide the necessary tools for many related subjects, such as stability and frequency response. State variable analysis and solution give us, in addition to mathematical elegance, a deep insight into the physical behavior of networks. The last chapter gives a brief introduction to linear, time-varying networks.

Throughout the text, the following features are used:

1. Problems relevant to a particular topic are listed by number next to the discussion of that topic. It is hoped that such an arrangement will make the learning of the material more systematic and more helpful.

2. Topics which are more advanced are marked with an asterisk (*) at the beginning of the appropriate section. They may be skipped during the first reading without loss of continuity. Later, the student and the instructor are encouraged to return and integrate them into the study.

3. Examples are marked clearly with a title ("Example 3") and with a square mark (□) at the end, all between two thick lines.

4. In Appendix F, selected solutions and hints to problems are given. The student should use these judiciously as a helpful tool, not as a substitute for thinking and

---

* See, for example, R. E. Thomas and A. J. Rosa, *Circuits and Signals*. New York: John Wiley & Sons, 1984, and D. F. Mix and N. M. Schmitt, *Circuit Analysis for Engineers: Continuous and Discrete-Time Systems*. New York: John Wiley & Sons, 1985.

working! A more complete Solutions Manual is available to instructors upon adoption.

5. References and bibliographical listings at the end of each chapter are kept to a minimum. Too many references may look impressive, but they tend to discourage, by their sheer numbers, even the most well-intended student. This is not to say, however, that a resourceful instructor should not encourage the better students to read additional material. The given lists are a good start in that direction.

I gratefully acknowledge the help of Robin Morel, Ireena Erteza, and Sabina Erteza, who helped me in the preparation of Appendix E. My secretary, Mrs. Joan Lillie, typed the final manuscript and managed the administrative aspects of my work. My editors at John Wiley & Sons, along with the staff, provided prompt, courteous, and efficient assistance throughout the stages of this project. My sincere thanks go to them.

*Albuquerque, New Mexico*                                                    **Shlomo Karni**
*August 1985*

# CONTENTS

# 1

# INTRODUCTION

In this chapter, we review the basic properties of network elements and the formulation of loop equations and node equations.

## 1-1  SYSTEMS AND NETWORKS

In its broadest sense, a *system* can be defined as a group, or collection, of components, each with its specific characteristics, interacting in some prescribed manner. So, an automobile with all its parts, a generating station supplying electric power to customers through transmission lines, the nerves of a living organism, a group of people with certain mutual interests—all are examples of systems. In fact, any conceivable group of entities, interrelated in some fashion can be called a system.

In particular, an electrical network is a system, consisting of *elements* (components) such as resistors, capacitors, voltage and current sources, transistors, diodes, etc. Our aim in this book is the analysis of electrical networks, that is the development of the mathematical relations among the variables that describe the behavior of the network.

## 1-2  SOME CHARACTERISTICS OF ELECTRICAL NETWORKS

In certain cases, it will be useful to represent a network by the classical "black box." See Fig. 1-1. It has, in general, $m$ inputs (excitations), $x_1(t)$, $x_2(t), \ldots, x_m(t)$ and $n$ outputs (responses), $y_1(t), y_2(t), \ldots, y_n(t)$.[†] Note that, in general, $m$ is not equal to $n$. We use the notation

$$S\{x_1, x_2, \ldots, x_m\} = \{y_1, y_2, y_3, \ldots, y_n\} \qquad (1\text{-}1a)$$

[†] These are general symbols; others are $e(t)$ for excitation and $r(t)$ for response. Specific excitations and responses will be denoted by appropriate symbols such as $i(t)$ for a current, $v(t)$ for a voltage, etc.

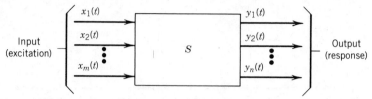

**FIGURE 1-1.** A "black box."

or

$$\{x_1, x_2, \ldots, x_m\} \xrightarrow{\ S\ } \{y_1, y_2, y_3, \ldots, y_n\} \tag{1-1b}$$

to mean, "the inputs $x_1, x_2, \ldots, x_m$ to network $S$ produce the outputs $y_1, y_2, \ldots, y_n$."
In network analysis, we deal in general with the following problems.

1. Given the network $S$, obtain the input-output relations in the mathematical form.
   In other words, find the explicit form of Eq. (1-1) for a given network.
2. Having obtained the equations for the given network, solve them to obtain the
   outputs corresponding to given inputs.
3. Discuss the mathematical properties of the equations of a network, and, hence,
   certain properties of the network.

A *continuous-time* network is characterized by inputs and outputs which are functions
of the *continuous variable t*. In a *discrete-time* network, they vary only at discrete values of
time. In the former case we designate these inputs and outputs as $x(t)$ and $y(t)$,
whereas in the latter case they are denoted by $x(n)$ and $y(n)$, where $n = 1, 2, 3, \ldots,$

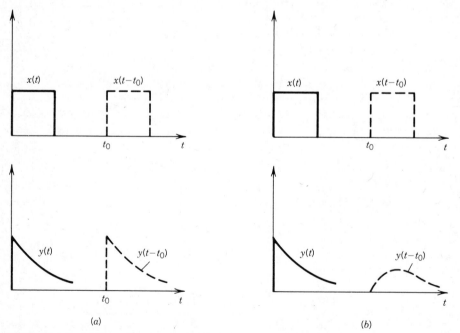

(a)                    (b)

**FIGURE 1-2.** (*a*) A constant network; (*b*) A time-varying network.

are the discrete points on the time axis. Continuous-time networks can be described, in general, by *differential equations*, and discrete-time networks by *difference equations*.

A *constant (time-invariant)* network is one whose response has a *shape* depending only on the *shape* of the excitation and not on the time of application of the excitation (see Fig. 1-2). Mathematically, this can be expressed as follows: with all initial conditions zero, if

$$S\{x(t)\} = y(t) \tag{1-2a}$$

then

$$S\{x(t - t_0)\} = y(t - t_0) \tag{1-2b}$$

where the notation of Eq. (1-1) is used. In other words: a time-shift $t_0$ in the input produces an identical time-shift in the output.

In a *time-varying* network, Eq. (1-2b) does not hold. Typically, the parameters of a time-invariant network will be constants, and those of a time-varying network will be functions of time.

## Example 1

A (hypothetical) network is described by the equation $v(t) = i(t/2)$, where $v$ is the output voltage and $i$ is the input current. Applying Eq. (1-2), we obtain

$$S\{i(t - t_0)\} = i[\tfrac{1}{2}(t - t_0)]$$

but on the other hand,

$$v(t - t_0) = i(\tfrac{1}{2}t - t_0)$$

obviously, $i(t/2 - t_0) \neq i[1/2(t - t_0)]$, and, therefore, the network is time-varying. $\square$

## Example 2

The output current of a certain network is given by $i(t) = [v(t)]^2$, where $v$ is the input voltage. With Eq. (1-2), we obtain

$$S\{v(t - t_0)\} = [v(t - t_0)]^2$$

on one hand, and

$$i(t - t_0) = [v(t - t_0)]^2$$

on the other. This network, then, is constant. $\square$

A *lumped* element has physical dimensions which do not affect its describing equation. More precisely, if $d$ is the largest dimension of the element, and $\lambda$ is the wavelength of the signal, then a lumped element satisfies

$$d \ll \lambda \tag{1-3a}$$

The wavelength $\lambda$ is given by

$$\lambda = \frac{c}{f} \qquad (1\text{-}3\text{b})$$

where $c = 3 \times 10^8$ m/s = velocity of electromagnetic waves (velocity of light), and $f$, in Hertz, is the frequency of the signal. *Total* differential equations describe a network made up of lumped elements. In a *distributed network*, variations with respect to, say, the length of the components *are* important; such a network is described by *partial* differential equations, involving space variation as well as time variation.

---

### Example 3

At audio frequencies, with $f = 1{,}000$ Hz, $\lambda = 3 \times 10^5$ meters $\approx 186$ miles. At typical microwave frequencies, $f = 10^{10}$ Hz, then $\lambda = 3$ cm $\approx 1.2$ inches. Thus, for audio circuits, all elements are considered lumped. In microwave circuits, elements such as waveguides are distributed and their describing equations will involve space variations as well as time variations.

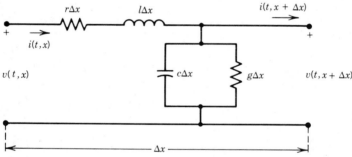

**FIGURE 1-3.** A model of a distributed network.

The model of the transmission line, Fig. 1-3, is that of a distributed network, since its equations are[1]

$$-\frac{\partial v(t, x)}{\partial x} = ri(t, x) + l\frac{\partial i(t, x)}{\partial t}$$

$$-\frac{\partial i(t, x)}{\partial x} = gv(t, x) + c\frac{\partial v(t, x)}{\partial t}$$

showing dependence on time and position; voltage and current at a given time will vary along the length of transmission line.   □

---

## 1-3   ELEMENTS AND SOURCES

It is both an amazing and a comforting fact that we can model, analyze, and design the most complicated electrical networks using only a few basic elements. Let us

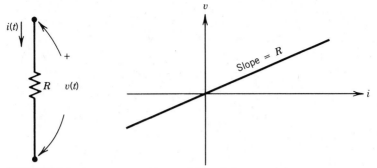

**FIGURE 1-4.** A constant resistor and its $v–i$ characteristic.

review their properties and introduce certain additional concepts associated with them.

## The Resistor

A lumped, time invariant, *linear* resistor is defined by its voltage-current relation

$$v(t) = Ri(t) \qquad (1\text{-}4a)$$

as shown in Fig. 1-4. The voltage $v(t)$ and the current $i(t)$ have their reference signs as indicated.[†] The units of $v(t)$ are *volts* (V), of $i(t)$ are *amperes* (A), and the resistance $R$ is in *ohms* ($\Omega$).

The inverse relationship, $i$ in terms of $v$, is, of course,

$$i(t) = \frac{1}{R} v(t) = Gv(t) \qquad (1\text{-}4b)$$

where $G$, the *conductance*, is in *mhos* ($\mho$).

The relationships in Eq. (1-4) indicate that the output of a resistor, either $v$ or $i$, depends, at any time $t = t_0$, only on the input, $i$ or $v$, respectively, *at that time $t = t_0$*, and not on past values of the input. Such an element is called *instantaneous (memoriless)*.

## The Capacitor

A constant, lumped, linear capacitor is defined by its charge–voltage relation

$$q(t) = Cv(t) \qquad (1\text{-}5)$$

as shown in Fig. 1-5. The charge $q(t)$ is measured in *coulombs* and the capacitance $C$ is in *farads* (F). In network analysis, we are interested in the relations between currents and voltages; therefore, we recall the basic relation between current and charge

$$i(t) = \frac{dq(t)}{dt} \qquad (1\text{-}6)$$

---

[†] Throughout this book, lower case letters will denote functions of time, for example, $v(t)$, $i(t)$. Sometimes the parenthetical $t$ may be omitted for convenience.

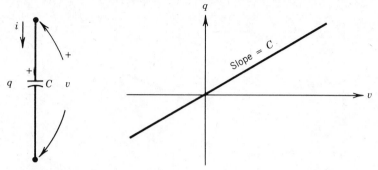

**FIGURE 1-5.** A constant capacitor and its $q-v$ characteristic.

and differentiate Eq. (1-5). The result is

$$\frac{dq(t)}{dt} = i(t) = C\frac{dv(t)}{dt} \tag{1-7a}$$

and it provides the $i-v$ relationship for the capacitor.

The inverse relationship, $v$ in terms of $i$, is obtained from Eq. (1-5) and by integrating Eq. (1-7a):

$$\frac{q(t)}{C} = v(t) = \frac{1}{C}\int_{-\infty}^{t} i(x)\,dx \tag{1-7b}$$

The integral of $i$, from $-\infty$ to any time $t$, represents the total charge on the capacitor at time $t$. The dummy variable $x$ is used in the integrand in order not to confuse it with $t$.

Unlike the resistor, the capacitor has a *memory*: past values of the input ($i$) also affect the output ($v$), as seen in Eq. (1-7b). Such an element, where the output at $t = t_0$ depends on present *and* on past values of the input, is called *dynamic*.

Finally, it will be convenient for us to rewrite Eq. (1-7b) by expressing the integal as a sum of two integrals:

$$\int_{-\infty}^{t} i(x)\,dx = \int_{-\infty}^{0} i(x)\,dx + \int_{0}^{t} i(x)\,dx = q(0) + \int_{0}^{t} i(x)\,dx \tag{1-8}$$

where $t = 0$ is some convenient *initial* time, and $q(0)$ is the initial charge (initial condition) on the capacitor. Then Eq. (1-7b) becomes

$$v(t) = \frac{q(0)}{C} + \frac{1}{C}\int_{0}^{t} i(x)\,dx \tag{1-9}$$

that is,

$$v(t) = v(0) + \frac{1}{C}\int_{0}^{t} i(x)\,dx \tag{1-10}$$

with $v(0)$ the initial voltage across the capacitor.

## The Inductor

A constant, lumped, linear inductor is defined by its flux-current relation

$$\phi(t) = Li(t) \tag{1-11}$$

as shown in Fig. 1-6. The flux $\phi(t)$ is measured in *webers* and the inductance $L$ in *henries* (H).

In order to obtain the $v-i$ relations, we recall that, according to experimental observations (of Faraday, Lenz, and others)

$$v(t) = \frac{d\phi(t)}{dt} \tag{1-12}$$

and differentiate Eq. (1-11)

$$\frac{d\phi(t)}{dt} = v(t) = L\frac{di(t)}{dt} \tag{1-13a}$$

This is the desired $v-i$ relationship for the inductor. It is interesting and instructive to recognize that Eqs. (1-7a) and (1-13a) are of identical form, but with voltage and current exchanging places; that is, for the capacitor

$$i_C(t) = C\frac{dv_C(t)}{dt} \tag{1-7a}$$

while for the inductor

$$v_L(t) = L\frac{di_L(t)}{dt} \tag{1-13a}$$

Here we have added the subscripts $C$ and $L$ for clarity. Such relationships, where elements obey the same equation in form but with $v$ and $i$ replaced, are called *duals*. The principle of duality will be very useful to us.

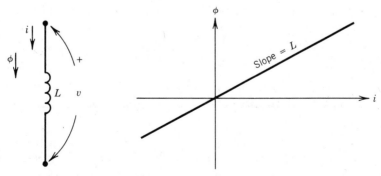

**FIGURE 1-6.** A constant inductor and its $\phi-i$ characteristic.

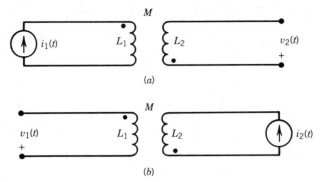

**FIGURE 1-7.** Mutually coupled inductors.

Following the development of the $v$–$i$ relation for a capacitor, and using duality, we obtain the $i$–$v$ relation for an inductor

$$i(t) = i(0) + \frac{1}{L} \int_0^t v(x)\, dx \qquad (1\text{-}13b)$$

**Probs. *1-1***
***1-2***
***1-3***
***1-4***

Two mutually coupled inductors are shown in Fig. 1-7. In Fig. 1-7$a$, $L_1$ carries a current $i_1(t)$ which creates a flux $\phi_1(t)$. Part of this flux links $L_2$ and induces in $L_2$ the voltage

$$v_2 = \pm M \frac{di_1}{dt} \qquad (1\text{-}13c)$$

where $M$ (in henries) is the mutual inductance. The polarity (reference) of $v_2$ depends on the relative sense of the windings of $L_1$ and $L_2$. This information is given usually by an equivalent notation: a dot marking is placed on one terminal of each inductor, such that a current entering one dot on an inductor induces a voltage referenced plus (+) at the corresponding dot on the second inductor. Therefore, in Fig. 1-7$a$, the induced voltage $v_2$ is referenced as shown. In Fig. 1-7$b$, the current $i_2$ induces a voltage $v_1$ whose magnitude is $M\, di_2/dt$. Since $i_2$ does not enter its dot, the induced voltage $v_1$ is negative on the other dot, or positive as shown.

If $L_2$ in Fig. 1-7$a$, or $L_1$ in Fig 1-7$b$, carries also a current, then the additional voltage due to it is added.

---

### *Example 4*

In the network shown in Fig. 1-8, we write by superposition the total $v_1$ and $v_2$ as

$$v_1 = L_1 \frac{di_1}{dt} + M \frac{di_2}{dt}$$

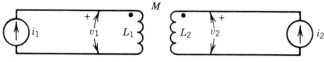

**FIGURE 1-8.** Example 4.

and

$$v_2 = M\frac{di_1}{dt} + L_2\frac{di_2}{dt}$$

Here $L_1 \, di_1/dt$ is the "self" voltage drop across $L_1$ due to $i_1$ alone, referenced as in Fig. 1-7. The induced voltage in $L_1$ is $M \, di_2/dt$, and its sign $(+)$ is determined by the dot marking: $i_2$, the inducing current, enters its own dot; hence the induced voltage is $(+)$ at the dot marking on $L_1$. In a similar way, the total voltage $v_2$ consists of the "self" term $L_2 \, di_2/dt$ and the induced term $+M \, di_1/dt$. □

For any two coupled inductors $L_a$ and $L_b$, it can be shown that

$$M_{ab} = k\sqrt{L_a L_b} \qquad (1\text{-}13d)$$

where $0 \le k \le 1$ is the *coefficient of coupling*. Also

$$M_{ab} = M_{ba} \qquad (1\text{-}13e)$$

## Voltage Sources

A voltage source is a source of electrical energy which has a prescribed voltage $v(t)$ across its terminals. A voltage source is specified completely by its reference mark and by its time function (waveform). Two common voltage sources are shown in Fig. 1-9: a constant (dc) voltage source, and a sinusoidal voltage source. In the first case $v(t) = 12\,\text{V}$, while in the second case $v(t) = V_m \sin \beta t$. The normal application of a voltage source to a network $N$ is through a switch which is *open* until, say, $t = 0$. At $t = 0$, the switch is *closed*. See Fig. 1-10.

There are two types of voltage sources: (a) an *independent* source, where the waveform of $v$ is only a function of time. The two previous examples are independent voltage sources; (b) a *dependent* (controlled) source, where the waveform of $v$ depends on (is controlled by) another voltage or current in the network. In Fig. 1-11 we see two such dependent voltage sources. The controller $v_1$, case (a), and $i_3$, case (b), must be shown specifically in order to define the sources $v_2$ and $v_4$. The multipliers $k$ and $r$ are, respectively, dimensionless and in ohms.

A current source is a source of electrical energy which has a prescribed current $i(t)$ through it. It is specified completely by its reference arrow and by its time function (waveform). In Fig. 1-12 we see the symbol for a current source, and two types of dependent current sources: $\alpha$ is dimensionless, and $g$ is in mhos.[†]

[†] Some authors use a circle for independent sources and a diamond shape for dependent sources.

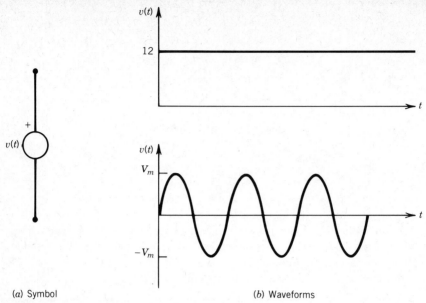

(a) Symbol                    (b) Waveforms

**FIGURE 1-9.** Voltage sources.

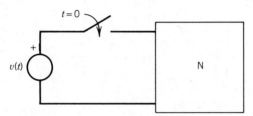

**FIGURE 1-10.** A voltage source applied to a network $N$.

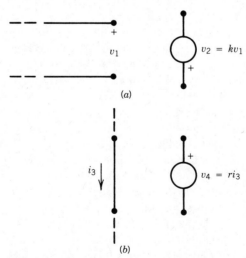

**FIGURE 1-11.** (a) A voltage-controlled voltage source; (b) A current-controlled voltage source.

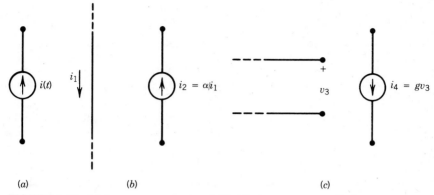

(a)  (b)  (c)

**FIGURE 1-12.**  (*a*) A current source (symbol); (*b*) A current-controlled current source; (*c*) A voltage-controlled current source.

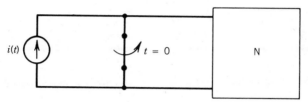

**FIGURE 1-13.**  A current source applied to a network *N*.

The normal application of a current source to a network *N* is through a switch which remains *closed* until $t = 0$ (the dual case of a voltage source!) At $t = 0$, the switch is *opened*, as in Fig. 1-13.  ***Prob. 1-5***

The *state* of a network is the minimal amount of information needed at $t = t_0$ so that, given an input for $t > t_0$, the output can be computed for all $t > t_0$.

---

## Example 5

The state of the network shown in Fig. 1-14a is the current (or flux) in the inductor. For, given this state at $t = t_0$, together with the excitation $v(t)$, $t > t_0$, the response of the network can be completely determined for all $t > t_0$. Similarly, the voltage (or

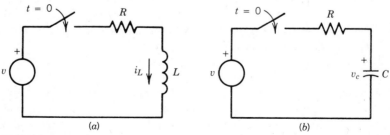

(a)  (b)

**FIGURE 1-14.**  Examples 5, 6, and 7.

charge) of the capacitor in Fig. 1-14$b$ is the state of that network. If $t_0 = 0$, we speak of the *initial state* (or initial conditions) of the network. □

---

The *zero-state response* is the output $y$ due to an input $x$ when the initial state of the network is zero, that is, all initial conditions are zero. Such a network is also said to be *at rest*, or *initially relaxed*.

The *zero-input response* of a network is the output $y$ due to an identically zero input, $x \equiv 0$. This output, then, is dependent on initial charges in some capacitors and/or initial currents in some inductors.

---

## Example 6

Consider the *RL* series circuit shown in Fig. 1-14$a$ and let $R = 1 \, \Omega$, $L = 1 \, H$. Also let $t_0 = 0$. If $v(t) = 1$ for $t > 0$, then the zero-state current (response) is found to be $1 - e^{-t}$, $t \geq 0$. □

---

## Example 7

Consider the *RC* network shown in Fig. 1-14$b$ with $R = 1 \, \Omega$, $C = 1 \, F$. Let $t_0 = 0$, and let the initial state of the network—the charge on $C$—be given, $Q_0$. The zero-input current is found to be $Q_0 e^{-t}$, $t \geq 0$. □

**Prob. 1-6**

---

## 1-4 LINEARITY AND SUPERPOSITION

A network described by

$$S\{x(t)\} = y(t) \tag{1-1}$$

is *linear* if

$$S\{a_1 x_1 + a_2 x_2\} = a_1 S\{x_1\} + a_2 S\{x_2\} = a_1 y_1 + a_2 y_2 \tag{1-14}$$

where $a_1$ and $a_2$ are any arbitrary constants. The property described by Eq. (1-14) is *the principle of superposition* and can be stated as follows: *the response of a linear system to a sum of excitations, each multiplied by an arbitrary constant, is the sum of the individual responses to each of these excitations, multiplied by the same constants.*

A linear network must satisfy Eq. (1-14) for the zero-input response and for the zero-state responses; that is, it must be zero-input linear [(Eq. 1-14) must hold for zero-input] and zero-state linear [(Eq. 1-14) must hold for zero-state]. In addition, the total response due to a given state and a given input must be the sum of the zero-state response and the zero-state response. In applying Eq. (1-14), we treat initial conditions as excitations.

Two corollaries result from the principle of superposition, Eq. (1-14):

1. *Additivity*   Since $a_1$ and $a_2$ are arbitrary, let $a_1 = a_2 = 1$. Then

$$S\{x_1 + x_2\} = S\{x_1\} + S\{x_2\} = y_1 + y_2 \qquad (1\text{-}15a)$$

2. *Homogeneity*   Let $a_1 \neq 0$ and $a_2 = 0$ to obtain

$$S\{a_1 x_1\} = a_1 S\{x_1\} = a_1 y_1 \qquad (1\text{-}15b)$$

Therefore, linearity implies both additivity and homogeneity. Either additivity or homogeneity is a necessary condition for linearity, and in some cases it might be easier to show that a network is nonlinear because it violates one or the other. A network is linear if, and only if, it is both additive and homogeneous.

---

### Example 8

Both networks shown in Fig. 1-1 are linear.   □

---

### Example 9

A semiconductor diode is described by the equation

$$i(t) = I(e^{40v} - 1)$$

This diode is nonlinear, since the zero-state output does not obey additivity,

$$I(e^{40(v_1+v_2)} - 1) \neq I(e^{40v_1} - 1) + I(e^{40v_2} - 1) \qquad □$$

*Probs. 1-7*
*1-8*
*1-9*

---

In summary: our main studies within the scope of this book will be of *lumped, linear* networks. Unless stated otherwise specifically, such networks will be also *continuous-time* and *constant*.

## 1-5   KIRCHHOFF'S LAWS: LOOP AND NODE EQUATIONS

In a typical problem in network analysis, we are given a number of elements that are interconnected in some fashion. We are asked to *solve* this network, that is, find all the currents and voltages in the network. To write the necessary equations we apply Kirchhoff's current law (KCL) and Kirchhoff's voltage law (KVL) to the given network; in addition, we use the equations which define the various elements in the network to obtain the necessary equations for solving the network.

It is hoped that these ideas are familiar to the student, along with their applications.[†] As an illustration and a review, let us consider the following examples.

---

### Example 10

For the network shown in Fig. 1-15, write the loop equations for $t > 0$. Element values are in ohms, henries, and farads. Also given is the initial state of the network. More precisely, these are the initial conditions *just before* $t = 0$. We designate these as occurring at $t = 0_-$. Therefore, the initial current in the inductor and the initial voltage across the capacitor are given as $i_3(0_-) = 4$ amperes and $v_4(0_-) = 1$ volt.

A "primitive" approach tells us that there are ten voltages and currents to be determined: $v_1, i_1, v_2, i_2, \ldots, v_5, i_5$. Therefore we need ten equations; of these, five define the elements:

$$v_1 = 1i_1 \qquad\qquad (v\text{–}i \text{ relation for a resistor})$$

$$v_2 = 2i_2 \qquad\qquad (v\text{–}i \text{ relation for a resistor})$$

$$v_3 = 3\frac{di_3}{dt} \qquad\qquad (v\text{–}i \text{ relation for an inductor})$$

$$v_4 = 1 + 4\int_{0_-}^{t} i_4(x)\, dx \qquad\qquad (v\text{–}i \text{ relation for a capacitor})$$

$$v_5 = 10e^{-t} \qquad\qquad (\text{a voltage source})$$

The remaining five relations must be obtained from Kirchhoff's laws. We write KCL as $\Sigma i = 0$ at a node, with a $(+)$ sign for currents leaving and a $(-)$ sign for currents entering that node. Similarly, and dually, KVL is written as $\Sigma v = 0$ around a loop, with a $(+)$ sign for a voltage drop and a $(-)$ for a voltage rise around that loop.

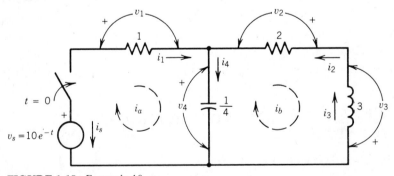

**FIGURE 1-15.** Example 10.

---

[†] A more detailed study is given in a later chapter.

Specifically here,

$$i_1 + i_5 = 0 \qquad \text{(KCL)}$$

$$-i_1 - i_2 + i_4 = 0 \qquad \text{(KCL)}$$

$$i_2 - i_3 = 0 \qquad \text{(KCL)}$$

$$v_1 + v_4 - v_5 = 0 \qquad \text{(KVL)}$$

$$-v_2 - v_3 - v_4 = 0 \qquad \text{(KVL)}$$

These are, then, the ten time-domain equations that must be solved.

In a more "sophisticated" way, we can define new variables, the *loop currents* $i_a(t)$ and $i_b(t)$. Their usefulness is obvious, since all the element currents are linear combinations of these, i.e., $i_1 = i_a; i_2 = -i_b; i_3 = -i_b; i_4 = i_a - i_b; i_5 = -i_a$. The previous three KCL equations are already implied in these relations, for example

$$0 = i_1 + i_5 = i_a - i_a$$

and the two equations of KVL in the time domain become here

$$1i_a + 1 + 4 \int_{0-}^{t} (i_a - i_b)\, dx - 10e^{-t} = 0$$

and

$$2i_b + 3\frac{di_b}{dt} - 1 - 4 \int_{0-}^{t} (i_a - i_b)\, dx = 0$$

It is certainly easier to solve these two loop equations than the "primitive" ten equations! In the following chapters, we will study such solutions in detail.  ☐

---

### Example 11

For the network shown, write the node equations for $t > 0$. Element values are in ohms, henries, and farads, with initial conditions as shown (assumed zero when not given). Note the dependent current source.

Rather than begin with the "primitive" approach, let us define two *node voltages*, $v_1(t)$ and $v_2(t)$. The two equations for $v_1$ and $v_2$ are obtained by applying KCL at each node and writing each current in terms of its $i-v$ relation. More specifically, at the node $v_1$, with $(+)$ for currents leaving and $(-)$ for currents entering, we have

$$0.1\frac{dv_1}{dt} + 4v_1 + 10 + 0.5\frac{d}{dt}(v_1 - v_2) = 0$$

and at the node $v_2$

$$-10 - 2(4v_1) - 1 + 3 \int_{0-}^{t} v_2(x)\, dx + 0.5\frac{d}{dt}(v_2 - v_1) = 0$$

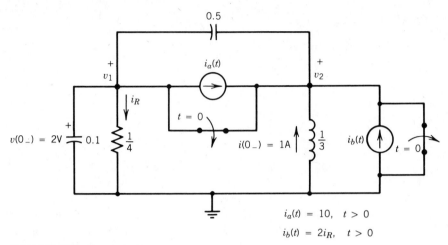

**FIGURE 1-16.** Example 11.

These are the two required equations, in the time domain. Be sure that every term is accounted for, with its sign and with its expression. Again, the *solution* of such equations will be studied fully in subsequent chapters. □

***Probs. 1-10***
***through***
***1-15***

## PROBLEMS

1-1 A linear, *time-varying* resistor is defined by the relationship

$$v(t) = R(t)i(t)$$

Draw the $v$–$i$ characteristics of such a resistor for various times, $t_1, t_2, t_3 \ldots$.

1-2 A linear, *time-varying* capacitor is defined by the relationship

$$q(t) = C(t)v(t)$$

a) Draw the $q$–$v$ characteristics of such a capacitor for various times, $t_1, t_2, t_3, \ldots$.

b) Derive the $i$–$v$ and the $v$–$i$ relations for this capacitor. Compare with a linear constant capacitor.

1-3 A linear, *time-varying* inductor is defined by the relationship

$$\phi(t) = L(t)i(t)$$

a) Draw the $\phi$–$i$ characteristics of such an inductor for various times $t_1, t_2, t_3, \ldots$.

b) Derive the $v$–$i$ and the $i$–$v$ relations for this inductor. Compare with a linear constant inductor.

1-4 For each of the circuits shown, write the expressions for the voltages.

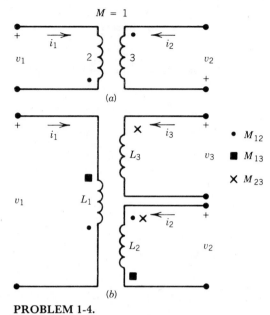

PROBLEM 1-4.

1-5 Explain fully why the normal position of the switch for a voltage source is *open* prior to application (and not closed)—see Fig. 1-10 in the text. Explain also why, for a current source, the switch is closed (and not open).

1-6 Derive fully the solutions given for the networks in Fig. 1-14 of this chapter. Specifically:

a) The zero-state current $i_L(t) = 1 - e^{-t}, t \geq 0$

b) The zero-input current $i_c(t) = Q_0 e^{-t}, t \geq 0$
*Hints*: The differential equations are

a) $\dfrac{di_L}{dt} + i_L = 1, \quad i_L(0_-) = 0$

b) $\dfrac{dv_c}{dt} + v_c = 0, \quad v_c(0_-) = Q_0/C$

1-7 Prove that the networks shown in Fig. 1-14 are linear.

1-8 Classify each system (continuous or discrete-time, constant or time-varying, linear or nonlinear, lumped or distributed) as applicable:

a) $2t\dfrac{dy}{dt} + 3y = x \qquad x \xrightarrow{\ S\ } y$

b) $y(n+1) - y(n) = 1$

c) $x(t) \xrightarrow{\ S\ } \dfrac{dx(t)}{dt} \qquad$ (a differentiator)

d) $x(t) \xrightarrow{\ S\ } \int_0^t x(\eta)\, d\eta$ (an integrator)

e) $x(t) \xrightarrow{\ S\ } Kx(t)$

(an amplifier $K > 1$)

(an attenuator $K < 1$)

1-9 A system is described by the given input-output curve. Is the system linear?

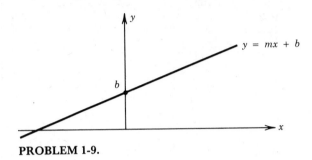

**PROBLEM 1-9.**

1-10 In writing a KCL equation, $\Sigma i = 0$, we adopted the convention of a $(+)$ sign for a current leaving the node, and a $(-)$ for a current entering. Can we switch and use a $(-)$ for entering and a $(+)$ for leaving? Explain fully.

1-11 Repeat Problem 1-10 for KVL, $\Sigma v = 0$, with its convention of signs.

1-12 The network shown is a model of a FET amplifier circuit. Write the node equation for it. All initial conditions are zero.

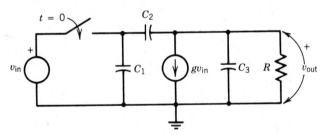

**PROBLEM 1-12.**

1-13 The network shown is called a "matching network" for a particular amplifier. Write the loop equations for it.

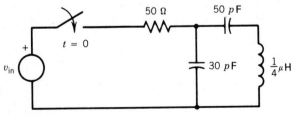

**PROBLEM 1-13.**

1-14 The op-amp circuit shown is a feedback amplifier.[†] Write the time-domain equation relating the output voltage $v_0(t)$ to the input voltage $v_{in}(t)$.

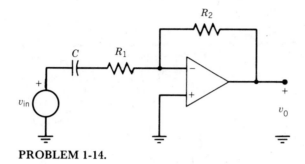

**PROBLEM 1-14.**

*Hint*: Write KVL around the input loop and KCL at the inverting $(-)$ terminal of the op-amp.

1-15 The network shown represents a certain transistor amplifier, with two dependent sources. Write the time domain equation relating $v_0$ to $v_{in}$.

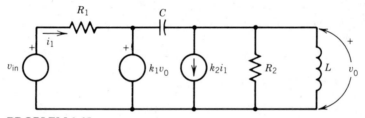

**PROBLEM 1-15.**

## REFERENCES

1. C. T. A. Johnk, *Engineering Electromagnetic Fields and Waves.* New York: John Wiley & Sons, Inc., 1975.

---

[†] For a review of op-amps, see Appendix B.

# 2

# THE LAPLACE TRANSFORM

In this chapter we introduce the Laplace transform, a powerful mathematical tool used in solving simultaneous loop equations or nodes equations which we formulated in the previous chapter. Prior to this, however, we have to study a few useful functions related to this topic.

## 2-1  THE UNIT STEP FUNCTION

The unit step function, $u(t)$, is defined as follows

$$u(t) = \begin{cases} 1 & t > 0 \\ 0 & t < 0 \end{cases} \tag{2-1}$$

and is shown in Fig. 2-1. This function, then, can be used to describe an electrical switch: in Fig. 2-2a, the voltage source $v(t) = 10e^{-t}$ is applied at $t = 0$ to the network $N$, by closing the switch. The same information is given concisely in Fig. 2-2b, where the voltage $v(t)$ is multiplied by $u(t)$. For $t < 0$, there is no voltage applied to $N$, since

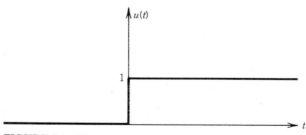

**FIGURE 2-1.** The unit step function.

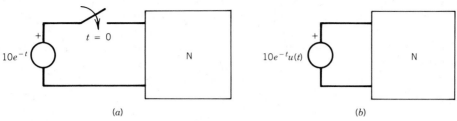

(a)                                (b)

**FIGURE 2-2.** The unit step function as an electrical switch.

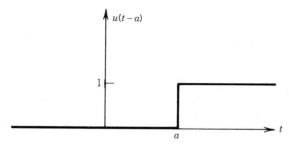

**FIGURE 2-3.** A shifted (delayed) unit step function.

$u(t) = 0$ then. After $t = 0$, the applied voltage is $10e^{-t}$, since $u(t) = 1$. Therefore, the operation of the switch is completely described by the multiplier $u(t)$.

A *time-shifted* unit step function is defined as follows

$$u(t - a) = \begin{cases} 1 & t > a \\ 0 & t < a \end{cases} \tag{2-2}$$

and shown in Fig. 2-3, where $a > 0$. This is a *delayed* unit step function, describing an electrical switch activated at $t = a$. In fact, a generalized definition of a unit step function is

$$u(\cdot) = \begin{cases} 1 & \cdot > 0 \\ 0 & \cdot < 0 \end{cases} \tag{2-3}$$

in other words, the unit step function is equal to one whenever its argument is positive; it equals to zero elsewhere. This general definition is easy to remember and it covers, of course, the two previous special cases.

---

## Example 1

Plot the following unit step functions

a)  $u(t + 2)$

b)  $-u(4 - t)$

c)  $u(-t - 3)$

We use the general definition, and obtain,

$$u(t + 2) = \begin{cases} 1 & \text{when } t + 2 > 0, \text{ or } t > -2 \\ 0 & \text{when } t + 2 < 0, \text{ or } t < -2 \end{cases}$$

$$-u(4 - t) = \begin{cases} -1 & \text{when } 4 - t > 0, \text{ or } t < 4 \\ 0 & \text{when } 4 - t < 0, \text{ or } t > 4 \end{cases}$$

$$u(-t - 3) = \begin{cases} 1 & \text{when } -t - 3 > 0, \text{ or } t < -3 \\ 0 & \text{when } -t - 3 < 0, \text{ or } t > -3 \end{cases}$$

These functions are shown in Fig. 2-4.

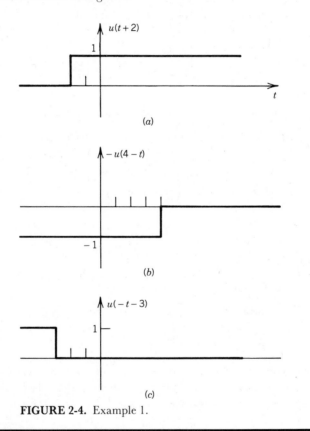

(a)

(b)

(c)

**FIGURE 2-4.** Example 1.

It will be also instructive to consider the unit step function as a limiting case of a *continuous* function. This is shown in Fig. 2-5, where we observe that

$$u(t - a) = \lim_{\varepsilon \to 0} g(t) \tag{2-4}$$

We can think of $g(t)$ as representing a real (non-ideal) switch where $\varepsilon$ is the length of time needed to activate the switch.

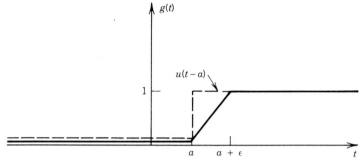

**FIGURE 2-5.** The unit step function as a limit.

There are other useful applications for the unit step function. It serves to describe many common waveforms.

## Example 2

Write the mathematical expression for the *gate function* shown in Fig. 2-6a. Such gate functions are frequently used in digital and computer circuits. Without the use of the unit step function, we could write

$$v(t) = \begin{cases} 0 & t < 0 \\ 10 & 0 < t < 2 \\ 0 & t > 2 \end{cases}$$

which is rather long and clumsy.

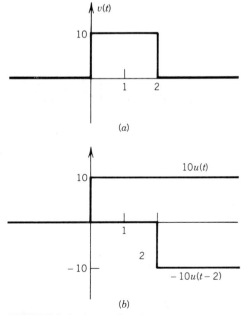

**FIGURE 2-6.** A gate function.

Using the unit step function, matters are much easier: we can think of $v(t)$ as the sum (superposition) of the two step functions shown in Fig. 2-6$b$. Therefore

$$v(t) = 10u(t) - 10u(t-2)$$

and this succinct expression says it all!   □

---

## Example 3

The *unit ramp function*, $r(t)$, is defined as follows

$$r(t) = tu(t) \tag{2-5}$$

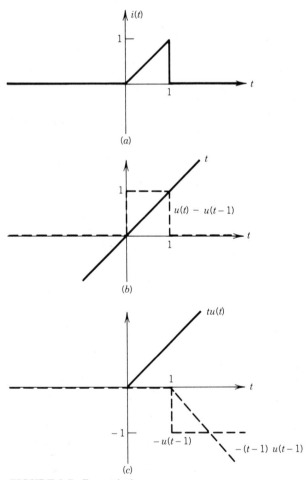

**FIGURE 2-7.** Example 3.

with similar definitions for a shifted (delayed) ramp. Find the expression for the triangular waveform $i(t)$, shown in Fig. 2-7a. The current $i(t)$ can be considered as the product of the two functions shown in Fig. 2-7b, that is,

$$i(t) = t[u(t) - u(t - 1)]$$

or, alternatively, as the sum of the three functions shown in Fig. 2-7c.

$$i(t) = tu(t) - (t - 1)u(t - 1) - u(t - 1).$$

These two expressions are identical, and it is a matter of convenience or preference in choosing to write the one or the other. The second form has the advantage of expressing a sum (superposition!) and of showing uniformity in the terms: $tu(t)$ contains only $t$, whereas $(t - 1)u(t - 1)$ is uniform in $(t - 1)$, and is, in fact, the function $tu(t)$ shifted by one unit to the right. ☐ ***Probs. 2-1***
***through***
***2-5***

---

Let us consider next the derivative of the unit step function. Such an operation is prompted not only by a mathematical urge, but by a simple circuit consideration: if we let a unit step voltage $v(t) = u(t)$ excite a capacitor (without any initial voltage across it), the current through it is given by $i(t) = C \, dv/dt = C \, du(t)/dt$. Since $u(t)$ is discontinuous, let us go back to the function $g(t)$ in Fig. 2-5. Its derivative is plotted in Fig. 2-8. Recall that $u(t - a)$ is obtained in the limit as $\varepsilon \to 0$ in $g(t)$. The derivative of $g(t)$, $g'(t)$, is a gate function between $t = a$ and $t = a + \varepsilon$, of height $1/\varepsilon$. The area under $g'(t)$ equals to 1, *independent of* $\varepsilon$.

If we proceed formally to the limit as $\varepsilon \to 0$, we obtain the function in Fig. 2-8b: an extremely high gate (height $\to \infty$ as $\varepsilon \to 0$) at $t = a$, with a narrow base (base $\to 0$ as $\varepsilon \to 0$), yet with an area equal to 1. This function is the derivative of the unit step function and is called the *unit impulse* function, or the *delta* function:

$$\delta(t - a) = \frac{d}{dt} u(t - a) \tag{2-6}$$

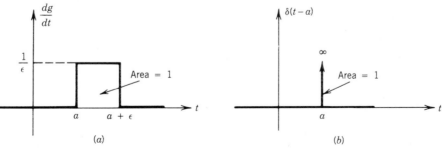

**FIGURE 2-8.** (*a*) The derivative of $g(t)$; (*b*) The limit as $\varepsilon \to 0$.

While this approach is not quite rigorous, there is a solid foundation for such functions in mathematical literature. Our main goal here is to establish the working tools that these functions provide for circuit analysis, particularly with the Laplace transform.

### Example 4

Let us consider in full detail the problem mentioned earlier: an uncharged capacitor, of $C$ farads, is excited at $t = 0$ with a unit step voltage (a 1-volt battery connected to it at $t = 0$). We have, therefore, $v(t) = u(t)$, and the current in the capacitor is

$$i(t) = C\frac{dv}{dt} = C\frac{du(t)}{dt} = C\delta(t)$$

and is shown in Fig. 2-9$b$: it is a unit impulse current, with an area of $C$ units. Does this make sense? Of course! The charge across the capacitor is

$$q(t) = Cv(t) = Cu(t)$$

and is a step function: no charge for $t < 0$, instantly becoming $C$ coulombs for $t > 0$. This charge is accompanied by a current $i(t) = dq/dt$, instantly "rushing in", and at $t = 0$ (in no time at all) depositing $C$ coulombs on the capacitor. The function $i(t) = C\delta(t)$ describes this current and the area under $i(t)$ equals to that charge.

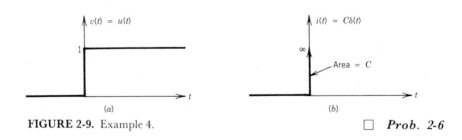

(a)                                    (b)

**FIGURE 2-9.** Example 4.                                □ **Prob. 2-6**

Let us repeat and summarize the properties of $\delta(t - a)$, as follows:

$$\delta(t - a) = 0 \quad t \neq a$$
$$\delta(t - a) \to \infty \quad t = a \tag{2-7}$$
$$\int_{-\infty}^{\infty} \delta(t - a)\, dt = 1.$$

The last integral expresses the area under the impulse function, which is sometimes called the *strength* of the impulse.

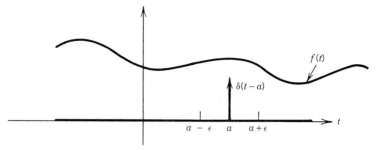

**FIGURE 2-10.** Sampling property.

Another important property of $\delta(t - a)$ is the so-called *sampling property*. Its derivation is as follows: consider the definite integral

$$I = \int_{-\infty}^{\infty} f(t)\delta(t - a) \, dt \tag{2-8}$$

which appears in many engineering problems (we will encounter it also). Here $f(t)$ is some given function, continuous at $t = a$. The situation is shown in Fig. 2-10. We wish to evaluate the integral $I$.

Since $\delta(t - a)$ is zero for $t < a$ and for $t > a$, the integrand $f(t)\delta(t - a)$ is also zero there and the area (the integral) is zero. Consequently, the limits on the integral can be changed to

$$I = \int_{a-\varepsilon}^{a+\varepsilon} f(t)\delta(t - a) \, dt \tag{2-9}$$

where $\varepsilon$ is a small (positive) interval before and after $t = a$. Now, since $f(t)$ is continuous at $t = a$, $f(a - \varepsilon) \simeq f(a) \simeq f(a + \varepsilon)$. In other words, over the range of integration in Eq. (2-9), $f(t)$ has a constant value $f(a)$. Then we write

$$I = \int_{a-\varepsilon}^{a+\varepsilon} f(a)\delta(t - a) \, dt = f(a) \int_{a-\varepsilon}^{a+\varepsilon} \delta(t - a) \, dt$$

$$= f(a) \cdot 1 = f(a) \tag{2-10}$$

where we made use of the fact that the area under $\delta(t - a)$ is unity. The final result is then: the integral $I$ in Eq. (2-8) is equal to $f(a)$, the value of $f(t)$ *at the point* $t = a$ where the impulse occurs. We say that, in Eq. (2-8), the impulse function *samples* only one value of $f(t)$, namely, the one at $t = a$; hence the term "sampling property" for the impulse function.

---

## Example 5

Evaluate the following integrals:

a) $\displaystyle\int_{-\infty}^{\infty} \cos\frac{\pi}{4} t \, \delta(t - 1) \, dt = \cos\frac{\pi}{4} t \bigg]_{t=1} = 0.707$

b) $\displaystyle\int_{0.4}^{10} e^{-t}\delta(t)\,dt = 0$

because the impulse occurs outside the limits of integration; over this entire range the integrand is zero and hence the area is also zero.

c) $\displaystyle\int_{-\infty}^{\infty} f(x)\delta(t - x)\,dx = f(t)$

where we use (without proof) the fact that $\delta(t - x) = \delta(x - t)$. This last integral will be very useful to us later. □ **Probs. 2-7**
**2-8**

## 2-2   THE LAPLACE TRANSFORM

What is a transform? We can say that, in general, a transform is an *indirect, but an easier* way of solving a problem. For example, the monetary system is a transform method of transacting business. The direct method, bartering, is difficult and limited. Instead, the value of goods is *transformed* into a monetary equivalent, and we use money to facilitate all transactions. Similarly, the Morse code is a transform: letters are transformed (coded) into dots and dashes, which makes it easier to send messages.

Among mathematical transforms, the phasor transform should be familiar to us. Sinusoidal voltages and currents are transformed (coded) into complex numbers. Integro-differential loop- or node-equations are transformed into simple *algebraic* equations; their solution is easy, and we obtain the transformed (coded) phasor solutions. An inverse transform (decoding), by inspection from a simple table of listings, gives us the time-domain answers.

The Laplace transform has the same features: although indirect, it is a simpler method; the transform solutions are obtained easily, and then we invert (decode) them to get the final time-domain solutions. Table 2-1 shows schematically the two approaches: the direct (difficult) and the transform methods.

The difficulties involved in the direct method should be stressed again: (a) given two or more simultaneous integro-differential equations, it is very difficult, if not practically impossible, to calculate the homogeneous solution, then the particular integral. (b) Even if done, the homogeneous solution will include arbitrary constants which must be carried through until a later step; then, and only then, the given initial conditions are used to evaluate them.

The Laplace transform converts the integro-differential equations into algebraic equations in the very first step, and, furthermore, the initial conditions are built into this step. Both parts of the solution (homogeneous and particular) are included. From here on, only algebraic steps are involved. The solution, in transform (coded) language, is then decoded (inverted) back into the time-domain by looking it up in appropriate tables.

**TABLE 2-1 Methods of Solving Integro-differential Equations**

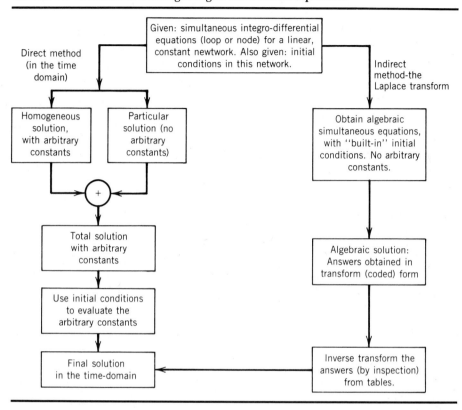

## Definition

Let $f(t)u(t)$ be either a voltage or a current, defined for $0 < t < \infty$. Its Laplace transform is

$$\mathscr{L}\{f(t)u(t)\} = \int_{0_-}^{\infty} f(t)u(t)e^{-st}\,dt = F(s) \tag{2-11}$$

We note that the definite integral is only a function of $s$, not of $t$, and we designate it by $F(s)$. The new variable $s$ is, in general, a complex number

$$s = \sigma + j\omega \tag{2-12}$$

and its units, as evidenced from $e^{-st}$, are in $s^{-1}$, or frequency. It is called the *complex frequency variable*. The time-domain function $f(t)u(t)$ is thus transformed into a frequency-domain function $F(s)$. The symbol $\mathscr{L}$ is read "the Laplace transform of . . . ." Conversely, we write

$$\mathscr{L}^{-1}F(s) = f(t)u(t) \tag{2-13}$$

where $\mathscr{L}^{-1}$ stands for "the inverse Laplace transform of . . . ."

Because of the upper limit $t \to \infty$ in Eq. (2-11), there is the question of the convergence (existence) of this integral. The sufficient conditions for this existence are: (a) the function $f(t)u(t)$ must be continuous, or, at most, have a finite number of finite discontinuities over the interval $0 \le t < \infty$; and, (b) the function $f(t)u(t)$ must be of *exponential order*, that is, if two real constants $M > 0$ and $c$ can be found such that

$$\lim_{t \to \infty} |f(t)| \le Me^{ct} \tag{2-14}$$

then the integral converges for all $\sigma > c$. Fortunately, the common functions in engineering satisfy these conditions.

---

## Example 6

Find $\mathscr{L}u(t)$ and the appropriate $\sigma$ for its existence. We write, by definition,

$$\mathscr{L}u(t) = \int_{0-}^{\infty} u(t)e^{-st}\, dt = \int_{0-}^{\infty} 1e^{-st}\, dt = \left. \frac{e^{-st}}{-s} \right]_{0-}^{\infty}$$

Applying Eq. (2-14), we find that $M = 1$ and $c = 0$. Consequently, we know that the value of the integral at the upper limit vanishes for $\sigma > 0$ and therefore

$$\mathscr{L}u(t) = \frac{1}{s} \qquad \sigma > 0$$

(Note: $|e^{-st}| = |e^{-(\sigma + j\omega)t}| = |e^{-\sigma t}||e^{-j\omega t}|$ and $|e^{-j\omega t}| \equiv 1$). □

---

This will be, then, our typical approach in finding the Laplace transform of functions:

a) Set up the defining integral
b) Simplify the integrand if possible
c) Evaluate the integral
d) Find $\sigma$ for which the upper limit vanishes [or for which Eq. (2-14) holds]
e) Tabulate the result for future use.

---

## Example 7

Find $\mathscr{L}e^{at}u(t)$, $a$ real. We write

$$\mathscr{L}e^{at}u(t) = \int_{0-}^{\infty} e^{at}u(t)e^{-st}\, dt$$

$$= \int_{0-}^{\infty} e^{(a-s)t}\, dt = \left. \frac{e^{(a-s)t}}{(a-s)} \right]_{0-}^{\infty}$$

Here $|e^{at}| \leq 1e^{at}$, showing $M = 1$ and $c = a$. Therefore, the evaluation of the integral gives

$$\mathscr{L} e^{at} u(t) = \frac{1}{s - a} \qquad \sigma > a \qquad \square \ \textbf{\textit{Probs. 2-10}}$$
$$\textbf{\textit{2-11}}$$

A short and useful table of Laplace transforms is given here, and is reproduced conveniently on the inside of the front cover:

**TABLE 2-2  A Short Table of Laplace Transforms**

| OPERATIONS | | |
|---|---|---|
| t-domain | s-domain | Note |
| 1.  $a_1 f_1(t) + a_2 f_2(t)$ | $a_1 F_1(s) + a_2 F_2(s)$ | Linearity |
| 2.  $\dfrac{df}{dt} = f'(t)$ | $sF(s) - f(0_-)$ | |
| $\vdots$ | | Differentiation in the t-domain |
| $\dfrac{d^n f}{dt^n}$ | $s^n F(s) - s^{n-1} f(0_-) - $ $s^{n-2} f'(0_-) - \cdots - f^{(n-1)}(0_-)$ | |
| 3.  $\displaystyle\int_0^t f(x)\, dx$ | $\dfrac{1}{s} F(s)$ | Integration in the t-domain |
| 4.  $(-t)^n f(t)$ | $\dfrac{d^n F(s)}{ds^n}$ | Differentiation in the s-domain |
| 5.  $f(t - a)u(t - a)$ | $F(s)e^{-as}$ | Shifting in the t-domain |
| 6.  $f(t)e^{-at}$ | $F(s + a)$ | Shifting in the s-domain |
| 7.  $f(at)$ | $\dfrac{1}{a} F\left(\dfrac{s}{a}\right)$ | Scaling |
| 8.  $\lim\limits_{t \to 0_+} f(t)$ | $\lim\limits_{s \to \infty} sF(s)$ | Initial value |
| 9.  $\lim\limits_{t \to \infty} f(t)$ | $\lim\limits_{s \to 0} sF(s)$ | Final value |
| 10.  $f_1(t) * f_2(t)$ | $F_1(s) \cdot F_2(s)$ | Convolution in the t-domain |

**Table 2-2** (*continued*)

| FUNCTIONS | |
| --- | --- |
| $f(t)u(t)$ | $F(s)$ |
| 1.  $u(t)$ | $\dfrac{1}{s}$ |
| 2.  $\delta(t)$ | $1$ |
| 3.  $\delta'(t)$ | $s$ |
| 4.  $t^n u(t),\ n = 0, 1, 2,\ldots$ | $\dfrac{n!}{s^{n+1}}$ |
| 5.  $e^{-bt}u(t)$ | $\dfrac{1}{s+b}$ |
| 6.  $t^n e^{-bt}u(t)$ | $\dfrac{n!}{(s+b)^{n+1}}$ |
| 7.  $\sin \beta t\, u(t)$ | $\dfrac{\beta}{s^2 + \beta^2}$ |
| 8.  $\cos \beta t\, u(t)$ | $\dfrac{s}{s^2 + \beta^2}$ |
| 9.  $t \cos \beta t\, u(t)$ | $\dfrac{s^2 - \beta^2}{(s^2 + \beta^2)^2}$ |
| 10.  $t \sin \beta t\, u(t)$ | $\dfrac{2\beta s}{(s^2 + \beta^2)^2}$ |
| 11.  $e^{-bt}\sin \beta t\, u(t)$ | $\dfrac{\beta}{(s+b)^2 + \beta^2}$ |
| 12.  $e^{-bt}\cos \beta t\, u(t)$ | $\dfrac{s+b}{(s+b)^2 + \beta^2}$ |
| 13.  $te^{-bt}\cos \beta t\, u(t)$ | $\dfrac{(s+b)^2 + \beta^2}{[(s+b)^2 + \beta^2]^2}$ |
| 14.  $te^{-bt}\sin \beta t\, u(t)$ | $\dfrac{2\beta(s+b)}{[(s+b)^2 + \beta^2]^2}$ |
| 15.  $\sinh \beta t\, u(t)$ | $\dfrac{\beta}{s^2 - \beta^2}$ |
| 16.  $\cosh \beta t\, u(t)$ | $\dfrac{s}{s^2 - \beta^2}$ |

**FIGURE 2-11.** The points $t = 0_-$, 0, and $0_+$.

A few comments are important at this stage:

1. The time functions that are considered start at $t = 0$, as evidenced by the multiplier $u(t)$. In other words, the past values (for $t < 0$) of $f(t)$ are of no importance; only the initial conditions of $f(t)$ are needed. Such a consideration fits our situation in circuit analysis: only initial values of capacitive voltages and inductive currents, and their future $(t > 0)$ values are needed for a complete description of the network's response. Luckily we don't need the entire past history $(-\infty < t < 0)$ of these voltages and currents!

2. While, at times, the multiplier $u(t)$ may seem redundant in Eq. (2-11), it will become crucial in other cases. For this reason, let us keep it there always, and discard it appropriately only during the evaluation of the integral.

3. The lower limit in the defining integral, Eq. (2-11), is $0_-$, that is "just before" $t = 0$, or, more precisely, "the limit as $t \rightarrow 0$ through negative values from the left." A different (and valid) definition uses $0_+$ as the lower limit. See Fig. 2-11, where the distances are enlarged in order to clarify these items. The choice of $0_-$ makes it easier to handle the impulse function $\delta(t)$. More on this subject will be discussed subsequently.

## 2-3   PROPERTIES OF THE LAPLACE TRANSFORM

Like other transforms, the Laplace transform has certain useful properties. Let us develop these.

### Linearity

For any constants $a_1$ and $a_2$,

$$\mathscr{L}\{a_1 f_1(t)u(t) + a_2 f_2(t)u(t)\} = a_1 F_1(s) + a_2 F_2(s) \qquad (2\text{-}15)$$

The proof follows directly from the defining integral, Eq. (2-11). This property states that a multiplication by a constant in the time domain corresponds to the same multiplication in the transform domain, and that the Laplace transform of a sum (or difference) of functions is the sum (or difference) of their transforms. (This property is *not* true, for example, in the logarithmic transform!).

### Time Differentiation

We will be interested in the Laplace transform of $df/dt$. By definition, we write

$$\mathscr{L}\frac{df}{dt} = \int_{0_-}^{\infty} \frac{df}{dt} e^{-st} dt \qquad (2\text{-}16)$$

and carry out this integration by parts, letting

$$u = e^{-st} \qquad\qquad dv = (df/dt)\, dt$$

$$\therefore\, du = -se^{-st}\, dt \qquad \therefore\, v = f(t)$$

The result is:

$$\mathscr{L}\frac{df}{dt} = f(t)e^{-st}\Big]_{0-}^{\infty} - \int_{0-}^{\infty} -se^{-st}f(t)\, dt = -f(0_-) + sF(s) \qquad (2\text{-}17)$$

where, by assumption on the exponential order of $f(t)$, the value of $f(t)e^{-st}$ at $t = \infty$ is zero. Consequently,

$$\mathscr{L}\frac{df}{dt} = sF(s) - f(0_-) \qquad (2\text{-}18)$$

where $F(s) = \mathscr{L}f(t)$.

This result fulfills two of the expected advantages of the Laplace transform: a differentiation (relatively difficult) is transformed into multiplication by $s$ (a simple algebraic step), and the initial condition of $f(t)$, $f(0_-)$, is built in.

We can extend this procedure to higher derivatives. For example,

$$\mathscr{L}\frac{d^2f}{dt^2} = \mathscr{L}\frac{d}{dt}\left(\frac{df}{dt}\right) = \mathscr{L}\frac{dg(t)}{dt} \qquad (2\text{-}19)$$

where we let temporarily $\dfrac{df}{dt} = g(t)$. In accordance with Eq. (2-18), we have

$$\mathscr{L}\frac{dg(t)}{dt} = sG(s) - g(0_-)$$

and since

$$G(s) = \mathscr{L}g(t) = \mathscr{L}\frac{df}{dt} = sF(s) - f(0_-)$$

$$g(0_-) = \left(\frac{df}{dt}\right)_{0-} = f'(0_-)$$

we have

$$\mathscr{L}\frac{d^2f}{dt^2} = s[sF(s) - f(0_-)] - f'(0_-) = s^2F(s) - sf(0_-) - f'(0_-) \qquad (2\text{-}20)$$

Again, double differentiation is transformed into multiplication by $s^2$, with the two natural initial conditions, $f(0_-)$ and $f'(0_-)$, built in. Extension to higher derivatives yields

$$\mathscr{L}\frac{d^nf}{dt^n} = s^nF(s) - s^{n-1}f(0_-) - s^{n-2}f'(0_-) - \cdots - f^{(n-1)}(0_-) \qquad (2\text{-}21)$$

**Probs. 2-13**
**2-14**

A complete preliminary example is in order here.

## Example 8

For the network shown in Fig. 2-12, the current source excites the RC parallel circuit for $t \geq 0$. The initial voltage of the capacitor is given. We write KCL at the node labelled $v(t)$ as

$$0.1 \frac{dv(t)}{dt} + 2v(t) = 10u(t)$$

a first-order differential equation. Taking the Laplace transform on both sides, we get:

$$0.1[sV(s) - 4] + 2V(s) = \frac{10}{s}$$

an algebraic equation in the unknown $V(s)$. We solve it algebraically by collecting terms first

$$(0.1s + 2)V(s) = \frac{10}{s} + 0.4$$

$$\therefore V(s) = \frac{10}{s(0.1s + 2)} + \frac{0.4}{0.1s + 2} = \frac{100}{s(s + 20)} + \frac{4}{s + 20}$$

The first fraction can be written as follows

$$\frac{100}{s(s + 20)} = \frac{5}{s} + \frac{-5}{s + 20}$$

which is easily verified by recombining the partial fractions (much more on this topic will be presented soon). Therefore

$$V(s) = \frac{5}{s} - \frac{1}{s + 20}$$

and, from our table of transforms, we read

$$v(t) - \mathcal{L}^{-1}V(s) = 5u(t) - e^{-20t}u(t) = (5 - e^{-20t})u(t).$$

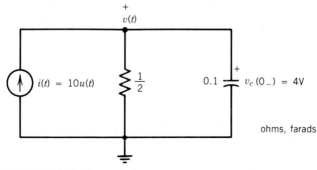

FIGURE 2-12. Example 8.

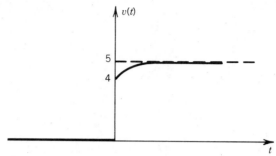

**FIGURE 2-13.** Total response for Example 8.

This total response is plotted in Fig. 2-13.

Let us verify this result by reworking it (for the last time?) via the classical time-domain method. In doing so, let us keep in mind the comparison chart of Table 2-1 and emphasize the corresponding steps:

First, we need the homogeneous solution to

$$0.1 \frac{dv_H}{dt} + 2v_H = 0$$

The assumed solution is of the form

$$v_H(t) = Ke^{st} \quad \therefore \frac{dv_H}{dt} = Kse^{st}$$

Substituting, we get

$$(0.1s + 2)Ke^{st} = 0$$

Since $Ke^{st} \neq 0$ (i.e., we want a nontrivial solution), then

$$0.1s + 2 = 0 \quad \therefore s = -20$$

The last equation is the *characteristic equation* of the network, and its root is the *characteristic value*. Notice how the characteristic equation appears almost as a "by-product" in the Laplace transform method: it is the multiplying factor of $V(s)$ just prior to the algebraic solution $V(s) = \ldots$.

Therefore, the homogeneous solution is

$$v_H(t) = Ke^{-20t}$$

Next, we need a particular solution to

$$0.1 \frac{dv_p}{dt} + 2v_p = 10 \quad t > 0$$

Since the righthand side is a constant, we assume

$$v_p = A \quad \text{a constant}$$

Substituting, we get

$$0.1(0) + 2A = 10 \quad \therefore A = 5$$

The total solution is therefore

$$v(t) = v_H + v_p = Ke^{-20t} + 5$$

Now we use the initial condition to find $K$

$$v(0_-) = 4 = Ke^0 + 5 \quad \therefore K = -1$$

Finally, then,

$$v(t) = -e^{-20t} + 5 \quad t > 0$$

in agreement with the solution obtained by the Laplace transform.
There is no doubt that the Laplace transform method is easier and faster! ☐

***Probs. 2-15***
***2-16***

## Impulse Functions in $f(t)$

A question might arise at this point (if not earlier) about Eq. (2-11): Since the Laplace transform could have been defined originally in Eq. (2-11) with $0_+$ as the lower limit, what happens if $f(0_-) \neq f(0_+)$? The answer is very simple. Either one of these choices ($0_-$ or $0_+$) in Eq. (2-11) is valid, provided we are consistent throughout all the following derivations. It is easier, however, to work with $0_-$ as the lower limit, because this choice *automatically* gives the correct answers even when $f(0_-) \neq f(0_+)$. The following example illustrates this point.

### *Example 9*

In the network shown in Example 8, the capacitor is uncharged until $t = 0$, at which time the current source $i(t) = \delta(t)$ is applied. The differential equation is

$$C\frac{dv}{dt} + \frac{1}{R}v = \delta(t)$$

Apply the Laplace transform to this equation, using $0_-$ as the lower limit in Eq. (2-11). We obtain

$$C[sV(s) - v(0_-)] + \frac{1}{R}V(s) = 1$$

and since $v(0_-) = 0$ (i.e., the capacitor is uncharged until $t = 0$), the transformed equation becomes

$$CsV(s) + \frac{1}{R}V(s) = 1$$

If we want to use the Laplace transform with $0_+$ as the lower limit in Eq. (2-11), we obtain

$$C[sV(s) - v(0_+)] + \frac{1}{R} V(s) = 0$$

The right-hand side of this equation is zero because

$$\int_{0_+}^{\infty} \delta(t) e^{-st} \, dt = 0$$

since the impulse $\delta(t)$ occurs outside the limits of the integral. We still need to evaluate $v(0_+)$. This must be done as a separate "subproblem", based on physical considerations. The current $i(t) = \delta(t)$ amounts to one unit of charge placed instantly on the capacitor. Therefore, $q(0_+) = 1$ and $v(0_+) = q(0_+)/C = 1/C$. Inserting this value into the transformed equation, we get

$$C\left[ sV(s) - \frac{1}{C} \right] + \frac{1}{R} V(s) = 0$$

or

$$CsV(s) + \frac{1}{R} V(s) = 1$$

precisely as in the previous calculation.  □

---

Since the evaluation of conditions at $t = 0_+$ is an additional task and may be quite involved, it is preferable to use the Laplace transform with $0_-$ as the lower limit. A similar situation occurs in $f'(t)$ when $f(t)$ has a step discontinuity at $t = 0$. Then $f'(t)$ will have the term $[f(0_+) - f(0_-)]\delta(t)$ in it. This impulse function will be included in the Laplace transform if $0_-$ is chosen as the lower limit. See Fig. 2-14.

### Time Integration

If we let

$$g(t) = \int_{0_-}^{t} f(x) \, dx \tag{2-22}$$

then

$$f(t) = \frac{dg}{dt}$$

and the previous results on differentiation may be applied. We obtain

$$F(s) = \mathscr{L}f(t) = sG(s) - g(0_-) \tag{2-23}$$

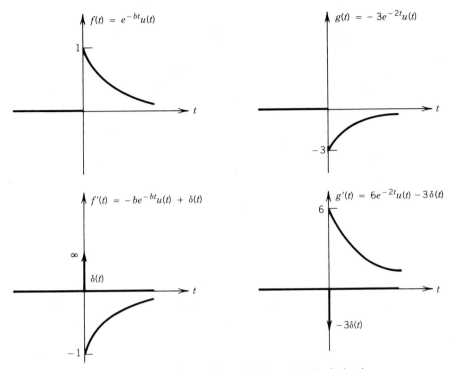

**FIGURE 2-14.** Two functions with step discontinuity and their derivatives.

However, $g(0_-) = 0$, as can be seen from Eq. (2-22). Therefore,

$$G(s) = \mathcal{L} \int_{0_-}^{t} f(x) \, dx = \frac{1}{s} F(s) \qquad (2\text{-}24)$$

Thus, integration in the time domain is transformed into division by $s$ in the frequency domain. It also makes sense: integration and differentiation are inverse operations, as are division and multiplication by $s$.

---

### Example 10

Solve the network shown in Fig. 2-15 by Laplace transform. We write KVL around the loop

$$2i(t) - 3 + \frac{1}{0.1} \int_{0_-}^{t} i(x) \, dx = 4e^{-t} u(t)$$

and take the Laplace transform

$$2I(s) - \frac{3}{s} + \frac{10}{s} I(s) = \frac{4}{s + 1}$$

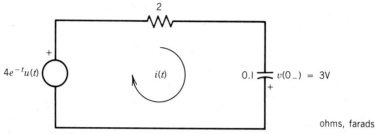

ohms, farads

**FIGURE 2-15.** Example 10.

Rearrange and solve algebraically

$$\left(2 + \frac{10}{s}\right)I(s) = \frac{4}{s+1} - \frac{3}{s}$$

$$\therefore I(s) = \frac{2s}{(s+1)(s+5)} - \frac{3/2}{s+5}$$

This is the transform answer. Its inversion to $i(t)$ will be discussed later. □

## Example 11

In the op-amp circuit shown in Fig. 2-16, calculate $V_0(s)$, then invert to get $v_0(t)$. The capacitor is initially uncharged. We write KCL at the inverting terminal $(-)$:

$$\frac{v_{in} - 0}{R} + C\frac{d}{dt}[v_0 - 0] = 0$$

Take the Laplace transform and solve for $V_0(s)$

$$\frac{1}{R}V_{in}(s) + CsV_0(s) = 0$$

$$\therefore V_0(s) = -\frac{1}{RC}\frac{V_{in}(s)}{s}$$

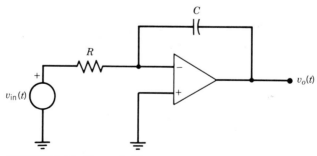

**FIGURE 2-16.** Example 11.

Therefore

$$v_0(t) = -\frac{1}{RC} \int_{0-}^{t} v_{in}(x)\, dx$$

This circuit, then, acts as an integrator of the input (with a scaled multiplier $-1/RC$).

☐ **Probs. 2-17**
**through**
**2-21**

## Complex Frequency Differentiation

The mathematical properties of $F(s)$ allow to differentiate Eq. (2.11) with respect to $s$ under the integral sign. Thus,

$$\frac{dF(s)}{ds} = \int_{0-}^{\infty} f(t)\frac{d}{ds}e^{-st}\, dt = \int_{0-}^{\infty} [-tf(t)]e^{-st}\, dt \qquad (2\text{-}25)$$

Therefore

$$\mathscr{L}\{-tf(t)\} = \frac{dF(s)}{ds} \qquad (2\text{-}26)$$

Continuing this process, we obtain

$$\mathscr{L}\{(-t)^n f(t)\} = \frac{d^n F(s)}{ds^n} \qquad (2\text{-}27)$$

This property is useful in obtaining new transforms.

*Example 12*

$$\mathscr{L}u(t) = \frac{1}{s}$$

therefore,

$$\mathscr{L}tu(t) = -\frac{d}{ds}\left(\frac{1}{s}\right) = \frac{1}{s^2}$$

$$\mathscr{L}t^2 u(t) = \frac{d^2}{ds^2}\left(\frac{1}{s}\right) = \frac{2}{s^3}$$

etc.                                                                                      ☐

**Example 13**

$$\mathscr{L} e^{-at} u(t) = \frac{1}{s + a} .$$

therefore,

$$\mathscr{L} t e^{-at} u(t) = -\frac{d}{ds} \left( \frac{1}{s + a} \right) = \frac{1}{(s + a)^2}$$

$$\mathscr{L} t^2 e^{-at} u(t) = \frac{d^2}{ds^2} \left( \frac{1}{s + a} \right) = \frac{2}{(s + a)^3}$$

etc.

□

**Probs. 2-22**
**2-23**

## Complex Frequency Integration

If we integrate Eq. (2-11) on both sides with respect to $s$ between the limits shown, the result is

$$\int_s^\infty F(s) \, ds = \int_s^\infty \left[ \int_{0-}^\infty f(t) e^{-st} \, dt \right] ds$$

$$= \int_{0-}^\infty f(t) \, dt \int_s^\infty e^{-st} \, ds = \int_{0-}^\infty \frac{f(t)}{t} e^{-st} \, dt \qquad (2\text{-}28)$$

Therefore,

$$\mathscr{L} \left\{ \frac{f(t)}{t} \right\} = \int_s^\infty F(s) \, ds \qquad (2\text{-}29)$$

**Example 14**

$$\mathscr{L} \sin t = \frac{1}{s^2 + 1}$$

therefore,

$$\mathscr{L} \frac{\sin t}{t} = \int_s^\infty \frac{1}{s^2 + 1} \, ds = \tan^{-1} s \Big]_s^\infty = \frac{\pi}{2} - \tan^{-1} s = \tan^{-1} \frac{1}{s}$$

While we are at it, let us establish here a well-known definite integral,

$$\int_0^\infty \frac{\sin t}{t} \, dt = \frac{\pi}{2}$$

To do this, write fully the Laplace transform of sin $t/t$

$$\mathcal{L} \frac{\sin t}{t} = \int_0^\infty \frac{\sin t}{t} e^{-st} \, dt = \tan^{-1} \frac{1}{s}$$

and let $s \to 0$ to obtain the desired result.   $\square$

## Time Shifting

In Fig. 2-17 we see several functions. The Laplace-transformable function is the one in Fig. 2-17b, obtained from the one in Fig. 2-17a by multiplication with $u(t)$. When this function is shifted to the right by $a$ units, its expression becomes $f(t - a)u(t - a)$. It is here that the use of the step function, both in $f(t)u(t)$ and in $f(t - a)u(t - a)$, is essential. Without $u(t)$, erroneous expressions might result for the shifting and for the Laplace transform.

By definition, we write

$$\mathcal{L}f(t - a)u(t - a) = \int_{0-}^\infty f(t - a)u(t - a)e^{-st} \, dt \qquad (2\text{-}30)$$

Since $u(t - a) = 0$ for $0 < t < a$, the lower limit can be changed to $a$,

$$\int_a^\infty f(t - a)e^{-st} \, dt$$

and the integral is evaluated by substituting a new variable $x = t - a$. The result is

$$\int_a^\infty f(t - a)e^{-st} \, dt = \int_0^\infty f(x)e^{-s(x+a)} \, dx = e^{-as} \int_0^\infty f(x)e^{-sx} \, dx = e^{-as}F(s)$$

Therefore,

$$\mathcal{L}f(t - a)u(t - a) = e^{-as}F(s) \qquad (2\text{-}31)$$

where $F(s) = \mathcal{L}f(t)u(t)$ is the Laplace transform of the nonshifted function

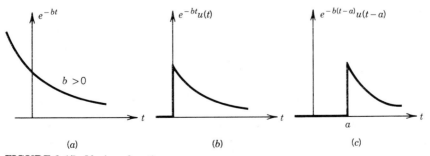

(a)                    (b)                    (c)

**FIGURE 2-17.**  Various functions.

---

### Example 15

Find the Laplace transform $F(s)$ of a periodic function $f(t)$ starting at $t = 0$, with a period $T$. See Fig. 2-18.

Consider the first cycle of this function as a new function, say $f_1(t)$, defined by

$$f_1(t) = \begin{cases} f(t) & 0 < t < T \\ 0 & \text{elsewhere} \end{cases}$$

Then we can write $f(t)$ in terms of $f_1(t)$ as follows:

$$f(t) = f_1(t) + f_1(t - T)u(t - T) + f_1(t - 2T)u(t - 2T) + \cdots$$

which states simply that $f(t)$ consists of the sum of the first cycle *plus* the same cycle shifted by one period, *plus* the same cycle shifted by two periods, etc. Therefore,

$$F(s) = F_1(s) + e^{-Ts}F_1(s) + e^{-2Ts}F_1(s) + \cdots = F_1(s)[1 + e^{-Ts} + e^{-2Ts} + \cdots]$$

$$= F_1(s)/(1 - e^{-Ts})$$

where the identity $1 + x + x^2 + \cdots = 1/(1 - x)$ was used in the last step. Thus, we have obtained in closed form the Laplace of a periodic function in terms of the Laplace transform of its first cycle which is

$$F_1(s) = \int_{0_-}^{\infty} f_1(t)e^{-st}\,dt = \int_{0_-}^{T} f(t)e^{-st}\,dt$$

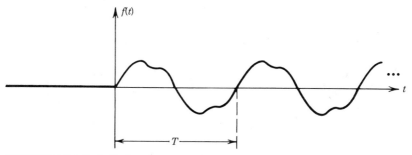

**FIGURE 2-18** Periodic function.

---

### s Shifting

We wish to evaluate $\mathcal{L}f(t)e^{-at}u(t)$ in terms of $F(s)$, the Laplace transform of $f(t)u(t)$. By a direct evaluation, we write

$$\mathcal{L}f(t)e^{-at}u(t) = \int_{0_-}^{\infty} f(t)e^{-at}e^{-st}\,dt = \int_{0_-}^{\infty} f(t)e^{-(s+a)t}\,dt = F(s + a) \quad (2\text{-}32)$$

Again, we see how a transform simplifies matters: in the time domain we have $f(t)e^{-at}$, a multiplication of $f(t)$ by the exponential function $e^{-at}$; in the s-domain, the new transform is simply the old $F$, with $s$ replaced by $(s + a)$.

**Example 16**

We know that

$$\mathcal{L} tu(t) = \frac{1}{s^2}$$

Therefore

$$\mathcal{L} te^{-at}u(t) = \frac{1}{(s + a)^2}$$

a result obtained earlier by a different property. $\square$

**Example 17**

Given $V(s) = \dfrac{2}{s^2 + 2s + 10}$, find $v(t)$. We complete the square in the denominator

$$\frac{2}{s^2 + 2s + 10} = \frac{2}{(s + 1)^2 + 9}$$

and recognize immediately $F(s + 1)$. Therefore,

$$v(t) = e^{-t}\mathcal{L}^{-1}\frac{2}{s^2 + 9} = \frac{2}{3}e^{-t}\sin 3t\, u(t) \qquad \square$$

**Scaling of $t$ or of $s$**

If we use a *scaled* time variable, given by $at$, where $a > 0$, then, by definition

$$\mathcal{L} f(at) = \int_{0-}^{\infty} f(at)e^{-st}\, dt \qquad (2\text{-}33)$$

We let $x = at$, $dx = a\, dt$, and obtain

$$\mathcal{L} f(at) = \frac{1}{a}\int_{0-}^{\infty} f(x)e^{-(s/a)x}\, dx = \frac{1}{a}F\left(\frac{s}{a}\right)$$

This result shows also the effect of scaling the frequency variable from $s$ to $s/a$, namely

$$\mathcal{L} af(at) = F\left(\frac{s}{a}\right) \qquad (2\text{-}34)$$

***Example 18***

$$\mathscr{L} \sin t = \frac{1}{s^2 + 1}$$

therefore,

$$\mathscr{L} \sin \beta t = \frac{1}{\beta} \frac{1}{\left(\dfrac{s}{\beta}\right)^2 + 1} = \frac{\beta}{s^2 + \beta^2}$$

□

***Probs. 2-25
through
2-29***

We have now enough tools of the Laplace transform to begin its applications to circuit analysis. This is done in the next chapters.

## PROBLEMS

2-1  Write the expression for $i(t)$ as shown, using step functions

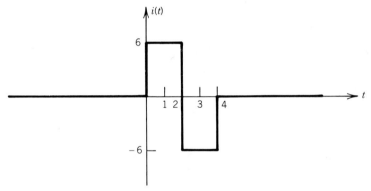

**PROBLEM 2-1.**

2-2  Write the expression for $v(t)$ as shown

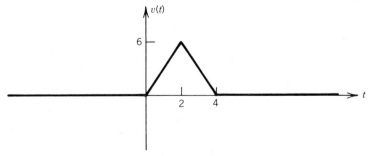

**PROBLEM 2-2.**

2-3 Write the expression (an infinite series) for $v(t)$ as shown. Build it slowly, by first writing the expressions for $v_1$, $v_2$, $v_3$, . . . .

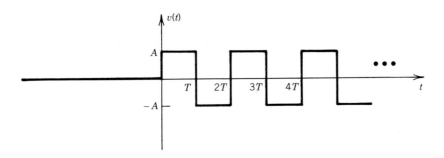

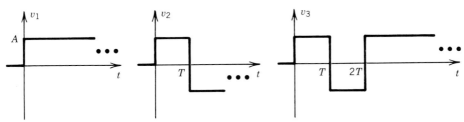

**PROBLEM 2-3.**

2-4 Higher-order functions may be obtained by integrating the unit step function.

  a) Show that the *unit ramp* function, defined by Eq. (2-5) in the text, can be written also as

$$r(t) = \int_0^t u(x)\ dx$$

  b) Sketch the *unit parabola* function, defined as

$$p(t) = \int_0^t r(x)\ dx$$

  and write its explicit expression.

2-5 Plot carefully each function:

  a) $(t - 1)u(t)$      b) $(t + 1)u(t)$

  c) $tu(t - 1)$      d) $tu(t + 1)$

  e) $(t - 1)u(t - 2)$      f) $(t + 1)u(t + 2)$

  g) $(t - 2)u(t - 1)$      h) $(t + 2)u(t + 1)$

2-6 Describe fully, in mathematical and in physical terms, the following experiments:

  a) an inductor ($L$ henries), without any initial current through it, is excited at $t = 0$ by a unit step current source. What is the voltage? the flux?

b) A mass $m$, initially at rest, is excited at $t = 0$ by a force $K$, applied instantly. What is the acceleration? the momentum?

2-7 By plotting the curves for several increasing values of $\lambda$, show that

$$u(t) = \lim_{\lambda \to \infty} \left[ \frac{1}{2} + \frac{1}{\pi} \tan^{-1}(\lambda t) \right]$$

2-8 Higher derivatives of the unit step function can be obtained as limiting cases. Consider the function $p(t)$ shown in the figure. As $\varepsilon \to 0$ we have

$$\lim_{\varepsilon \to 0} p(t) = \delta(t - a)$$

Draw carefully $dp(t)/dt$, and obtain the properties of the derivative of $\delta(t - a)$, called the *doublet*:

$$\lim_{\varepsilon \to 0} \frac{dp(t)}{dt} = \delta'(t - a) = \frac{d}{dt} \delta(t - a)$$

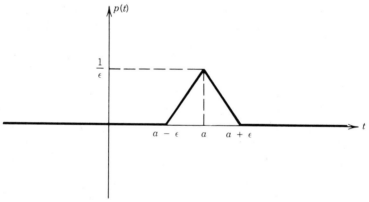

**PROBLEM 2-8.**

2-9 Discuss the following entities as transforms. Mention the direct method, the difficulties in it, and the ease of the indirect (transform) method in each case:

a) The letters of the alphabet
b) The (spoken and written) language
c) The binary code for computers
d) Logarithms

2-10 Derive in full detail the Laplace transform of each function, including conditions on $\sigma$. Compare your results with Table 2-2 in the text. When necessary, use integration by parts:

a) $e^{j\beta t} u(t)$
b) $A$, a constant
c) $u(t - a),\ a > 0$

d) $\delta(t - a)$ (*Hint*: sampling property)

e) $tu(t)$

f) $\sin \beta t\, u(t)$

g) $\cos \beta t\, u(t)$

h) $te^{at}u(t)$

i) $e^{at} \sin \beta t\, u(t)$

j) $e^{at} \cos \beta t\, u(t)$

2-11 To illustrate the sufficient, but not necessary, condition on the convergence of $f(t)u(t)$, consider the function $t^{-1/2}u(t)$.

a) Show that it has an *infinite* discontinuity.

b) Set up and evaluate the integral defining its Laplace transform. In doing so, you'll need first to change variables to show that

$$\int_{0_-}^{\infty} t^{-1/2}e^{-st}\, dt = \frac{2}{\sqrt{s}} \int_{0_-}^{\infty} e^{-x^2}\, dx$$

Then, to evaluate $J = \int_{0_-}^{\infty} e^{-x^2}\, dx$, write

$$J^2 = \int_{0_-}^{\infty} e^{-x^2}\, dx \int_{0_-}^{\infty} e^{-y^2}\, dy = \int_{0_-}^{\infty} \int_{0_-}^{\infty} e^{-(x^2+y^2)}\, dx\, dy$$

and change to polar coordinates

$$x^2 + y^2 = r^2$$
$$dx\, dy = r\, dr\, d\theta$$

to show that

$$J = \sqrt{\pi}/2$$

2-12 Does the function $e^{t^2}u(t)$ have a Laplace transform? Why?

2-13 Use the property of $\mathscr{L}(df/dt)$ to obtain $\mathscr{L}\delta(t)$, with $f(t) = u(t)$.

2-14 Extend Problem 2.13 to $\mathscr{L}\delta'(t)$, $\mathscr{L}\delta''(t)$, $\ldots \mathscr{L}\delta^{(n)}(t)$, where

$$\delta^n(t) = \frac{d^n\delta(t)}{dt^n}.$$

2-15 Solve the given *RL* circuit by the following steps:

a) Write the loop equation;

b) Laplace transform it;

c) Solve algebraically for $I(s)$;

d) Reduce $I(s)$ to terms listed in our table;

e) Invert $I(s)$ using the table;

f) Plot $i(t)$

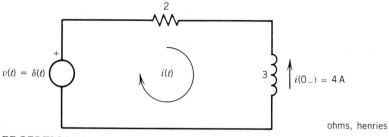

**PROBLEM 2-15.**

2-16 (a) Derive the following transforms by using the property of $\mathcal{L}(df/dt)$.
(b) Verify by actually differentiating first, then transforming:

a) $\mathcal{L}\dfrac{d}{dt}[tu(t)]$

b) $\mathcal{L}\dfrac{d}{dt}[\sin\beta t\, u(t)]$

c) $\mathcal{L}\dfrac{d}{dt}[e^{at}u(t)]$

d) $\mathcal{L}\dfrac{d}{dt}[\cos\beta t\, u(t)]$

2-17 (a) Derive the following transforms, by using the property of $\mathcal{L}\{\int_{0_-}^{t} f(x)\,dx\}$.
(b) Verify by integrating first.

a) $\mathcal{L}\left\{\displaystyle\int_{0_-}^{t} \delta(x)\,dx\right\}$

b) $\mathcal{L}\left\{\displaystyle\int_{0_-}^{t} u(x)\,dx\right\}$

c) $\mathcal{L}\left\{\displaystyle\int_{0_-}^{t} e^{ax}\,dx\right\}$

d) $\mathcal{L}\left\{\displaystyle\int_{0_-}^{t} \cos\beta x\,dx\right\}$

2-18 In the op-amp circuit shown, calculate the transform output voltage $V_0(s)$, then invert $V_0(s)$ to obtain $v_0(t)$. The capacitor is initially uncharged.

2-19 Calculate $V_0(s)$ in the op-amp circuit shown. The capacitor is initially uncharged.

2-20 A certain linear system obeys the following differential equation

$$\frac{d^2y}{dt^2} + 7\frac{dy}{dt} + 12y = 4e^{-t}u(t)$$

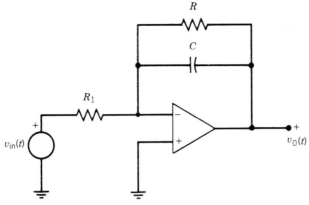

**PROBLEM 2-18.**

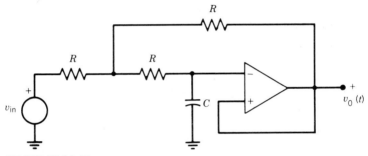

**PROBLEM 2-19.**

with the initial conditions $y(0_-) = -1$, $y'(0_-) = 2$.

a) Solve for $Y(s)$.

b) Identify the characteristic equation and find the characteristic values.

2-21 Repeat Problem 2-20 for

$$\frac{d^2y}{dt^2} + 4\frac{dy}{dt} + 20y = 2u(t)$$

with

$$y(0_-) = 0, \qquad y'(0_-) = -1$$

2-22 a) From $\mathcal{L} \sin \beta t\, u(t) = \dfrac{\beta}{s^2 + \beta^2}$ obtain $\mathcal{L}t \sin \beta t\, u(t)$

b) From $\mathcal{L} \cos \beta t\, u(t) = \dfrac{s}{s^2 + \beta^2}$ obtain $\mathcal{L}t \cos \beta t\, u(t)$

2-23 From $\mathcal{L}e^{-at}u(t) = \dfrac{1}{s + a}$ obtain $\mathcal{L}te^{-at}u(t)$. Plot both time functions.

2-24  Given $I(s) = \dfrac{1}{s(s+1)} e^{-2s}$. Find $i(t)$ by using the integration and time-shifting properties. Plot $i(t)$.

2-25  Obtain the Laplace transform of the half-wave rectified sinusoid shown.

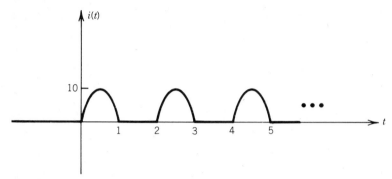

**PROBLEM 2-25.**

2-26  Given

$$V(s) = \frac{2s - 3}{s^2 + 4s + 20}$$

find $v(t)$. *Hint*: complete the square and don't forget $(s + a)$ in the numerator also.

2-27  Obtain

a)  $\mathscr{L} t e^{-2t} \cos 4t\, u(t)$   from $\mathscr{L} \cos 4t\, u(t)$
b)  $\mathscr{L} t^2 e^{-t} u(t)$       from $\mathscr{L} e^{-t} u(t)$

by using properties of the Laplace transform.

2-28  Using Table 2-2, and properties of the Laplace transform, find $\mathscr{L}^{-1}$ of:

a)  $3/(s + 2)^4$
b)  $4e^{-3s}/(s + 1)^2$

2-29  When $f(t)$ and $F(s)$ contain a parameter, say $b$, which is independent of $t$ and $s$, we can perform in Eq. 2-11 valid operations with respect to $b$. For example, take $\partial/\partial b$ in the transform of $\sin bt\, u(t)$ to obtain a new time function and its Laplace transform.

# 3

# LAPLACE TRANSFORM IN CIRCUIT ANALYSIS

In this chapter, we apply the Laplace transform to loop- and node-equations, and develop several important concepts related to this analysis.

## 3-1 LOOP AND NODE EQUATIONS

Let us apply the Laplace transform to two examples.

### Example 1

This is the same problem as in Example 9 of Chapter 1, repeated here for convenience. The two loop equations for the unknown loop currents $i_a$ and $i_b$ are[†]:

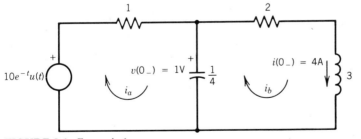

**FIGURE 3-1.** Example 1.

KVL loop a:

$$1i_a + 1 + 4\int_{0_-}^{t} (i_a - i_b)\, dx = 10e^{-t}u(t)$$

---

[†] Recall: lower case $i$ and $v$ denote time functions, $i(t)$ and $v(t)$. Capital letters $I$ and $V$ are transforms, $I(s)$ and $V(s)$.

53

KVL loop b:

$$2i_b + 3\frac{di_b}{dt} - 1 + 4\int_{0-}^{t} (i_b - i_a)\, dx = 0$$

We take the Laplace transform of these equations:

$$I_a + \frac{1}{s} + \frac{4}{s}(I_a - I_b) = \frac{10}{s+1}$$

$$2I_b + 3(sI_b - 4) - \frac{1}{s} + \frac{4}{s}(I_b - I_a) = 0$$

and upon rearrangement in matrix form

$$\begin{bmatrix} 1 + \dfrac{4}{s} & -\dfrac{4}{s} \\[2ex] -\dfrac{4}{s} & 3s + 2 + \dfrac{4}{s} \end{bmatrix} \begin{bmatrix} I_a \\ I_b \end{bmatrix} = \begin{bmatrix} \dfrac{10}{s+1} \\[2ex] 0 \end{bmatrix} + \begin{bmatrix} -\dfrac{1}{s} \\[2ex] 12 + \dfrac{1}{s} \end{bmatrix}$$

These two simultaneous algebraic equations can be solved for $I_a$ and $I_b$, for example by determinants, using Cramer's rule. The results are

$$I_a = \frac{\begin{vmatrix} \dfrac{10}{s+1} - \dfrac{1}{s} & -\dfrac{4}{s} \\[2ex] 12 + \dfrac{1}{s} & 3s + 2 + \dfrac{4}{s} \end{vmatrix}}{\begin{vmatrix} 1 + \dfrac{4}{s} & -\dfrac{4}{s} \\[2ex] -\dfrac{4}{s} & 3s + 2 + \dfrac{4}{s} \end{vmatrix}} \qquad I_b = \frac{\begin{vmatrix} 1 + \dfrac{4}{s} & \dfrac{10}{s+1} - \dfrac{1}{s} \\[2ex] -\dfrac{4}{s} & 12 + \dfrac{1}{s} \end{vmatrix}}{\begin{vmatrix} 1 + \dfrac{4}{s} & -\dfrac{4}{s} \\[2ex] -\dfrac{4}{s} & 3s + 2 + \dfrac{4}{s} \end{vmatrix}} \qquad \square$$

Before getting too involved in algebra, let us make several important observations:

1. The determinant in the denominator is the factor which multiplied $I_a$ and $I_b$ prior to their solution. As shown in Example 8 in Chapter 2, this is the *characteristic polynomial* of the network; its coefficients depend entirely on the element values of the network—hence its name. When equated to zero, it yields the *characteristic equation* and the *characteristic values*. Here we get (verify!)

$$3s^2 + 14s + 12 = 0$$

$$\therefore s_1 = -1.131 \quad \therefore s_2 = -3.535$$

2. If we recall how the characteristic equation was obtained via the classical method, we can draw the following conclusion: the two simultaneous integro-differential

equations can be reduced to a single second-order differential equation in $i_a$ (or in $i_b$)

$$3\frac{d^2 i_a}{dt^2} + 14\frac{d i_a}{dt} + 12 i_a = \cdots$$

Obviously, we won't go that way—after all, the Laplace transform is easier. Nonetheless, we learn from this step an important lesson: the single differential equation is of order 2, which is also the order of the characteristic equation. As a supporting proof, we know that a second-order differential equation requires two independent initial conditions for its solution; they are here $v_C(0_-) = 1$ and $i_L(0_-) = 4$.

In summary: the *order of a network*, $n$, is equal to the order of its characteristic equation, which also equals the order of its single equivalent differential equation, which equals the number of the independently specified initial conditions. The latter is obvious *by inspection*: the number of capacitors $n_C$ plus the number of inductors $n_L$

$$n = n_C + n_L \tag{3-1}$$

(A slight modification of this relation will be given later.)

3. The righthand side of the two transformed loop equations contains two types of terms:

(a) $10/(s + 1)$, the transform of the voltage source, grouped into the first matrix on the righthand side, and

(b) $1/s$ and 12, due to the initial state (initial conditions), grouped into the second matrix.

Consequently (as expected!), the total response, $I_a$ or $I_b$, will consist of two parts: the zero-input response, and the zero-state response. The zero-input response is obtained by ignoring the voltage source, that is, by omitting the term $10/(s + 1)$ from the equations; the zero-state response, as the name implies, is obtained by setting the initial state to zero, that is, by dropping the terms $1/s$ and 12 from the equations.

With these observations, we expand the determinant solution for $I_a$ as follows:

$$I_a = \underbrace{\frac{\dfrac{10}{s+1}\left(3s + 2 + \dfrac{4}{s}\right)}{\dfrac{3s^2 + 14s + 12}{s}}}_{\text{zero-state response}} + \underbrace{\frac{-\dfrac{1}{s}\left(3s + 2 + \dfrac{4}{s}\right) + \dfrac{4}{s}\left(12 + \dfrac{1}{s}\right)}{\dfrac{3s^2 + 14s + 12}{s}}}_{\text{zero-input response}}$$

After clearing all fractions, we obtain

$$I_a(s) = \frac{30s^2 + 20s + 40}{(s + 1)(3s^2 + 14s + 12)} + \frac{-3s + 46}{3s^2 + 14s + 12}$$

This is the transform answer, ready to be inverted into $i_a(t)$.
A similar development holds for $I_b(s)$.                                           ☐

### Example 2

This example is a continuation of Example 10 in Chapter 1. We make the immediate observation about the order of the network, $n = 3$, thus expecting a cubic characteristic equation and 3 characteristic values.

The Laplace transform of the two node equations yields

$$0.1(sV_1 - 2) + 4V_1 + \frac{10}{s} + 0.5s(V_1 - V_2) = 0$$

$$-\frac{10}{s} - 8V_1 - \frac{1}{s} + \frac{3}{s}V_2 + 0.5s(V_2 - V_1) = 0$$

In matrix form, these equations are

$$\begin{bmatrix} 0.6s + 4 & -0.5s \\ -8 - 0.5s & 0.5s + \dfrac{3}{s} \end{bmatrix} \begin{bmatrix} V_1 \\ V_2 \end{bmatrix} = \begin{bmatrix} -\dfrac{10}{s} \\ \dfrac{10}{s} \end{bmatrix} + \begin{bmatrix} 0.2 \\ \dfrac{1}{s} \end{bmatrix}$$

On the right hand side, the first matrix accounts for the zero-state response, while the second one accounts for the zero-input response. The completion of this example is left as a problem (Problem 3-3).    $\square$    **Probs. 3-1**
**through**
**3-5**

## *3-2   TRANSFORM DIAGRAMS—LOOP EQUATIONS

Next, we would like to derive a systematic way of writing directly the transformed loop or node equations, ready for solution (by determinants or matrix inversion). Let us consider loop equations first.

### The Resistor

A resistor has the $v-i$ relationship as in Eq. (1-2), repeated here for convenience:

$$v_R(t) = Ri_R(t) \tag{3-2}$$

The Laplace transform of this equation yields

$$V_R(s) = RI_R(s) \tag{3-3}$$

with no initial conditions involved; the resistor is an instantaneous, memoryless element.

Equation (3-3) has the general form of $V(s) = Z(s)I(s)$, where $Z(s)$ is the *impedance* of an element, and describes the *transform diagram* of the resistor as shown in Fig. 3.2.

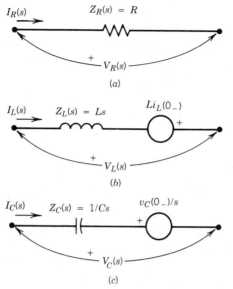

**FIGURE 3-2.** Transform diagrams for loop analysis.

## The Inductor

An inductor is defined by

$$v_L(t) = L\frac{di_L(t)}{dt} \tag{3-4}$$

Transforming this expression, we obtain[†]:

$$V_L(s) = L[sI_L(s) - i_L(0_-)] = LsI_L(s) - Li_L(0_-) \tag{3-5}$$

We recognize that the right-hand side consists of a drop across the impedance $Z_L(s) = Ls$ and of an "initial-condition" source, $Li_L(0_-)$, with a polarity as shown in the transform diagram, Fig. 3-2. It should be noted that $i_L(0_-)$ is a number which can be positive or negative, as illustrated in the previous examples.

## The Capacitor

The capacitor is described by the $v$–$i$ relationship

$$v_C(t) = \frac{1}{C}\int_{0_-}^{t} i_C(x)\,dx + v_C(0_-) \tag{3-6}$$

Transforming, we obtain

$$V_C(s) = \frac{1}{Cs}I_C(s) + \frac{v_C(0_-)}{s} \tag{3-7}$$

[†] As pointed out in Chapter 2, it is easier (but not essential) to work with $t = 0_-$ rather than $0_+$. The student should review carefully Section 2-3 in that chapter.

Again, we notice that, in the transform diagram, the impedance of the capacitor $Z_C(s) = 1/Cs$ accounts for the first term on the right-hand side of Eq. (3-7); the "initial-condition" source $v_C(0_-)/s$ is added in series, with a polarity as shown. Note that $v_C(0_-)$ is a number, positive or negative (positive if the given polarity of $v_C(0)$ — or of the initial charge $q_C(0_-)$ — agrees with the chosen reference polarity of $v_C(t)$, negative otherwise).

## Loop Equations

Given a network to be analyzed by loop equations, we shall proceed to write these directly in their Laplace form. The steps are:

a) Draw the transform diagram.

b) Write Kirchhoff's voltage law around each of the chosen loops.

The details are illustrated in an example.

---

### Example 3

In the network shown in Fig. 3-3$a$, element values are in ohms, henries, and farads. Initial conditions are in amperes and volts. The three loop currents $i_1(t)$, $i_2(t)$ and $i_3(t)$ are chosen in the clockwise direction. The transform diagram is shown in Fig. 3-3$b$, according to the rules studied previously. Special attention should be paid to the polarity marks on the "initial-condition" sources.

Since the network is linear, we shall *use superposition to write the loop equations*. Specifically, we shall *sum up all the voltage drops around each loop caused by the various loop currents*.

Around the first loop (the loop traced by $I_1$): the voltage drops caused by $I_1$ alone, with $I_2 = I_3 = 0$, are across the resistor ($Z_R = 2$) and the inductor ($Z_L = s$), that is, $I_1(s + 2)$. Next, consider $I_2$ alone ($I_1 = I_3 = 0$): it causes a drop in the first loop across the inductor ($Z_L = s$), given by $-sI_2$. Notice the negative sign, since this is a negative drop (a rise) in the direction of the first loop. Finally, $I_3$ alone causes a drop in the first loop across the resistor ($Z_R = 2$), given by $-2I_3$ (again, a negative drop in the direction traced around the first loop). By superposition, then, the *total* voltage drop around the first loop is $(s + 2)I_1 - sI_2 - 2I_3$, and it must be equal to the *total* voltage rise of the sources around this loop, namely $2/(s + 1) - 2$. The first loop equation is therefore

$$(s + 2)I_1 - sI_2 - 2I_3 = \frac{2}{s + 1} - 2$$

Around the second loop, $I_1$ alone causes a drop of $-sI_1$. The loop current $I_2$, alone, causes a drop of $I_2 (s + 3 + 1/2s)$, and $I_3$ alone causes no drop at all. The *total* voltage rise of the sources around the second loop is $[(10/s) + (2/s) + 2]$. Therefore, the second loop equation reads

$$-sI_1 + \left(s + 3 + \frac{1}{2s}\right) I_2 + 0I_3 = \frac{12}{s} + 2$$

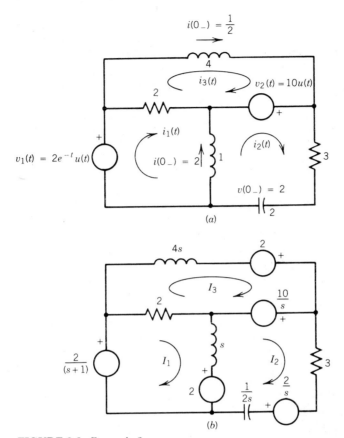

**FIGURE 3-3.** Example 3.

Similarly, around the third loop: Current $I_1$ *alone* causes a drop of $-2I_1$, current $I_2$ *alone* causes no drop, and current $I_3$ *alone* causes a drop of $(4s + 2)I_3$. The total rise due to sources around the third loop is $(2 - 10/s)$. The third loop equation reads

$$-2I_1 + 0I_2 + (4s + 2)I_3 = 2 - \frac{10}{s}$$

These loop equations can be written in matrix form as follows:

$$\begin{bmatrix} s + 2 & -s & -2 \\ -s & s + 3 + \dfrac{1}{2s} & 0 \\ -2 & 0 & 4s + 2 \end{bmatrix} \begin{bmatrix} I_1 \\ I_2 \\ I_3 \end{bmatrix} = \begin{bmatrix} \dfrac{2}{s+1} \\ \dfrac{10}{s} \\ \dfrac{-10}{s} \end{bmatrix} + \begin{bmatrix} -2 \\ \dfrac{2}{s} + 2 \\ 2 \end{bmatrix}$$

and are ready for solution. The two matrices on the right account for the zero-state and the zero-input responses, respectively.

Let us emphasize again: After drawing the transform diagram, consider one loop at a time and use superposition to obtain the total voltage drop around each loop. This is done by letting $I_1 \neq 0, I_2 = I_3 = \cdots = 0$ and finding the voltage drops due to $I_1$ in that loop, then letting $I_2 \neq 0, I_1 = I_3 = \cdots = 0$ and finding the voltage drops due to $I_2$ in that loop, etc. This total drop must be equal to the total rise of voltage sources around the loop. This method has the additional advantage of yielding the desired equations in their final form, without the need to collect or rearrange terms. After some practice, the student should be able to write the equations directly by inspection of the transform diagram.  □

## *3-3  TRANSFORM DIAGRAMS—NODE EQUATIONS

On a dual basis, we can derive the transform diagrams from the $i$–$v$ relations of the elements. Take the Laplace transform to obtain:

$$I_R(s) = GV_R(s) \qquad (\text{where } G = 1/R) \tag{3-8}$$

$$I_L(s) = \frac{1}{Ls} V_L(s) + \frac{i_L(0_-)}{s} \tag{3-9}$$

and

$$I_C(s) = CsV_C(s) - Cv(0_-) = CsV_C(s) - q(0_-) \tag{3-10}$$

where each equation exhibits the *admittance* of each element, $Y_R(s) = G$, $Y_L(s) = 1/Ls$, $Y_C(s) = Cs$. In addition, there are "initial-condition" sources. The transform diagrams are shown in Fig. 3-4. The student is urged to verify carefully the reference directions and the polarities in the transform diagrams as related to Eqs. (3-8), (3-9), and (3-10). Here, too, $i_L(0_-)$ is a number, positive or negative; so is $q(0_-)$. The previous remarks on their agreement with the chosen references apply here also.

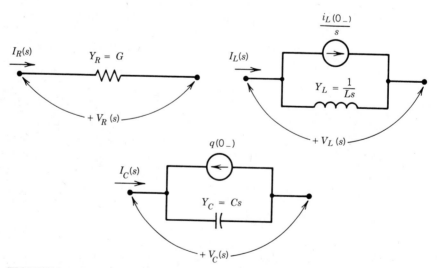

**FIGURE 3-4.** Transform diagrams for node analysis.

## Node Equations

Having decided on node analysis, we proceed with the following steps:

a)  Draw the transform diagram.

b)  Write Kirchhoff's current law at each node.

---

### *Example 4*

Consider the network shown in Fig. 3-5a with its transform diagram in Fig. 3-5b. Because of linearity, we can *use superposition to write the node equations by summing up all the currents leaving each node.*

At the first node (labelled with $V_1$): Let $V_1 \neq 0$ and all other node voltages zero, $V_2 = 0$. This in effect grounds the second node. There are three currents *leaving* this node, namely $\frac{1}{2}V_1$ (through one resistor), $\frac{1}{4}V_1$ (through the other resistor) and

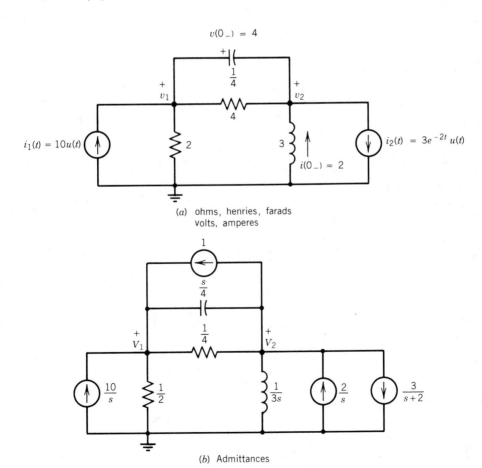

(a)  ohms, henries, farads
volts, amperes

(b)  Admittances

**FIGURE 3-5.** Example 4.

$(s/4) V_1$ through the capacitor. Next consider $V_2 \neq 0$ alone, while $V_1 = 0$: A current given by $[-\frac{1}{4} V_2 - (s/4) V_2]$ *leaves* the first node due to $V_2 \neq 0$. Notice the negative sign! By superposition, then, the *total* current *leaving* the first node is

$$\left(\frac{1}{2} + \frac{1}{4} + \frac{s}{4}\right) V_1 - \left(\frac{1}{4} + \frac{s}{4}\right) V_2$$

and this must be equal to the *total* current *entering* the first node from sources connected to this node, namely $(10/s) + 1$. The first node equation reads therefore

$$\left(\frac{3}{4} + \frac{s}{4}\right) V_1 - \left(\frac{1}{4} + \frac{s}{4}\right) V_2 = \frac{10}{s} + 1$$

At the second node (labelled with $V_2$): Let $V_1 \neq 0$ and $V_2 = 0$. The current that *leaves* this node then is $[-\frac{1}{4} V_1 - (s/4) V_1]$. Next, consider $V_2 \neq 0$ and $V_1 = 0$. The current leaving node two now is $[(1/3s) V_2 + \frac{1}{4} V_2 + (s/4) V_2]$. By superposition, then, the *total* current *leaving* node two is

$$-\left(\frac{1}{4} + \frac{s}{4}\right) V_1 + \left(\frac{1}{3s} + \frac{1}{4} + \frac{s}{4}\right) V_2$$

and must be equal to the *total* current *entering* node 2 from the sources, that is, $(2/s) - [3/(s + 2)] - 1$ (notice the negative sign on two of the sources: in order to enter this node, their references must be reversed, that is, a minus sign must be assigned). The second node equation reads

$$-\left(\frac{1}{4} + \frac{s}{4}\right) V_1 + \left(\frac{1}{3s} + \frac{1}{4} + \frac{s}{4}\right) V_2 = \frac{2}{s} - \frac{3}{s + 2} - 1$$

The two node equations are already in their final form. If desired, they can be rewritten in matrix notation as

$$\begin{bmatrix} \dfrac{3}{4} + \dfrac{s}{4} & -\dfrac{1}{4} - \dfrac{s}{4} \\ -\dfrac{1}{4} - \dfrac{s}{4} & \dfrac{1}{3s} + \dfrac{1}{4} + \dfrac{s}{4} \end{bmatrix} \begin{bmatrix} V_1 \\ V_2 \end{bmatrix} = \begin{bmatrix} \dfrac{10}{s} \\ -\dfrac{3}{s + 2} \end{bmatrix} + \begin{bmatrix} 1 \\ \dfrac{2}{s} - 1 \end{bmatrix}$$

Briefly, then: in the transform diagram consider one node at a time and use superposition. Let $V_1 \neq 0$, $V_2 = V_3 = \cdots = 0$ (ground nodes 2, 3, ...), and find the currents that leave that node due to $V_1$ alone. Then let $V_2 \neq 0$ and $V_1 = V_3 = \cdots = 0$ (ground) and find the currents that leave that node due to $V_2$. Repeat for $V_3 \neq 0$, $V_1 = V_2 = \cdots = 0$. The total of these currents must be equal to the total current sources entering that node. Repeat the entire procedure for every node. The advantages of this method are identical to those listed already: the method provides insight into a basic principle, i.e., superposition, in addition to yielding directly the required equations in their final form.                    □

## 3-4 SUMMARY

### Loop Analysis

1. Replace every resistor with an impedance $R$.

2. Replace every inductor with an impedance $Ls$ in series with an "initial condition" voltage source. This source is of value $Li(0_-)$, and its polarity is such that the given initial current leaves the positive terminal.

3. Replace every capacitor with an impedance $1/Cs$ in series with an "initial condition" voltage source. This source is of value $v_C(0_-)/s$, and its polarity is the given polarity of $v_C(0_-)$.

4. Replace every given excitation (source) by its transform.

5. Draw the transform loop currents and apply superposition to each loop, summing all the drops in the passive elements. On the right-hand side of the equation, in each loop, will appear the total of all sources around that loop (positive when a rise, negative when a drop in the direction of the loop).

6. The resulting equations will have the general matrix form

$$\mathbf{ZI} = \mathbf{E} \tag{3-11}$$

where

$\mathbf{Z}$ is a square matrix of the impedances, the *loop impedance matrix*,

$\mathbf{I}$ is the column matrix (vector) of the unknown transform loop currents,

$\mathbf{E}$ is a column matrix (vector) of all the sources around each loop, both actual sources (accounting for zero-state response) and "initial condition" sources (accounting for zero-input response).

It should be noted that Eq. (3-11) is a generalized form of *Ohm's law*.

### Node Analysis

1. Replace every resistor with an admittance $G = 1/R$.

2. Replace every inductor (without mutual coupling) with an admittance $1/Ls$ in parallel with an "initial condition" current source. This source is of value $i_L(0_-)/s$ and its direction is the given direction of $i_L(0_-)$.

3. Replace every capacitor with an admittance $Cs$ in parallel with an "initial condition" current source. This source is of value $q_C(0_-) = Cv_C(0_-)$, and its direction is such that the current leaves the positive terminal of the given $v_C(0_-)$.

4. Replace every given excitation (source) by its transform.

5. Label the transform node voltages and apply superposition to each node, summing all the currents that leave that node through the passive elements. On the right-hand side of the equation, in each node, will appear all the sources that enter this node (positive when directed to the node, negative when directed away from it).

6. The resulting equations will have the general matrix form

$$\mathbf{YV} = \mathbf{J} \tag{3-12}$$

where

> **Y** is a square matrix of the admittances, the *node admittance matrix*,
>
> **V** is a column matrix (vector) of the unknown transform node voltages,
>
> **J** is a column matrix (vector) of all the sources at each node, zero-state sources and zero-input sources.

Equation (3-12) is a generalized form of *Ohm's law*.

Finally, a word about networks with dependent sources: when these are present (as in Example 2 in this chapter), superposition still holds for the linear network, but we must be careful. The first four steps listed here for the transform analysis are still valid. After that, it is best to write KVL around each loop or KCL at each node individually, not by superposition. Among the final (and perhaps unexpected) results are those dependent sources *not* appearing on the right-hand side, as well as the possible non-symmetry of **Z** ($z_{pq} \neq z_{qp}$) or **Y** ($y_{jk} \neq y_{kj}$).    ***Probs. 3-6***
***through***
***3-14***

## 3-5 RATIONAL FUNCTIONS: POLES AND ZEROS

We are ready now for the final step in the Laplace transform: having calculated an unknown $V(s)$ or $I(s)$, find $\mathcal{L}^{-1}V(s) = v(t)$ or $\mathcal{L}^{-1}I(s) = i(t)$. Let us designate by $F(s)$ either $V(s)$ or $I(s)$ and use this generic notation.

From our experience in the previous chapters we recognize that, after clearing all fractions, $F(s)$ takes on the form

$$F(s) = K \frac{s^m + a_1 s^{m-1} + \cdots + a_{m-1}s + a_m}{s^n + b_1 s^{n-1} + \cdots + b_{n-1}s + b_n} = K \frac{N(s)}{D(s)} \tag{3-13}$$

i.e., a ratio of two polynomials, $N(s)$ and $D(s)$, with integer powers of $s$. Such a ratio is called a *rational function in s*. The coefficients $K$, $a_p$, and $b_q$ are real because they are obtained from only addition, subtraction, multiplication or division of the coefficients in the loop or node equations. The coefficient $K$ is a factor which conveniently makes the coefficients of $s^m$ and $s^n$ unity. We wish to expand $F(s)$ in *partial fractions*, listed in our Laplace transform table, in order to be able to write by inspection $f(t) = \mathcal{L}^{-1}F(s)$. Some ideas of partial fractions are, no doubt, familiar to us from integral calculus. Here we will review and expand them. Prior to doing it, however, we need a few additional terms and definitions.

Let the roots of $N(s) = 0$ be designated by $z_1, z_2, \ldots, z_m$, and the roots of $D(s) = 0$ by $p_1, p_2, \ldots, p_n$.[†] Since all the $a$'s and $b$'s are real, we conclude that all the $z$'s and $p$'s are either *real* or *complex conjugate*. Then we write

$$F(s) = K \frac{(s - z_1)(s - z_2) \cdots (s - z_m)}{(s - p_1)(s - p_2) \cdots (s - p_n)} \tag{3-14}$$

---

[†] If $m$ or $n \geq 3$, there are several computer programs which find these roots.

and we assume that this rational function is in its lowest form, that is, there are no common factors between $N(s)$ and $D(s)$.

The roots of $N(s) = 0$, that is, $s = z_1, \ldots, z_m$, are called the *zeros* of $F(s)$, since

$$F(z_k) = 0 \tag{3-15}$$

The roots of $D(s) = 0$ are the *poles* of $F(s)$, a pole being a value of $s$ for which $F(s)$ approaches infinity

$$\lim_{s \to p_k} F(s) \to \infty \tag{3-16}$$

## Example 5

Find the poles and zeros of

$$F(s) = 3.2 \, \frac{s - 1}{s^2 + 4s + 20}$$

There are two complex conjugate poles: $p_1 = -2 + j4$, $p_2 = -2 - j4 = p_1^*$. There is one zero at $z_1 = 1$. Another zero is at $z_2 = \infty$, since, as $s \to \infty$, $F(s) = 0$ (by l'Hôpital's rule). Notice that factors of $N(s)$ or $D(s)$, $(s - z_k)$ or $(s - p_k)$, account for the *finite* zeros or poles. Any zeros or poles at $s = \infty$ are due only to the difference of powers $(m - n)$. □

## Example 6

The rational function

$$F(s) = \frac{(s + 2)(s - 3)}{(s + 4)}$$

has two zeros, $z_1 = -2$, and $z_2 = 3$, and two poles, $p_1 = -4$, $p_2 = \infty$. □

## Example 7

The rational function

$$F(s) = 1.6 \, \frac{s^2 + s - 4}{s^2 + 2s + 6}$$

has only *finite* poles and zeros. At $s = \infty$ the function has the value of 1.6 (by l'Hôpital's rule). □

An additional distinction for poles and zeros is their multiplicity. If $z_1 \neq z_2 \neq \ldots$, then these zeros are *distinct* (or *simple*). If $z_1 = z_2 = \ldots = z_p \neq z_{p+1}, \ldots$, then $z_1$ is a *multiple* zero of multiplicity $p$. Typically, simple zeros yield distinct factors in $N(s)$,

$$N(s) = (s - z_1)(s - z_2) \ldots \tag{3-17}$$

while multiple zeros yield

$$N(s) = (s - z_1)^p (s - z_{p+1}) \ldots \tag{3-18}$$

The same distinction applies to poles. **Prob. 3-15**

Finally, it is often convenient to show the location of the poles and zeros in the complex $s$-plane. A zero is marked with a "**o**" and a poles with a "**x**". The multiplicity ($> 1$) is placed in parentheses next to the zero or the pole.

---

## Example 8

The pole-zero pattern of

$$F(s) = \frac{(s - 2)^3}{(s^2 + 1)(s + 4)}$$

is shown in Fig. 3-6.

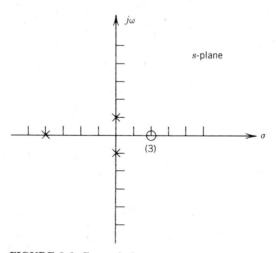

**FIGURE 3-6.** Example 8.

---

## 3-6 PARTIAL FRACTION EXPANSION

Our problem now is: given $F(s)$ as in Eq. (3-13), expand it in partial fractions. The procedure follows several steps in a precise, algorithmic fashion.

First, $F(s)$ will be assumed to be a *proper* rational function, that is $m < n$, the degree of the numerator $N(s)$ is less than that of the denominator $D(s)$. If this is not the case, longhand division must be performed first. The longhand division is carried out until, in the remainder, the (new) numerator is of a lower degree than $D(s)$.

### Example 9

Find $v(t)$ if

$$V(s) = \frac{2s - 1}{s + 4}$$

Here $m = n = 1$ and $V(s)$ is improper. A longhand division yields

$$V(s) = 2 + \frac{-9}{s + 4}$$

and consequently

$$v(t) = 2\delta(t) - 9e^{-4t}u(t)$$

and we completed the problem entirely.   ☐

### Example 10

Let

$$I(s) = \frac{2s^4 - 3s + 1}{s^3 + 2s^2 + 10s}$$

Here, longhand division yields

$$I(s) = 2s - 4 + \frac{-12s^2 + 10s + 1}{s^3 + 2s^2 + 10s}$$

and so, term by term,

$$i(t) = 2\delta'(t) - 4\delta(t) + \mathcal{L}^{-1}I_1(s)$$

where $I_1(s)$ is a *proper* rational function, to be considered next.   ☐

With $F(s)$ a proper rational function, we assume that its poles are simple and multiple. Typically, we write

$$F(s) = K\frac{N(s)}{D(s)} = \frac{KN(s)}{(s - p_1)(s - p_2)\cdots(s - p_k)(s - p_i)^r} \tag{3-19}$$

to exhibit simple poles at $p_1, p_2, \ldots, p_k$, and a pole of multiplicity $r$ at $p_i$. Then the partial fraction of $F(s)$ will be

$$F(s) = \frac{A_1}{s - p_1} + \frac{A_2}{s - p_2} + \cdots + \frac{A_k}{s - p_k} +$$
$$\frac{A_{i,r}}{(s - p_i)^r} + \frac{A_{i,r-1}}{(s - p_i)^{r-1}} + \cdots + \frac{A_{i,1}}{s - p_i} \tag{3-20}$$

We make two observations: (a) if indeed we bring the righthand side of Eq. (3-20) under a common denominator, it will be $D(s)$ as in Eq. (3-19). Therefore, the assumed form of the expansion in Eq. (3-20) is valid; (b) the multiple pole gives rise to the term $A_{i,r}/(s - p_i)^r$ as well as to all the lower powers $(s - p_i)^{r-1}, \ldots, (s - p_i)$. These must be assumed to be present. We wish to find all the coefficients in the numerators, $A_1, A_2, \ldots$. (In passing, let us mention that, for simple poles, these coefficients are called residues.) The key to the several methods for calculating them lies in the fact that Eq. (3-20) is an *identity*, true for any value of $s$ and for any valid mathematical operations performed on it.

For convenience, we discuss separately various cases:

### 1   Simple Poles Only

Here

$$F(s) = \frac{N(s)}{(s - p_1)(s - p_2) \cdots (s - p_n)} = \frac{A_1}{s - p_1} + \frac{A_2}{s - p_1} + \cdots + \frac{A_n}{s - p_n} \tag{3-21}$$

and $p_1 \neq p_2 \neq \cdots \neq p_n$. One method (a universal one) to calculate the residues is to substitute convenient values for $s$ in Eq. (3-21) and thus obtain enough equations for the unknown residues.[†]

---

### Example 11

Let

$$F(s) = \frac{2s^2 + 2s - 6}{(s + 1)(s + 2)(s + 4)} = \frac{A_1}{s + 1} + \frac{A_2}{s + 2} + \frac{A_3}{s + 4}$$

For $s = 0$ we get

$$-\frac{6}{8} = A_1 + \frac{A_2}{2} + \frac{A_3}{4}$$

For $s = 1$ we get

$$-\frac{2}{30} = \frac{A_1}{2} + \frac{A_2}{3} + \frac{A_3}{5}$$

---

[†] See Appendix E for the solution of such equations.

For $s = 2$ we get

$$\frac{6}{72} = \frac{A_1}{3} + \frac{A_2}{4} + \frac{A_3}{6}$$

These three equations yield $A_1 = -2$, $A_2 = 1$, $A_3 = 3$ and consequently

$$f(t) = (-2e^{-t} + e^{-2t} + 3e^{-4t})u(t) \qquad\qquad \square$$

This method is straightforward, if somewhat tedious. It works in *all* cases. The only precaution here is not to use values of $s$ which are poles.

A faster and more elegant method is the following. Multiply the identity (3-21) by $(s - p_1)$:

$$F(s)(s - p_1) = \frac{N(s)}{(s - p_1)(s - p_2) \cdots}(s - p_1)$$

$$= A_1 + \frac{A_2}{s - p_2}(s - p_1) + \cdots + \frac{A_n}{s - p_n}(s - p_1) \qquad (3\text{-}22)$$

The idea is to "free" the residue $A_1$ from its denominator. Notice how $(s - p_1)$ cancels on the lefthand side due to this multiplication. Now let $s = p_1$ in Eq. (3-22). The result is $A_1$:

$$F(s)(s - p_1)]_{s = p_1} = A_1 \qquad (3\text{-}23)$$

since all the other terms on the righthand side vanish when $s = p_1$. A similar approach will yield $A_2, A_3, \ldots, A_n$. In general, then,

$$F(s)(s - p_i)]_{s = p_i} = A_i \qquad (3\text{-}24)$$

In practice, this method is quite fast: we "cover up" the factor $(s - p_i)$ in the denominator of $F(s)$ and evaluate the rest for $s = p_i$.

## Example 12

Let us repeat $F(s)$ from the previous example, using the "cover up" method: To evaluate $A_1$, we calculate

$$A_1 = \frac{2s^2 + 2s - 6}{(s + 2)(s + 4)}\bigg]_{s = -1} = \frac{-6}{3} = -2$$

Similarly,

$$A_2 = \frac{2s^2 + 2s - 6}{(s + 1)(s + 4)}\bigg]_{s = -2} = \frac{-2}{-2} = 1$$

and

$$A_3 = \frac{2s^2 + 2s - 6}{(s + 1)(s + 2)}\bigg]_{s = -4} = \frac{18}{6} = 3$$

as before.

☐ **Probs. 3-17**
**3-18**

## 2   Multiple Poles: The Classical Method

Rather than memorize lengthy formulas, it is best to "derive" each case separately. The amount of effort is reduced considerably, and so is the chance of mistakes. For this reason, the general formulas are given only later, and the method is outlined by means of an example.

### Example 13

$$F(s) = \frac{s}{(s + 1)(s + 3)^2} = \frac{A_1}{(s + 1)} + \frac{A_{2,2}}{(s + 3)^2} + \frac{A_{2,1}}{s + 3}$$

$A_1$ is found as before:

$$A_1 = F(s)(s + 1)]_{s = -1} = -\frac{1}{4}$$

Recall: the expression for $F(s)$ is an identity, and therefore anything valid can be done to it. For example, let us "free" $A_{2,2}$ by multiplying throughout by $(s + 3)^2$. We get

$$\frac{s}{s + 1} = \frac{A_1}{s + 1}(s + 3)^2 + A_{2,2} + A_{2,1}(s + 3)$$

If we now let $s = -3$, we obtain $3/2 = A_{2,2}$, since all the other terms on the right vanish.

In general, then, the coefficient $A_{i,r}$ in Eq. (3-20) is found exactly as in the case of a simple pole,

$$A_{i,r} = F(s)(s - p_i)^r]_{s = p_i} \tag{3-25}$$

What to do about $A_{i,r-1}$, $A_{i,r-2}$, etc.? In our example, how to "free" $A_{2,1}$? If we reconsider the identity obtained after the multiplication by $(s + 3)^2$, we note that we have a term $A_{2,1}(s + 3)$. To "free" $A_{2,1}$, let us differentiate with respect to $s$. This yields

$$\frac{d}{ds}\left(\frac{s}{s + 1}\right) = \frac{d}{ds}\left[A_1\frac{(s + 3)^2}{s + 1}\right] + A_{2,1}$$

and letting $s = -3$ will reduce to zero all the terms on the right, except $A_{2,1}$.

Therefore

$$\frac{d}{ds}\left(\frac{s}{s+1}\right)\Bigg]_{s=-3} = \frac{1}{(s+1)^2}\Bigg]_{s=-3} = \frac{1}{4} = A_{2,1} \qquad \square$$

Thus, in general, $A_{p,r-1}$, $A_{p,r-2}$, etc., can be obtained by successive differentiations, after multiplying Eq. (3-20) by $(s - p_i)^r$. The expressions for these coefficients are summarized as follows: Let

$$F_i(s) = F(s)(s - p_i)^r \tag{3-26}$$

then

$$A_{i,r} = F_i(s)]_{s=p_i}$$

$$A_{i,r-1} = \frac{1}{1!}\frac{dF_i(s)}{ds}\Bigg]_{s=p_i}$$

$$\vdots \tag{3-27}$$

$$A_{i,r-k} = \frac{1}{k!}\frac{d^k F_i(s)}{ds^k}\Bigg]_{s=p_i}$$

Again, the direct substitution into these equations is discouraged. It is easier to work out every case separately, as in the given example.

Finally, the universal method may be used: substitute convenient values of $s$ in Eq. (3-20) to obtain enough linear equations for the $A$'s, then solve these equations.

### 3  Multiple Poles: Expansion via Simple Poles

This method[1] makes use of the partial fraction expansion with simple poles. Let us take the previous example.

***Example 14***

Write $F(s)$ as

$$F(s) = \frac{1}{s+3}\left[\frac{s}{(s+1)(s+3)}\right]$$

factoring out the term that causes the multiple pole, and leaving inside the square brackets a function with simple poles only. Now we expand inside the square brackets:

$$F(s) = \frac{1}{s+3}\left[\frac{-1/2}{s+1} + \frac{3/2}{s+3}\right]$$

and bring back in the term $(s + 3)$:

$$F(s) = \frac{-1/2}{(s + 1)(s + 3)} + \frac{3/2}{(s + 3)^2}$$

The first term contains, again, only simply poles. Expand it to get the final result

$$F(s) = \frac{-1/4}{s + 1} + \frac{1/4}{s + 3} + \frac{3/2}{(s + 3)^2} \qquad \square$$

---

**Example 15**

Expand

$$F(s) = \frac{1}{(s + 1)^3 (s + 2)}$$

$$F(s) = \frac{1}{(s + 1)^2}\left[\frac{1}{(s + 1)(s + 2)}\right] = \frac{1}{(s + 1)^2}\left[\frac{1}{s + 1} + \frac{-1}{s + 2}\right]$$

$$= \frac{1}{s + 1}\left[\frac{1}{(s + 1)^2} + \frac{-1}{(s + 1)(s + 2)}\right]$$

$$= \frac{1}{s + 1}\left[\frac{1}{(s + 1)^2} + \frac{-1}{s + 1} + \frac{1}{s + 2}\right]$$

$$= \frac{1}{(s + 1)^3} - \frac{1}{(s + 1)^2} + \frac{1}{(s + 1)(s + 2)}$$

$$= \frac{1}{(s + 1)^3} - \frac{1}{(s + 1)^2} + \frac{1}{s + 1} - \frac{1}{s + 2}$$

Here, $(s + 1)^2$ was pulled out, then returned into the square brackets, one step at a time.[†]    $\square$    ***Probs. 3-19***
***3-20***

---

## 4    Complex Conjugate Poles

In this case it is always advisable to complete the square in the denominator.

---

**Example 16**

$$F(s) = \frac{s - 2}{s^2 + 2s + 10}$$

[†] See Appendix E.

The denominator is $(s + 1)^2 + 9$, so we write

$$F(s) = \frac{s + 1 - 3}{(s + 1)^2 + 3^2} = \frac{s + 1}{(s + 1)^2 + 3^2} + \frac{-3}{(s + 1)^2 + 3^2}$$

and, using the time shifting property, we write immediately

$$f(t) = (e^{-t}\cos 3t - e^{-t}\sin 3t)u(t)$$

*Note:* The "cover up" method, for simple poles only, could be used here also. However, this would involve manipulations with complex numbers—a tedious labor, prone to errors.   □

---

**Example 17**

$$F(s) = \frac{3s + 1}{(s^2 + 2s + 10)(s + 4)}$$

Here we write

$$F(s) = \frac{K_1 s + K_2}{s^2 + 2s + 10} + \frac{A}{s + 4}$$

The residue $A$ is calculated by the "cover up" method, $A = -11/18$. To calculate $K_1$ and $K_2$ it is best to use the universal method with, say, $s = 0$ and $s = 1$ as two convenient values. The final result is

$$F(s) = \frac{0.611s + 1.778}{(s + 1)^2 + 9} + \frac{-0.611}{s + 4}$$

In order to invert, rewrite the first term as a sum:

$$F(s) = \frac{0.611(s + 1)}{(s + 1)^2 + 9} + \frac{1.778 - 0.611}{(s + 1)^2 + 9} + \frac{-0.611}{s + 4}$$

The idea is to create (legally!) the term $(s + 1)$ in the numerator, adding and subtracting 0.611 for this purpose. Finally, then,

$$f(t) = (0.611e^{-t}\cos 3t + 0.389e^{-t}\sin 3t - 0.611e^{-4t})u(t)$$

This is a relatively shorter approach by comparison to the "cover up" method which would involve here the manipulation of complex numbers.   □   **Prob. 3-21**

---

There are other useful relationships in partial fraction expansions. One of these might be mentioned specifically: in Eq. (3-19) and its expansion, Eq. (3-20), the residues obey the following relations

$$\sum \text{all the residues} = 0 \qquad \text{if } n > m + 1 \qquad (3\text{-}28)$$

$$\sum \text{all the residues} = K \qquad \text{if } n = m + 1 \qquad (3\text{-}29)$$

(Recall: a residue is the multiplier of $1/(s - p_i)$ *only*, whether $p_i$ is a simple or a multiple pole.) These relationships can serve as a final check on the partial fraction expansion, or may help to determine one residue if we know all the other residues.

Several extensive tables of Laplace transforms exist. We list in the references two[2,3] such accessible books.

## *5 The Initial Value and Final Value Theorems

Two very useful theorems relate numerical values of $f(t)$ and $F(s)$. The *initial value theorem* states that

$$f(0_+) = \lim_{s \to \infty} sF(s) \qquad (3\text{-}30)$$

where $f(0_+)$ is the value of $f(t)$ at $t = 0_+$. The proof of this theorem takes into consideration two cases: (a) $f(t)$ is continuous at $t = 0$, and, (b) $f(t)$ has a step discontinuity at $t = 0$.

**Case (a)** Here $f(0_-) = f(0_+)$. Write the defining integral of the Laplace transform of $df/dt$:

$$\int_{0_-}^{\infty} f'(t)e^{-st}dt = sF(s) - f(0_-) \qquad (3\text{-}31)$$

Now take the limit as $s \to \infty$ in Eq. (3-31): on the lefthand side the integrand itself becomes zero, and therefore

$$0 = \lim_{s \to \infty} sF(s) - f(0_-) \qquad (3\text{-}32)$$

Consequently

$$f(0_-) = f(0_+) = \lim_{s \to \infty} sF(s)$$

**Case (b)** Here $f(0_-) \neq f(0_+)$ and the derivative $f'(t)$ will contain an impulse of strength $[f(0_+) - f(0_-)]$ at $t = 0$. Specifically, we write $f(t)$ as

$$f(t) = f_1(t) + [f(0_+) - f(0_-)]u(t) \qquad (3\text{-}33)$$

a sum of $f_1(t)$, a continuous function at $t = 0$, plus a step function discontinuity. The derivative $f'(t)$ is therefore

$$f'(t) = f_1'(t) + [f(0_+) - f(0_-)]\delta(t) \qquad (3\text{-}34)$$

From Eq. (3-32) we know that

$$f_1(0_-) = f_1(0_+) = \lim_{s \to \infty} sF_1(s) \qquad (3\text{-}35)$$

On the other hand, the Laplace transform of Eq. (3-34) yields

$$sF(s) - f(0_-) = sF_1(s) - f_1(0_-) + f(0_+) - f(0_-) \qquad (3\text{-}36)$$

or

$$sF(s) = sF_1(s) + f(0_+) - f(0_-) \qquad (3\text{-}37)$$

Take the limit as $s \to \infty$ in Eq. (3-37) and use Eq. (3-35) to obtain

$$f(0_+) = \lim_{s \to \infty} sF(s) \qquad (3\text{-}38)$$

---

### Example 18

Given

$$V(s) = \frac{1}{s}$$

then

$$\lim_{s \to \infty} sF(s) = 1 = v(0_+)$$

which is correct since $v(t) = u(t)$. □

---

### Example 19

Given

$$I(s) = \frac{1}{s(s + 1)}$$

what is $i(0_+)$? We write immediately

$$i(0_+) = \lim_{s \to \infty} sI(s) = \lim_{s \to \infty} \frac{1}{s + 1} = 0$$

As a check $i(t) = (1 - e^{-t})u(t)$ and therefore $i(0_+) = 0$. □

---

### Example 20

$$F(s) = 1$$

$$\therefore \lim_{s \to \infty} sF(s) = \infty$$

but the value of $f(t) = \delta(t)$ at $t = 0_+$ is zero. The initial value theorem is not applicable here, because $f(t)$ has a "sharper" discontinuity than a step discontinuity.

□

The initial value theorem can be used repeatedly to evaluate $f(0_+)$, $f'(0_+)$, $f''(0_+)$, . . .

**Example 21**

The charge in a certain network was computed as

$$Q(s) = \frac{2s^2 + 3}{s(s^2 + 2s + 10)}$$

Its initial value is

$$q(0_+) = \lim_{s \to \infty} sQ(s) = \lim_{s \to \infty} \frac{2s^2 + 3}{s^2 + 2s + 10} = 2 \text{ coulombs}$$

For the first derivative we write

$$\mathcal{L}q'(t) = sQ(s) - q(0_+) = \frac{-6s - 20}{s^2 + 2s + 10}$$

and therefore

$$i(0_+) = q'(0_+) = \lim_{s \to \infty} s\left(\frac{-6s - 20}{s^2 + 2s + 10}\right) = -6 \text{ amps}$$

The student is urged to verify these results by actually finding $q(t)$, $i(t) = dq/dt$ and evaluating them at $t = 0_+$.   □   **Prob. 3-22**

The *final value theorem* states

$$\lim_{t \to \infty} f(t) = \lim_{s \to 0} sF(s) \tag{3-39}$$

provided $sF(s)$ has no poles on the $j\omega$ axis or inside the right half of the $s$-plane. This restriction essentially guarantees the existence of the limit.

The formal proof of this theorem starts, again, with the integral defining the Laplace transform of $f'(t)$. We take the limit of it, as $s \to 0$

$$\lim_{s \to 0} \int_{0_-}^{\infty} f'(t)e^{-st} \, dt = \lim_{s \to 0} sF(s) - f(0_-) \tag{3-40}$$

The lefthand side of this equation yields

$$\int_{0_-}^{\infty} f'(t)\, dt = \lim_{t \to \infty} f(t) - f(0_-) \tag{3-41}$$

Equating the righthand sides of Eq. (3-40) and (3-41) gives the desired result, Eq. (3-39).

***Example 22***

$$F(s) = \frac{1}{s + a}$$

$$\therefore \lim_{t \to \infty} f(t) = \lim_{s \to 0} \frac{s}{s + a} = 0,$$

and, indeed

$$\lim_{t \to \infty} f(t) = \lim_{t \to \infty} e^{-at} = 0. \qquad \square$$

***Example 23***

$$F(s) = \frac{\omega}{s^2 + \omega^2}.$$

So, it would seem that

$$\lim_{t \to \infty} f(t) = \lim_{s \to 0} s \frac{\omega}{s^2 + \omega^2} = 0?$$

But the function $sF(s)$ has two poles on the $j\omega$ axis, so we cannot apply the final value theorem here. Indeed, here $f(t) = \sin \omega t$, and its limit for $t \to \infty$ is undefined.

$\square$   ***Probs. 3-23***
***3-24***

In summary: the initial and final value theorems provide a useful way to determine from $F(s)$ the values of $f(t)$ at $t = 0_+$ and $t \to \infty$, without actually calculating $f(t) = \mathcal{L}^{-1}F(s)$. It is interesting to note that $f(t)$ at the origin ($t = 0_+$) behaves as $sF(s)$ at infinity ($s \to \infty$); and dually, $f(t)$ behaves at infinity ($t \to \infty$) as $sF(s)$ at the origin ($s = 0$).

## PROBLEMS

3-1 For the network shown:

   a) Determine its order $n$.

   b) Write the loop equations in the time-domain.

   c) Laplace transform part $(b)$.

   d) Solve algebraically for the unknowns, separating the answers into the zero-state and zero-input parts. Note the dependent source.

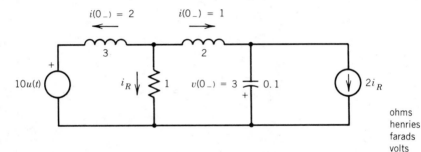

**PROBLEM 3-1.**

3-2 Repeat Prob. 3-1 for the network shown, with zero initial state.

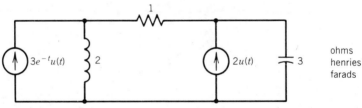

**PROBLEM 3-2.**

3-3 Complete Example 2 in this chapter:

   a) Find the characteristic equation and the characteristic values.

   b) Solve for $V_1(s)$ and $V_2(s)$, showing clearly the zero-state and zero-input parts.

3-4 For the network shown in Problem 3-1:

   a) Write the node equations.

   b) Laplace transform the node equations.

   c) Solve algebraically for the unknowns, separating into the zero-state and the zero-input parts.

   d) Obtain the characteristic equation and the characteristic values using the loop analysis formulation, and using the node analysis. (The equation and the values must be the same, of course!)

3-5 Repeat Problem 3-4 for the network in Problem 3-2.

3-6 Check carefully the dimension (units) of the various terms in Eqs. (3-5), (3-7), (3-9), and (3-12). As a start, check the dimensions of $I(s)$ and $V(s)$, using the defining integral of the Laplace transform.

3-7 Derive the transform diagram of a pair of mutually coupled inductors ($L_1$, $L_2$, and $M$) for loop analysis, as shown.

   a) Write the $v$–$i$ relationships.

   b) Transform these equations (do not forget initial conditions!)

   c) Draw the transform diagram, with impedances and "initial condition" sources, to fit these equations. Obtain two possible transform diagrams, one uncoupled and the other in the form of a "$T$", as shown

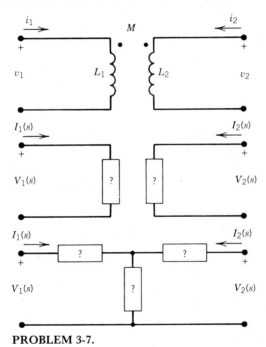

**PROBLEM 3-7.**

3-8 For the network shown, draw the transform diagram and, from it, write the loop equations in matrix form (element values are in ohms, henries, and farads, initial conditions in amperes and volts). Note the dependent source!

3-9 Repeat Problem 3-8 for the network shown. See Problem 3-7.

3-10 For the network shown, draw the transform diagram and, from it, write the node equations in matrix form.

3-11 For the circuit shown, calculate the voltage gain $V_{out}(s)/V_{in}(s)$, with all initial conditions zero. *Hint*: write two KCL equations, one each at the inverting ($-$) terminals of the op amps.

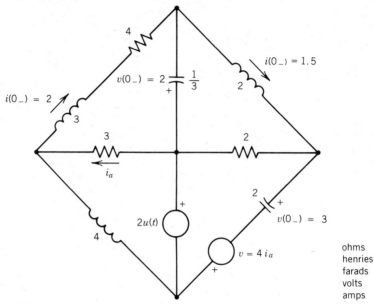

$4$

$v(0_-) = 2 \;=\!\!\!\!= \dfrac{1}{3}$

$i(0_-) = 1.5$

$i(0_-) = 2$

$3$

$2$

$3$

$i_a$

$2$

$2u(t)$

$2$

$v(0_-) = 3$

$4$

$v = 4\,i_a$

ohms
henries
farads
volts
amps

**PROBLEM 3-8.**

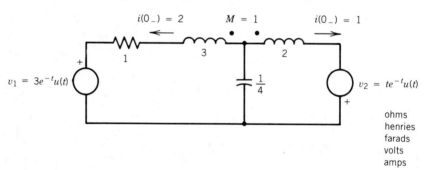

$i(0_-) = 2$      $M = 1$      $i(0_-) = 1$

$1$          $3$          $2$

$v_1 = 3e^{-t}u(t)$

$\dfrac{1}{4}$

$v_2 = te^{-t}u(t)$

ohms
henries
farads
volts
amps

**PROBLEM 3-9.**

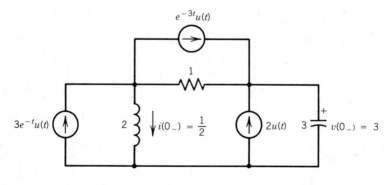

$e^{-3t}u(t)$

$1$

$3e^{-t}u(t)$      $2$      $i(0_-) = \dfrac{1}{2}$      $2u(t)$      $3$      $v(0_-) = 3$

ohms
henries
farads
volts
amps

**PROBLEM 3-10.**

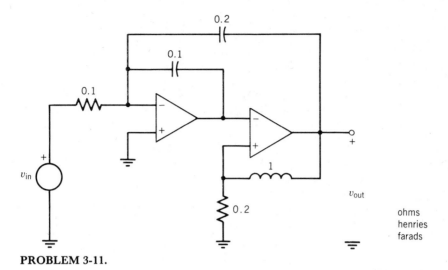

**PROBLEM 3-11.**

3-12 The circuit shown is a model of a field-effect transistor (FET) amplifier. Find the voltage gain $V_o(s)/V_{in}(s)$. Use the following typical values:

$$r_{in} = 10k\Omega \quad R_0 = 15k\Omega \quad C_1 = 50pF \quad C_2 = 10pF \quad g = 0.004$$

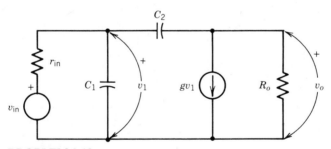

**PROBLEM 3-12.**

3-13 Draw the transform diagram, then write the loop equations in matrix form for the network shown. Notice the dependent source.

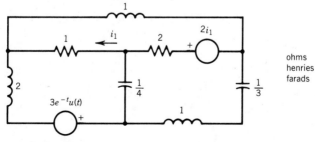

**PROBLEM 3-13.**

**3-14** In the figure is shown a linear equivalent circuit of a transistor amplifier. Use transform diagram techniques to find the *sinusoidal current gain*, defined by

$$\alpha(s)\Big]_{s=j\omega} = \frac{I_{out}(s)}{I_{in}(s)}\Big]_{s=j\omega}$$

Here "*b*," "*c*," "*e*" denote the base, collector, and emitter, respectively. All other parameters are given.

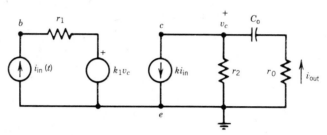

**PROBLEM 3-14.**

**3-15** For each rational function: (a) Arrange in the form of Eq. (3-13), that is $KN(s)/D(s)$; (b) Plot the pole-zero pattern.

a) $V(s) = \dfrac{2s - 3}{2s^2 + s + 4}$

b) $I(s) = \dfrac{3s^2 + 4s - 10}{10s^2 + 4s + 3}$

c) $V(s) = \dfrac{s^4 - 2s^2 - 35}{2s(s^2 + 7s + 10)}$

d) $I(s) = \dfrac{-3 + 4s}{2s^3 + 8s^2 + 10s}$

e) $V(s) = \dfrac{2.4}{s(s^2 + 6s + 6)}$

**3-16** For each of the following rational functions:

a) Calculate the poles and zeros (including any at $s = \infty$).

b) Write the function in its factored form, as in Eq. (3-14).

c) On the *s*-plane, mark for each function its poles and zeros.

1. $V(s) = \dfrac{3s^3 - 2s + 10}{(s^2 + 20s + 36)(s + 1)}$

2. $I(s) = \dfrac{4s - 1}{s^4 + 3s^3 + 12s^2 + 20s + 10}$

3. $V(s) = \dfrac{s^2 + 1}{(s + 3)(s + 10)(s^2 + 4s + 20)}$

4. $I(s) = \dfrac{2s^3 - 10}{s^3 + 8s^2 + 7s}$

5. $V(s) = \dfrac{s - 1}{s^4 + 13s^2 + 36}$

6. $I(s) = \dfrac{1 - 2s}{(s^3 + 3s^2 + 3s + 1)(s + 2)}$

7. $V(s) = \dfrac{3s + 10}{(s + 1)^2(s + 6)}$

8. $I(s) = \dfrac{12s^2 - 5s + 4}{(s + 2)(s^3 + 10s^2 + 20s + 40)}$

9. $V(s) = \dfrac{4 - 3s}{(s + 2)^3(s + 4)^2}$

10. $I(s) = \dfrac{s^2 + 10s - 25}{(s + 1)(s^2 + 10)}$

11. $V(s) = \dfrac{2s - 2}{s^3 + 7s^2 + 33s - 41}$

12. $I(s) = \dfrac{3s^3 + 18s^2 + 118s - 79}{s^4 + 12s^3 + 113s^2 + 484s + 1640}$

13. $V(s) = \dfrac{s^3 + 4s^2 - 80s}{s^4 + 2s^3 + 110s^2 + 200s + 1000}$

3-17 Invert each function, using partial fractions:

a) $I(s) = 3.25 \dfrac{s - 1}{s(s + 4)(s + 10)}$

b) $V(s) = \dfrac{1}{(s + 1)(s + 3)(s + 6)}$

c) $I(s) = 2.4 \dfrac{-s + 10}{(s + 2)(s + 4)}$

d) $V(s) = \dfrac{s^2 - s + 7}{(s^2 + 4s + 3)(s + 8)}$

3-18 a) Prove that a residue (for simple poles only!) can be also calculated as

$$A_i = \left. \frac{N(s)}{D'(s)} \right]_{s = p_i}$$

instead of Eq. (3-24). Hence $D'(s) = dD(s)/ds$. *Hint:* Write $D(s) = (s - p_i)D_a(s)$, as a partial factorization of $D(s)$.

b) Repeat Problem 3-17, using this method.

3-19   Invert each function

a)   $F_1(s) = \dfrac{-3}{s^2(s + 1)}$

b)   $F_2(s) = \dfrac{s^3 + 6s^2 + 8s + 2}{(s + 1)^2(s + 2)}$

3-20   Invert the functions 6 and 9 in Problem 3-16.

3-21   Invert the functions 1, 3, 5, and 10 in Problem 3-16.

3-22   Apply the initial value theorem to each function in Problems 3-17 and 3-19 and compare with the calculated value at $t = 0_+$ in that problem.

3-23   Apply the final value theorem to each function in Problem 3-16. Be sure to check first whether the final value theorem *can* be applied. Compare with the calculated value of the time function as $t \to \infty$.

3-24   An output voltage in a certain network is found as

$$V(s) = \frac{2s^3 + 10s^2 + s + 4}{s(s + 1)^2[(s + 2)^2 + 16]}$$

Without ever finding $v(t)$, write

a)   The waveform of $v(t)$, within arbitrary multipliers for amplitudes.

b)   The initial value of $v(t)$.

c)   The initial value $v'(0_+)$.

d)   The final value of $v(t)$.

## REFERENCES

1.  S. Karni and D. Etter, "An Algebraic-Recursive Algorithm for Partial Fraction Expansion with Multiple Poles," *IEEE Trans. on Education*, vol. 22, No. 1, pp. 25–27, February 1979. (See also appendix E in this book).
2.  M. Abramowitz and I. A. Stegun, *Handbook of Mathematical Functions*. Washington, D.C.: National Bureau of Standards no. 55, 1964, chapter 29. (Also published by Dover Publications).
3.  G. Doetsch, *Guide to the Applications of the Laplace and Z-Transforms*. New York: Van Nostrand Reinhold Company, 1971.

# 4

# EXTERNAL CHARACTERIZATION OF NETWORKS

As mentioned earlier, network characterization deals with the mathematical representation of input–output relationships. Here we shall discuss various ways of such representations when the network is given in its "black box" form, where only several terminals are accessible and are of interest.

## 4-1 N-PORT(N-P) AND N-TERMINAL (N-T) NETWORKS

A *port* in a network is a pair of terminals (among the many terminals that the network might have) for which the current entering one terminal is equal to the current leaving the other terminal. A port, labelled $k$, is shown in Fig. 4-1a, and we have for this port

$$I_k(s) = -I_{k'}(s) \tag{4-1}$$

This current equation is true for a port, because no matter what is connected to that port—a source or another passive element—the connection is in series; therefore the same current flows in these two terminals.

A 2-port network (2-P network) is shown in Fig. 4-1b where, quite often, port "1" (terminals 1–1') is the input port and port "2" (terminals 2–2') is the output port. Notice the reference direction adopted for the currents and the voltages.

By contrast, Eq. (4-1) need not be valid in an *N-terminal* network. Here, the terminals are not "paired off", and about all that can be said of all the terminal currents is that they must satisfy Kirchhoff's current law. For example, in the 4-terminal network (4-T network) shown in Fig. 4-1c, we have

$$I_1(s) + I_2(s) + I_3(s) + I_4(s) = 0 \tag{4-2}$$

and, without any further information, we cannot say, for example, that $I_3(s) = -I_4(s)$.

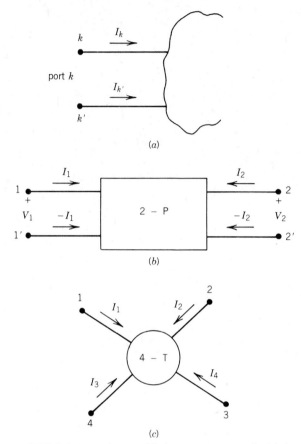

**FIGURE 4-1.** (*a*) A port; (*b*) A two-port network; (*c*) A four-terminal network.

   With either an N-P network or an N-T network, the representation by a "black box" means that we are interested only in the *external behavior* and *external characterization* at the *accessible* ports or terminals. Inside the "black box," the network may be very complex, but is of no immediate interest.

## 4-2   THE ONE-PORT (1-P) NETWORK

A one-port network is shown in Fig. 4.2. The accessible port variables are $V(s)$ and $I(s)$, while the remainder of the network (containing no independent sources) is enclosed in the proverbial "black box". Terminals 1–1′ are called the *driving-point* (*dp*) terminals. If the input is a current source $i(t)$, then the zero-state response is $v(t)$, and we have

$$V(s) = Z_{dp}(s)I(s) \tag{4-3}$$

where $Z_{dp}(s)$ is the *driving-point impedance* seen by the source at 1–1′. Simple examples

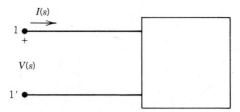

**FIGURE 4-2.** A one-port network.

of driving-point impedances for one ports are

$$Z_{dp}(s) = R \quad Z_{dp}(s) = Ls \quad Z_{dp}(s) = \frac{1}{Cs} \tag{4-4}$$

for, respectively, a one-port resistor, inductor, and capacitor. These were used in Chapter 3 for loop analysis.

Alternately, if the input is a voltage source $v(t)$, then the zero-state response is $i(t)$ and

$$I(s) = Y_{dp}(s)V(s) \tag{4-5}$$

where $Y_{dp}(s)$ is the *driving-point admittance* of the one-port network. For the single elements listed previously

$$Y_{dp}(s) = G = 1/R \quad Y_{dp}(s) = \frac{1}{Ls} \quad Y_{dp}(s) = Cs \tag{4-6}$$

and, in general, for the one-port

$$Y_{dp}(s) = Z_{dp}^{-1}(s) \tag{4-7}$$

An important observation is made here: Eqs. (4-3) and (4-5) are special cases of the general form

$$R(s) = H(s)E(s) \tag{4-8}$$

which relates the zero-state response, $R(s)$, to the excitation (input) $E(s)$ through $H(s)$, an appropriate *network function*, characteristic of the network. We will return to this relationship in greater detail in later chapters.

Some familiar methods of calculating driving-point functions are discussed next.

## Series Connection

One-port networks are connected in series if the same current flows through each one. See Fig. 4-3. It is easy to calculate the equivalent total driving-point impedance for n

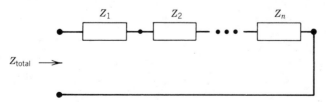

**FIGURE 4-3.** Series connection of one-port networks.

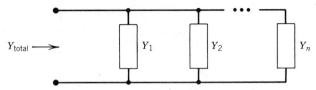

**FIGURE 4-4.** Parallel connection of one-port networks.

impedances connected in series

$$Z_{total}(s) = \sum_{k=1}^{n} Z_k(s) \tag{4-9}$$

## Parallel Connection

One-port networks are connected in parallel if the same voltage appears across each one. See Fig. 4-4. For such a connection, the equivalent total admittance is given by

$$Y_{total}(s) = \sum_{k=1}^{n} Y_k(s) \tag{4-10}$$

### Example 1

Find the driving-point admittance of the network shown in Fig. 4-5. The $RC$ branch is a series connection, therefore

$$Z_{RC} = \frac{10}{s} + 2 = \frac{2s + 10}{s}$$

The parallel connection with the inductor yields

$$Y_{RCL} = \frac{s}{2s + 10} + \frac{2}{s} = \frac{s^2 + 4s + 20}{2s^2 + 10s}$$

The single resistor is in series, so

$$Z_{total} = 3 + \frac{2s^2 + 10s}{s^2 + 4s + 20} = \frac{5s^2 + 22s + 60}{s^2 + 4s + 20}$$

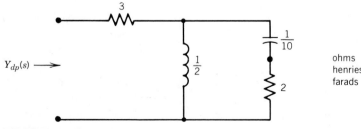

ohms
henries
farads

**FIGURE 4-5.** Example 1.

and

$$\frac{1}{Z_{total}} = Y_{total} = \frac{s^2 + 4s + 20}{5s^2 + 22s + 60}$$

All these steps can be summarized by inspection as follows

$$Y_{total} = \cfrac{1}{3 + \cfrac{1}{\cfrac{2}{s} + \cfrac{1}{\cfrac{10}{s} + 2}}}$$

This form is called a *continued fraction,* and upon clearing the denominators it is reduced to the final form of a rational function.   ☐ *Probs.* **4-1**
                                                                                                     *through*
                                                                                                     **4-5**

## Ladder Networks

Another important structure, quite common and useful in analysis and design, is the *ladder* network, Fig. 4-6. Let us designate the series arms of the ladder by their impedances. $Z_1, Z_3, \ldots$ and the shunt arms by their admittances, $Y_2, Y_4, \ldots$ Then we can write the total driving-point impedance at $1-1'$ as the continued fraction :

$$Z_{dp}(s) = Z_1 + \cfrac{1}{Y_2 + \cfrac{1}{Z_3 + \ddots}} \tag{4-11}$$

In the general case, with non-series-parallel structures or when dependent sources are present, there are no short cuts. The driving-point function may be found by applying the appropriate input ($I$ or $V$) at the port, and calculating the zero-state output ($V$ or $I$).

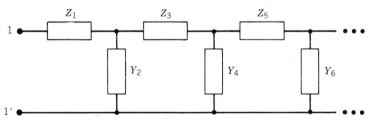

**FIGURE 4-6.** A ladder network.

## Example 2

Find $Y_{dp}(s)$ for the "bridged-T" 1-P shown in Fig. 4-7. We apply $V(s)$, in accordance with Eqs. (4-5) and (4-8). This is shown in dotted lines. Loop analysis yields the

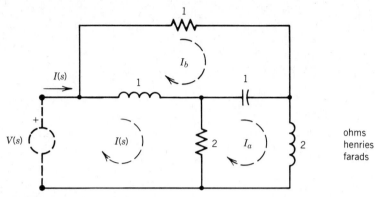

**FIGURE 4-7.** Example 2.

equation $\mathbf{ZI} = \mathbf{E}$ (as in the previous chapter), that is,

$$
\begin{bmatrix}
s + 2 & -2 & -s \\
-2 & 2s + 2 + \dfrac{1}{s} & -\dfrac{1}{s} \\
-s & -\dfrac{1}{s} & s + 1 + \dfrac{1}{s}
\end{bmatrix}
\begin{bmatrix}
I \\
I_a \\
I_b
\end{bmatrix}
=
\begin{bmatrix}
V \\
0 \\
0
\end{bmatrix}
$$

Solving for $I(s)$ only (recall: $I_a$ and $I_b$ are inaccessible!) we get

$$
I(s) = \frac{2s^3 + 4s^2 + 5s + 3}{6s^3 + 8s^2 + 5s + 2} V(s)
$$

exhibiting therefore

$$
Y_{dp}(s) = \frac{2s^3 + 4s^2 + 5s + 3}{6s^3 + 8s^2 + 5s + 2} \qquad \square
$$

## Example 3

Calculate the driving-point impedance of the 1-P shown. Note the dependent source! The input current $I$ is the controller for the dependent source. KCL at the top node

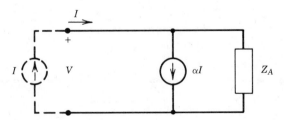

**FIGURE 4-8.** Example 3.

yields the current through $Z_A$ as $(1 - \alpha)I$. Therefore

$$V = Z_A(1 - \alpha)I$$

and

$$Z_{dp} = Z_A(1 - \alpha)$$

It is interesting to note that, for $\alpha = 2$, we get $Z_{dp} = -Z_A$, i.e., a *negative impedance converter*, changing $Z_A$ to $-Z_A$. ☐ **Prob. 4-6**

---

At this point, one comment is in order: it is not crucial whether we use $i(t)$ or $v(t)$ as the source for calculating *dp* functions. The reason is that $Z_{dp}^{-1}(s) = Y_{dp}(s)$. In later discussions, the general input–output relations, Eq. (4-8), will dictate precisely and uniquely the nature of the input ($i$ or $v$) and of the output ($i$ or $v$). Then we will not have a choice. It is good practice, therefore, to use the same accurate distinction in calculating driving-point functions.

## 4-3 THE 2-P NETWORK AND ITS OPEN-CIRCUIT PARAMETERS

The next and most common N-port is a 2-port, as shown in Fig. 4-9. Since we have no access to the inside of the "black box", we assume that it contains *no independent sources* (except at the ports); this means also no "initial condition" sources. Let us excite the 2-P with two current sources; the zero state responses will be the two port voltages. As usual, we write $I = I(s)$, $V = V(s)$ for brevity.

Since the 2-P is linear, superposition can be used, and we say that $V_1$ is caused in part by $I_1$ alone, and in part by $I_2$ alone. Therefore, by superposition,

$$V_1 = z_{11}I_1 + z_{12}I_2 \tag{4-12a}$$

where $z_{11}$ and $z_{12}$ are appropriate multipliers of $I_1$ and $I_2$; they are necessarily impedances (in ohms).

Similarly, $V_2$ is the superposition of the partial voltages caused by $I_1$ and $I_2$

$$V_2 = z_{21}I_1 + z_{22}I_2 \tag{4-12b}$$

Although Eqs. (4-12a,b) resemble the loop equations discussed in Chapter 3, there is a profound difference between them. Equations (4-12) describe voltage-current relationships *only* at the accessible ports, while inside the 2-port network there

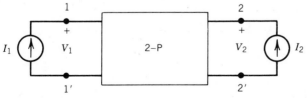

**FIGURE 4-9.** A 2-port (2-P) network.

may be many more loops. The following examples will illustrate this and other related points.

## Example 4

Derive the 2-P description, Eqs. (4-12), for the network shown in Fig. 4-10. To simplify the algebra, we are using an all-resistive network here. The two ports are identified with their voltages and their currents. We write (by superposition) the three loop equations

$$3I_1 + 0I_2 + 1I_3 = V_1$$

$$0I_1 + 5I_2 - 1I_3 = V_2$$

$$1I_1 - 1I_2 + 5I_3 = 0$$

The current $I_3$ is not a port current and is an "undesired" variable; we eliminate it using the third equation

$$I_3 = \frac{1}{5}(I_2 - I_1)$$

and substitute into the first two equations. The result is

$$\frac{14}{5}I_1 + \frac{1}{5}I_2 = V_1$$

$$\frac{1}{5}I_1 + \frac{24}{5}I_2 = V_2$$

These are the two port equations, Eqs. (4-12), as required.

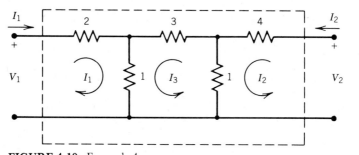

**FIGURE 4-10.** Example 4.

## Example 5

Derive the impedance parameters, Eqs. (4-12), for the 2-P shown in Fig. 4-11. Notice the dependent source: it is allowed. Only independent sources are not allowed inside

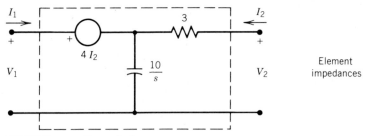

**FIGURE 4-11.** Example 5.

the 2-P. We write by inspection the two KVL equations, for each port:

$$- V_1 + 4I_2 + \frac{10}{s} (I_1 + I_2) = 0$$

$$- V_2 + 3I_2 + \frac{10}{s} (I_1 + I_2) = 0$$

A simple rearrangement of these equations yields the final result

$$\frac{10}{s} I_1 + \left( \frac{10}{s} + 4 \right) I_2 = V_1$$

$$\frac{10}{s} I_1 + \left( \frac{10}{s} + 3 \right) I_2 = V_2$$

and the parameters

$$z_{11} = 10/s = z_{21}; z_{12} = 10/s + 4; z_{22} = 10/s + 3 \qquad \square$$

The approach used in these examples is quite general, and does not require any memorization. It can be stated in two steps:

1. *Write valid equations (KVL and/or KCL) for the 2-P.*
2. *By elimination of undesired variables and by algebraic rearrangement, get the port equations in the desired form.*

We will elaborate on this approach (called here the *informal approach*) throughout this chapter. Indeed, it is reassuring to know that that's all! The method is general and applicable to *all* linear two-port networks.

A more formal approach to these parameters is as follows: consider Eqs. (4-12) again

$$V_1 = z_{11}I_1 + z_{12}I_2 \qquad (4\text{-}12\text{a})$$

$$V_2 = z_{21}I_1 + z_{22}I_2 \qquad (4\text{-}12\text{b})$$

If we set $I_1 = 0$ we get

$$z_{12} = \left. \frac{V_1}{I_2} \right]_{I_1 = 0} \qquad (4\text{-}13\text{a})$$

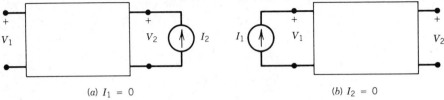

(a) $I_1 = 0$                    (b) $I_2 = 0$

**FIGURE 4-12.** A 2-P network with single inputs.

and

$$z_{22} = \frac{V_2}{I_2}\bigg]_{I_1 = 0} \tag{4-13b}$$

Next, set $I_2 = 0$ to get

$$z_{11} = \frac{V_1}{I_1}\bigg]_{I_2 = 0} \tag{4-13c}$$

and

$$z_{21} = \frac{V_2}{I_1}\bigg]_{I_2 = 0} \tag{4-13d}$$

These two experiments are shown in Fig. 4-12. What is happening here is, again, superposition: in Fig. 4-12a, only $I_2$ is acting and the partial responses $V_1$ and $V_2$ are then related to $z_{12}$ and $z_{22}$ by Eqs. (4-13a) and (4-13b). Then, with $I_1$ alone, Fig. 4-12b shows the partial responses $V_1$ and $V_2$, and their relations to $z_{11}$ and $z_{21}$ by Eqs. (4-13c) and (4-13d).

These equations indicate the nature of the $z$'s. Each one is measured, or computed, under *open-circuit conditions* (either port 1 or port 2 is open-circuited.) Therefore, we call them the *open-circuit impedance parameters*. To emphasize this, Eq. (4-12) will be rewritten in matrix form as

$$\mathbf{V} = \mathbf{Z}_{oc}\mathbf{I} \tag{4-14}$$

where the subscript oc means "open-circuit". The two parameters $z_{11}$ and $z_{22}$ are called *driving-point impedances* because each relates a voltage to a current at the same port; the other two, $z_{12}$ and $z_{21}$, are *transfer impedances*, since each relates a voltage output at one port to a current input at another port. We note that Eq. (4-14) is a matrix version of Eq. (4-8), relating a zero-state output ($\mathbf{V}$ here) to an input ($\mathbf{I}$) via an appropriate network function ($\mathbf{Z}_{oc}$). Such identifications (what input, what output, and the correct network functions) will be essential in our discussions now.

---

### Example 6

Let us rework Example 5 via the formal approach. The two separate experiments are shown: For Fig. 4-13a we write

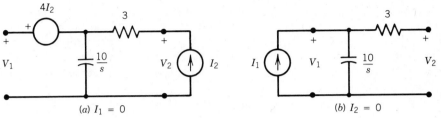

(a) $I_1 = 0$                                      (b) $I_2 = 0$

Element impedances

**FIGURE 4-13.**  Example 6.

$$V_1 = 4I_2 + \frac{10}{s}I_2 = \left(4 + \frac{10}{s}\right)I_2 \qquad \therefore z_{12} = 4 + \frac{10}{s}$$

and

$$V_2 = \left(3 + \frac{10}{s}\right)I_2 \qquad \therefore z_{22} = 3 + \frac{10}{s}$$

In Fig. 4-13$b$ notice that the dependent source $4I_2$ is zero since $I_2 = 0$. We write then

$$V_1 = \frac{10}{s}I_1 \qquad \therefore z_{11} = \frac{10}{s}$$

and

$$V_2 = \frac{10}{s}I_1 \qquad \therefore z_{21} = \frac{10}{s}$$

as before.                                                                 □

---

Recall that in Eq. (4-13) $z_{21}$ is the ratio of the response $V_2$ to an excitation $I_1$, while $z_{12}$ is the ratio of the response $V_1$ to an excitation $I_2$. If these ratios are not equal, this means that the two voltage responses will be unequal for $I_1 = I_2$. This network is *nonreciprocal* and is shown in Fig. 4-14.

If $z_{21}$ were equal to $z_{12}$, the two voltage responses would be equal. Such a network is *reciprocal*; that is, it obeys the *reciprocity theorem* which can be stated as follows: The ratio of a transform response (current or voltage) to an excitation transform (voltage or current) remains the same if we exchange the places of the excitation and the responses. Or, in other words: let $I_1(s)$ be the only excitation at port

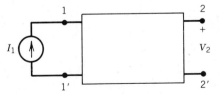

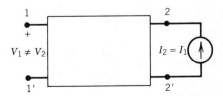

**FIGURE 4-14.**  A nonreciprocal network.

1 of a network, producing the response $V_2(s)$ at port 2. Next, remove the source $I_1$ (i.e., open-circuit port 1) and place it across the terminals of port 2, with the tail of the current reference at the positive reference of the former $V_2$. The response $V_1(s)$ across the open-circuited terminals of port 1 will be equal to the former $V_2$. Note that reciprocity applies only to ratios of voltage to current or current to voltage, but not to voltage ratios nor to current ratios.

Networks containing only resistors, inductors (possibly coupled), and capacitors are reciprocal. A loop analysis of such a network—as studied in the previous chapter—leads to the set of equations $\mathbf{ZI} = \mathbf{E}$. Here $\mathbf{Z}$ will be symmetric, that is, $z_{jk} = z_{kj}$. □ *Probs. 4-7*
*4-8*

## 4-4 A NEAT METHOD: MATRIX PARTITIONING

### Example 7

Find $\mathbf{Z}_{oc}$ for tne 2-P network shown in Fig. 4-15.

Although we can solve this problem exactly as we did earlier, let us show a systematic approach which is quite general and which requires no memorizations. We write Kirchhoff's voltage laws around the indicated four loops. In loops 1 and 2, the loop currents will be $I_1$ and $I_2$, respectively. In loops 3 and 4, we choose arbitrary directions for $I_3$ and $I_4$. Our ultimate result must not contain $I_3$ and $I_4$, of course.

$$12I_1 + 7I_2 - I_3 + 4I_4 = V_1 \quad \text{(around loop 1)}$$
$$7I_1 + 15I_2 + 3I_3 - 5I_4 = V_2 \quad \text{(around loop 2)}$$
$$-I_1 + 3I_2 + 6I_3 + 0I_4 = 0 \quad \text{(around loop 3)}$$
$$4I_1 - 5I_2 + 0I_3 + 15I_4 = 0 \quad \text{(around loop 4)}$$

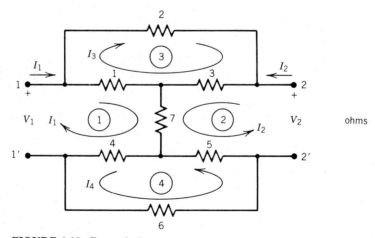

**FIGURE 4-15.** Example 7.

We could solve the last two equations for $I_3$ and $I_4$ in terms of $I_1$ and $I_2$, then substitute into the first and second equations. Let us do essentially the same thing by using *matrix partitioning* (see also appendix A).

Write the four equations in matrix form, partitioned as indicated:

$$\begin{bmatrix} 12 & 7 & -1 & 4 \\ 7 & 15 & 3 & -5 \\ \hline -1 & 3 & 6 & 0 \\ 4 & -5 & 0 & 15 \end{bmatrix} \begin{bmatrix} I_1 \\ I_2 \\ \hline I_3 \\ I_4 \end{bmatrix} = \begin{bmatrix} V_1 \\ V_2 \\ \hline 0 \\ 0 \end{bmatrix}$$

For convenience, let us designate the following submatrices

$$\mathbf{W}_1 = \begin{bmatrix} 12 & 7 \\ 7 & 15 \end{bmatrix} \quad \mathbf{W}_2 = \begin{bmatrix} -1 & 4 \\ 3 & -5 \end{bmatrix}$$

$$\mathbf{W}_3 = \begin{bmatrix} -1 & 3 \\ 4 & -5 \end{bmatrix} \quad \mathbf{W}_4 = \begin{bmatrix} 6 & 0 \\ 0 & 15 \end{bmatrix}$$

$$\mathbf{I}_a = \begin{bmatrix} I_1 \\ I_2 \end{bmatrix} \quad \mathbf{I}_b = \begin{bmatrix} I_3 \\ I_4 \end{bmatrix} \quad \mathbf{V}_a = \begin{bmatrix} V_1 \\ V_2 \end{bmatrix} \quad \mathbf{0} = \begin{bmatrix} 0 \\ 0 \end{bmatrix}$$

Our matrix equation is then

$$\begin{bmatrix} \mathbf{W}_1 & \mathbf{W}_2 \\ \mathbf{W}_3 & \mathbf{W}_4 \end{bmatrix} \begin{bmatrix} \mathbf{I}_a \\ \mathbf{I}_b \end{bmatrix} = \begin{bmatrix} \mathbf{V}_a \\ \mathbf{0} \end{bmatrix}$$

which yields two matrix equations as follows:

$$\mathbf{W}_1\mathbf{I}_a + \mathbf{W}_2\mathbf{I}_b = \mathbf{V}_a$$

and

$$\mathbf{W}_3\mathbf{I}_a + \mathbf{W}_4\mathbf{I}_b = \mathbf{0}$$

Recall that we want to eliminate the variables $I_3$ and $I_4$. This means that, between these two matrix equations, $\mathbf{I}_b$ must be eliminated. Since matrices are involved, we should be careful in such matters as inversion and pre- or post-multiplication. From the second equation, we write $\mathbf{W}_4\mathbf{I}_b = -\mathbf{W}_3\mathbf{I}_a$ and since $\mathbf{W}_4$ is nonsingular (verify this!), we can premultiply by its inverse to obtain

$$\mathbf{I}_b = -\mathbf{W}_4^{-1}\mathbf{W}_3\mathbf{I}_a$$

This expression for $\mathbf{I}_b$ is inserted into the first equation:

$$\mathbf{W}_1\mathbf{I}_a - \mathbf{W}_2\mathbf{W}_4^{-1}\mathbf{W}_3\mathbf{I}_a = \mathbf{V}_a$$

or finally

$$(\mathbf{W}_1 - \mathbf{W}_2\mathbf{W}_4^{-1}\mathbf{W}_3)\mathbf{I}_a = \mathbf{V}_a$$

This is the desired result. The numerical evaluations give

$$\mathbf{W}_4^{-1} = \begin{bmatrix} \dfrac{1}{6} & 0 \\ 0 & \dfrac{1}{15} \end{bmatrix} \qquad \mathbf{W}_2\mathbf{W}_4^{-1}\mathbf{W}_3 = \begin{bmatrix} \dfrac{37}{30} & -\dfrac{11}{6} \\ -\dfrac{11}{6} & \dfrac{19}{6} \end{bmatrix}$$

$$\mathbf{W}_1 - \mathbf{W}_2\mathbf{W}_4^{-1}\mathbf{W}_3 = \begin{bmatrix} \dfrac{323}{30} & \dfrac{53}{6} \\ \dfrac{53}{6} & \dfrac{71}{6} \end{bmatrix}$$

So

$$\begin{bmatrix} \dfrac{323}{30} & \dfrac{53}{6} \\ \dfrac{53}{6} & \dfrac{71}{6} \end{bmatrix} \begin{bmatrix} I_1 \\ I_2 \end{bmatrix} = \begin{bmatrix} V_1 \\ V_2 \end{bmatrix}$$

What we did here, essentially, is to eliminate $\mathbf{I}_b$ (containing the two undesired variables $I_3$ and $I_4$) between the two matrix equations. In a similar way, the submatrix $\mathbf{I}_b$ may contain more than two undesired (internal) currents, while $\mathbf{I}_a$ retains only the two port currents. □ ***Probs. 4-9***
***4-10***

## 4-5 THE -P NETWORK AND ITS SHORT-CIRCUIT PARAMETERS

In a dual fashion to the previous discussion, we wish to express $I_1$ and $I_2$ in terms of $V_1$ and $V_2$. Again, let us use superposition, with voltage sources as inputs, and currents as responses. We have, for Fig. 4-16, the following relationships

$$I_1 = y_{11}V_1 + y_{12}V_2 \tag{4-15a}$$

$$I_2 = y_{21}V_1 + y_{22}V_2 \tag{4-15b}$$

where the $y$'s are admittances, multiplying the respective inputs to contribute the partial responses to the total $I_1$ and $I_2$.

The informal approach is illustrated by an example.

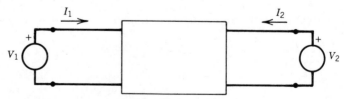

**FIGURE 4-16.** A 2-P with voltage inputs.

## Example 8

Calculate the $y$'s for the 2-P shown in Fig. 4-17. We write by inspection KCL for $I_1$ and for $I_2$

$$I_1 = 2V_1 + \alpha I_2$$

$$I_2 = \frac{3}{s} V_2 - \alpha I_2$$

Rearrange these equations into the desired form, using the second one in the first

$$I_1 = 2V_1 + \frac{3\alpha}{(\alpha + 1)s} V_2$$

and

$$I_2 = \frac{3}{(\alpha + 1)s} V_2$$

Therefore

$$y_{11} = 2, \qquad y_{12} = \frac{3\alpha}{(\alpha + 1)s}; \qquad y_{21} = 0; \qquad y_{22} = \frac{3}{(\alpha + 1)s}$$

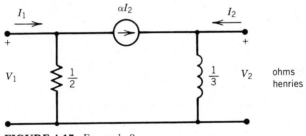

**FIGURE 4-17.**  Example 8.  □

The formal approach is, again, based on superposition. With $V_1$ alone as an input ($V_2 = 0$), we obtain from Eq. (4-15)

$$y_{11} = \frac{I_1}{V_1}\bigg]_{V_2=0} \qquad y_{21} = \frac{I_2}{V_1}\bigg]_{V_2=0} \qquad (4\text{-}16a)$$

and with $V_2$ alone ($V_1 = 0$) we have

$$y_{12} = \frac{I_1}{V_2}\bigg]_{V_1=0} \qquad y_{22} = \frac{I_2}{V_2}\bigg]_{V_1=0} \qquad (4\text{-}16b)$$

These two experiments are shown in Fig. 4-18.

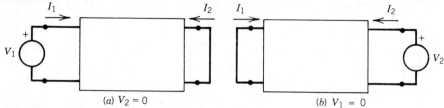

(a) $V_2 = 0$                                    (b) $V_1 = 0$

**FIGURE 4-18.**   A 2-P network with single voltage inputs.

These equations indicate how to calculate (or measure) the $y$'s. Each one is a *short circuit admittance*. To emphasize this, we write Eq. (4-15) in matrix form

$$I = Y_{sc}V \tag{4-17}$$

where the subscript sc means "short circuit." The two parameters $y_{11}$ and $y_{22}$ are *driving-point* short circuit admittances, while $y_{12}$ and $y_{21}$ are *transfer* in nature. Again, we identify in Eq. (4-17) the input ($V$), the zero-state output ($I$) and the appropriate network function ($Y_{sc}$).

The complete duality of $Z_{oc}$ and $Y_{sc}$ is recognized from the nature of the inputs (currents in one, voltages in the other), the outputs (voltages, currents), and conditions of measurement (oc, sc). Finally, we note that the matrix relations, Eqs. (4-14) and (4-17), are reciprocal, i.e.,

$$Z_{oc}^{-1} = Y_{sc} \tag{4-18}$$

and

$$Y_{sc}^{-1} = Z_{oc} \tag{4-19}$$

provided, of course, the inverses exist.

---

### Example 9

The sc parameters of the 2-P shown in Fig. 4-19 are (verify!).

$$Y_{sc} = \begin{bmatrix} \dfrac{1}{Z_A} & -\dfrac{1}{Z_A} \\[2mm] -\dfrac{1}{Z_A} & \dfrac{1}{Z_A} \end{bmatrix}$$

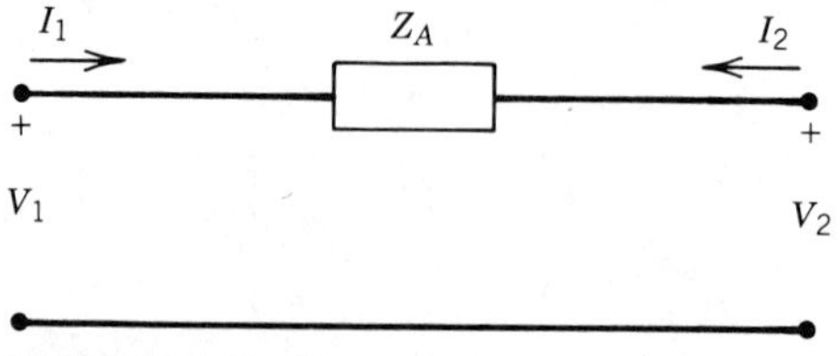

**FIGURE 4-19.**   Example 9.

This matrix is singular: therefore $\mathbf{Z}_{oc}$ does not exist. It is also easy to verify physically (electrically) why this 2-P has no $z$'s: under oc conditions, for example $I_2 = 0$, the current source $I_1$ is violating Kirchhoff's current law since it flows into an open circuit!

This example illustrates an important reason for the several descriptions of 2-P networks: one (or more) of the descriptions may not exist for a particular network; we will use then another one.

☐ *Probs. 4-13 through 4-19; 4-26 4-40*

## 4-6 THE 2-P NETWORK AND ITS HYBRID PARAMETERS

Let the 2-P be excited by $I_1$ at port 1 and by $V_2$ at port 2, as shown in Fig. 4-20. The respective outputs are then $V_1$ and $I_2$. We expect, therefore, the following relationships

$$V_1 = h_{11}I_1 + h_{12}V_2$$
$$I_2 = h_{21}I_1 + h_{22}V_2$$

(4-20)

Here the parameters $\mathbf{h}$ have mixed (hybrid) units, that is $h_{11}$ is an impedance, $h_{12}$ is dimensionless, $h_{21}$ is dimensionless, and $h_{22}$ is an admittance. They are called the *hybrid parameters*. As always, they can be derived informally by rearranging any two equations relating the $V$'s and the $I$'s. Alternately, they can be derived formally, using superposition.

The formal derivation yields

$$h_{11} = \frac{V_1}{I_1}\bigg]_{V_2=0} \qquad h_{12} = \frac{V_1}{V_2}\bigg]_{I_1=0}$$
$$h_{21} = \frac{I_2}{I_1}\bigg]_{V_2=0} \qquad h_{22} = \frac{I_2}{V_2}\bigg]_{I_1=0}$$

(4-21)

and we see that the conditions for the two experiments are also mixed (hybrid): one experiment is done under a short-circuit condition ($V_2 = 0$), the other under an open-circuit condition ($I_1 = 0$).

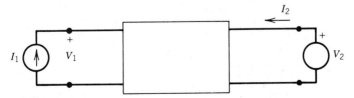

**FIGURE 4-20.** Hybrid inputs and outputs.

*Example 10*

A transistor model, quite useful for many applications, is shown in Fig. 4-21. There are two dependent sources. We write KVL for port 1

$$V_1 = 4000I_1 + 2 \times 10^{-4}V_2$$

and KCL for port 2

$$I_2 = 150I_1 + 10^{-4}V_2$$

thus obtaining immediately the hybrid parameters

$$h_{11} = 4000 \qquad h_{12} = 2 \times 10^{-4}$$
$$h_{21} = 150 \qquad h_{22} = 10^{-4}$$

In fact, the particular form of this model is based on its hybrid equations.

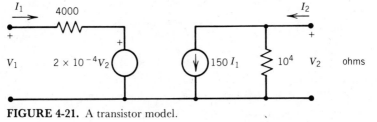

**FIGURE 4-21.** A transistor model.

The *inverse hybrid* parameters serve, as their name implies, to describe the 2-P as follows

$$I_1 = g_{11}V_1 + g_{12}I_2$$
$$V_2 = g_{21}V_1 + g_{22}I_2 \qquad (4\text{-}22)$$

Here the inputs are $V_1$ and $I_2$, and the outputs are $I_1$ and $V_2$. Formally and by superposition, we have

$$g_{11} = \frac{I_1}{V_1}\bigg]_{I_2=0} \qquad g_{12} = \frac{I_1}{I_2}\bigg]_{V_1=0}$$
$$g_{21} = \frac{V_2}{V_1}\bigg]_{I_2=0} \qquad g_{22} = \frac{V_2}{I_2}\bigg]_{V_1=0} \qquad (4\text{-}23)$$

also, when the inverse exists,

$$\begin{bmatrix} h_{11} & h_{12} \\ h_{21} & h_{22} \end{bmatrix}^{-1} = \begin{bmatrix} g_{11} & g_{12} \\ g_{21} & g_{22} \end{bmatrix} \qquad (4\text{-}24)$$

## 4-7   THE 2-P NETWORK AND ITS CHAIN PARAMETERS

The *chain* (or *transmission*) parameters were historically the earliest. Consequently, they differ from all the others in two respects: (1) they do not relate outputs to inputs as in Eq. (4-8), and (2) the reference direction of $I_2$ was then opposite to the one used now. The equations for the chain parameters simply relate the left-side variables $(V_1, I_1)$ to the right-side ones $(V_2, I_2)$, as shown in Fig. 4-22.

$$\begin{bmatrix} V_1 \\ I_1 \end{bmatrix} = \begin{bmatrix} A & B \\ C & D \end{bmatrix} \begin{bmatrix} V_2 \\ -I_2 \end{bmatrix} \tag{4-25}$$

that is,

$$\begin{aligned} V_1 &= AV_2 + B(-I_2) \\ I_1 &= CV_2 + D(-I_2) \end{aligned} \tag{4-26}$$

Although it is possible to obtain the chain parameters by using carefully the formal method, identifying the inputs and the other port's condition (oc or sc), it is by far easier to use the informal approach.

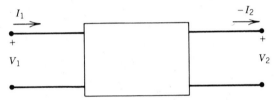

**FIGURE 4-22.**  Chain parameter variables.

---

## Example 11

Calculate the chain parameters for the 2-P shown in Fig. 4-23. We write directly

$$V_1 = \left( s + \frac{2}{s} \right) I_1 + \frac{2}{s} I_2$$

$$V_2 = \frac{2}{s} I_1 + \frac{2}{s} I_2$$

Rearrange these equations so that $V_1$ and $I_1$ are on the left side; the second equation yields

$$I_1 = \frac{s}{2} V_2 - I_2$$

that is,

$$C = \frac{s}{2} \qquad D = 1$$

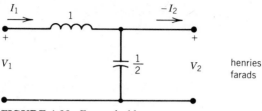

**FIGURE 4-23.** Example 11.

and when substituted into the first equation, it yields

$$V_1 = \left(\frac{s^2}{2} + 1\right) V_2 - sI_2$$

that is,

$$A = \frac{s^2}{2} + 1 \qquad B = s \qquad \qquad \square$$

The *reverse transmission* parameters describe $V_2$ and $-I_2$ in terms of $V_1$ and $I_1$,

$$\begin{bmatrix} V_2 \\ -I_2 \end{bmatrix} = \begin{bmatrix} E & F \\ G & H \end{bmatrix} \begin{bmatrix} V_1 \\ I_1 \end{bmatrix} \qquad (4\text{-}27)$$

and obviously

$$\begin{bmatrix} E & F \\ G & H \end{bmatrix} = \begin{bmatrix} A & B \\ C & D \end{bmatrix}^{-1} \qquad (4\text{-}28)$$

when the inverse exists. The reverse transmission parameters are seldom used.

The six different sets of these parameters are related algebraically, of course. Table 4.1 summarizes these relationships, where the symbol $\Delta$ means "the determinant of..." Thus,

$$\Delta z = z_{11}z_{22} - z_{12}z_{21} \qquad (4\text{-}29\text{a})$$

$$\Delta h = h_{11}h_{22} - h_{12}h_{21} \qquad (4\text{-}29\text{b})$$

$$\Delta A = AD - BC \qquad (4\text{-}29\text{c})$$

and so on. Elements in the same position are equal; for example $y_{12} = -\Delta A/B$; $h_{11} = 1/y_{11}$; $g_{22} = \Delta z/z_{11}$.

**Probs. 4-20**
**4-21**
**4-22**

## *4-8  THE N-T NETWORK AND THE INDEFINITE ADMITTANCE MATRIX

As mentioned in Section 4-1, the N-terminal (N-T) network, shown in Fig. 4-24, is characterized by Kirchhoff's current law

$$\sum_{k=1}^{n} I_k = 0 \qquad (4\text{-}30)$$

**TABLE 4.1  2-P Parameters**

| | $Z_{oc}$ | | $Y_{sc}$ | | $h$ | | $g$ | | $A$ | |
|---|---|---|---|---|---|---|---|---|---|---|
| **$Z_{oc}$** | $z_{11}$ | $z_{12}$ | $\dfrac{y_{22}}{\Delta y}$ | $\dfrac{-y_{12}}{\Delta y}$ | $\dfrac{\Delta h}{h_{22}}$ | $\dfrac{h_{12}}{h_{22}}$ | $\dfrac{1}{g_{11}}$ | $\dfrac{-g_{12}}{g_{11}}$ | $\dfrac{A}{C}$ | $\dfrac{\Delta A}{C}$ |
| | $z_{21}$ | $z_{22}$ | $\dfrac{-y_{21}}{\Delta y}$ | $\dfrac{y_{11}}{\Delta y}$ | $\dfrac{-h_{21}}{h_{22}}$ | $\dfrac{1}{h_{22}}$ | $\dfrac{g_{21}}{g_{11}}$ | $\dfrac{\Delta g}{g_{11}}$ | $\dfrac{1}{C}$ | $\dfrac{D}{C}$ |
| **$Y_{sc}$** | $\dfrac{z_{22}}{\Delta z}$ | $\dfrac{-z_{12}}{\Delta z}$ | $y_{11}$ | $y_{12}$ | $\dfrac{1}{h_{11}}$ | $\dfrac{-h_{12}}{h_{11}}$ | $\dfrac{\Delta g}{g_{22}}$ | $\dfrac{g_{12}}{g_{22}}$ | $\dfrac{D}{B}$ | $\dfrac{-\Delta A}{B}$ |
| | $\dfrac{-z_{21}}{\Delta z}$ | $\dfrac{z_{11}}{\Delta z}$ | $y_{21}$ | $y_{22}$ | $\dfrac{h_{21}}{h_{11}}$ | $\dfrac{\Delta h}{h_{11}}$ | $\dfrac{-g_{21}}{g_{22}}$ | $\dfrac{1}{g_{22}}$ | $\dfrac{-1}{B}$ | $\dfrac{A}{B}$ |
| **$h$** | $\dfrac{\Delta z}{z_{22}}$ | $\dfrac{z_{12}}{z_{22}}$ | $\dfrac{1}{y_{11}}$ | $\dfrac{-y_{12}}{y_{11}}$ | $h_{11}$ | $h_{12}$ | $\dfrac{g_{22}}{\Delta g}$ | $\dfrac{-g_{12}}{\Delta g}$ | $\dfrac{B}{D}$ | $\dfrac{\Delta A}{D}$ |
| | $\dfrac{-z_{21}}{z_{22}}$ | $\dfrac{1}{z_{22}}$ | $\dfrac{y_{21}}{y_{11}}$ | $\dfrac{\Delta y}{y_{11}}$ | $h_{21}$ | $h_{22}$ | $\dfrac{-g_{21}}{\Delta g}$ | $\dfrac{g_{11}}{\Delta g}$ | $\dfrac{-1}{D}$ | $\dfrac{C}{D}$ |
| **$g$** | $\dfrac{1}{z_{11}}$ | $\dfrac{-z_{12}}{z_{11}}$ | $\dfrac{\Delta y}{y_{22}}$ | $\dfrac{y_{12}}{y_{22}}$ | $\dfrac{h_{22}}{\Delta h}$ | $\dfrac{-h_{12}}{\Delta h}$ | $g_{11}$ | $g_{12}$ | $\dfrac{C}{A}$ | $\dfrac{-\Delta A}{A}$ |
| | $\dfrac{z_{21}}{z_{11}}$ | $\dfrac{\Delta z}{z_{11}}$ | $\dfrac{-y_{21}}{y_{22}}$ | $\dfrac{1}{y_{22}}$ | $\dfrac{-h_{21}}{\Delta h}$ | $\dfrac{h_{11}}{\Delta h}$ | $g_{21}$ | $g_{22}$ | $\dfrac{1}{A}$ | $\dfrac{B}{A}$ |
| **$A$** | $\dfrac{z_{11}}{z_{21}}$ | $\dfrac{\Delta z}{z_{21}}$ | $\dfrac{-y_{22}}{y_{21}}$ | $\dfrac{-1}{y_{21}}$ | $\dfrac{-\Delta h}{h_{21}}$ | $\dfrac{-h_{11}}{h_{21}}$ | $\dfrac{1}{g_{21}}$ | $\dfrac{g_{22}}{g_{21}}$ | $A$ | $B$ |
| | $\dfrac{1}{z_{21}}$ | $\dfrac{z_{22}}{z_{21}}$ | $\dfrac{-\Delta y}{y_{21}}$ | $\dfrac{-y_{11}}{y_{21}}$ | $\dfrac{-h_{22}}{h_{21}}$ | $\dfrac{-1}{h_{21}}$ | $\dfrac{g_{11}}{g_{21}}$ | $\dfrac{\Delta g}{g_{21}}$ | $C$ | $D$ |

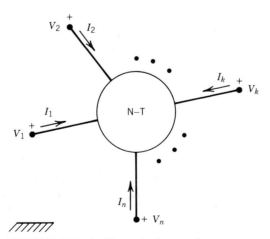

**FIGURE 4-24.**  An N-terminal network.

Without any additional information, we *cannot* "pair off" two terminals and say, for example, that $I_j = -I_{j+1}$ (in a port network this *is* true at port $j$).

To analyze the N-T network, we shall write Kirchhoff's current law, Eq. (4-30). First, let us choose the reference (datum) point of zero potential arbitrarily outside the N-T network. The voltage sources connected between the $n$ terminals and this datum are $V_1(s), V_2(s), \ldots, V_n(s)$. See Fig. 4-25. For a linear N-T network, superposition applies and each current is the sum of the partial contributions due to the $n$ sources. That is,

$$
\begin{aligned}
I_1 &= y_{11}V_1 + y_{12}V_2 + \cdots + y_{1n}V_n \\
I_2 &= y_{21}V_1 + y_{22}V_2 + \cdots + y_{2n}V_n \\
&\ \ \vdots \\
I_n &= y_{n1}V_1 + y_{n2}V_2 + \cdots + y_{nn}V_n
\end{aligned}
\tag{4-31}
$$

Although these equations resemble the short-circuit parameters, there is a distinct basic difference: in the present case, the datum (reference) node is arbitrary and outside the network. We write Eq. (4-31) in matrix form as

$$
\mathbf{I} = \mathbf{Y}_i \mathbf{V} \tag{4-32}
$$

where $\mathbf{Y}_i$ is the *indefinite admittance matrix*. The word "indefinite" serves to remind us that the reference node for voltages is arbitrarily outside the network. Formally, a typical element $y_{jk}$ of $\mathbf{Y}_i$ is given by

$$
y_{jk} = \left. \frac{I_j}{V_k} \right]_{V_1 = V_2 = \cdots = V_{k-1} = V_{k+1} = \cdots = V_n = 0} \tag{4-33}
$$

for $j = k$ or $j \neq k$. In words: connect all terminals, except terminal $k$, to the datum. Excite terminal $k$ with a voltage $V_k$ and measure (or compute) the current in terminal $j$. The ratio $I_j/V_k$ under these conditions yields $y_{jk}$. As in previous cases, a straightforward analysis is usually easier than the formal application of Eq. (4-33). Two examples will illustrate in this idea.

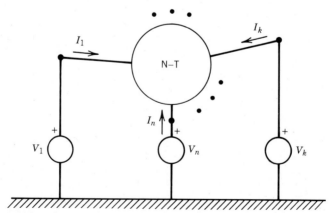

**FIGURE 4-25.** An outside datum for the N-terminal (N-T) network.

## *Example 12*

Find $\mathbf{Y}_i$ for the 3-T network shown in Fig. 4-26. No independent sources (in particular, no initial conditions) exist. The external sources are shown in dotted lines, for convenience. Write Kirchhoff's current law at each node: At node 1, $I_1$ (entering) is equal to the current (leaving) through the resistor plus the current (leaving) through the inductor, that is, $I_1 = (V_1 - V_2)/2 + (V_1 - V_3)/3s$. At node 2, in a similar fashion, we have $I_2 = (V_2 - V_1)/2 + 4s(V_2 - V_3)$; and at node 3, $I_3 = (V_3 - V_1)/3s + 4s(V_3 - V_2)$. Rearranging these equations, we get

$$
\begin{bmatrix} I_1 \\ I_2 \\ I_3 \end{bmatrix} = \begin{bmatrix} \dfrac{1}{2} + \dfrac{1}{3s} & -\dfrac{1}{2} & -\dfrac{1}{3s} \\ -\dfrac{1}{2} & \dfrac{1}{2} + 4s & -4s \\ -\dfrac{1}{3s} & -4s & \dfrac{1}{3s} + 4s \end{bmatrix} \begin{bmatrix} V_1 \\ V_2 \\ V_3 \end{bmatrix}
$$

that is, $\mathbf{I} = \mathbf{Y}_i\mathbf{V}$. Notice that these equations can be written by inspection.

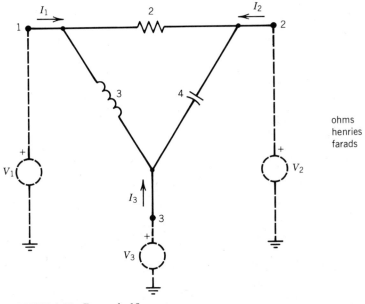

ohms
henries
farads

**FIGURE 4-26.** Example 12.

## *Example 13*

The circuit shown in Fig. 4-27 is a model of a transistor (FET) amplifier. There is a voltage-dependent current source.

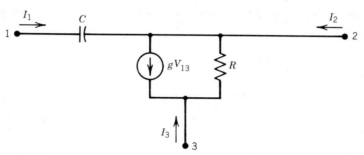

**FIGURE 4-27.** Example 13.

To calculate $\mathbf{Y}_i$, we write KCL at each node:

$$I_1 = Cs(V_1 - V_2)$$

$$I_2 = \frac{1}{R}(V_2 - V_3) + Cs(V_2 - V_1) + g(V_1 - V_3)$$

$$I_3 = \frac{1}{R}(V_3 - V_2) - g(V_1 - V_3)$$

From these equations we write

$$
\begin{bmatrix} I_1 \\ I_2 \\ I_3 \end{bmatrix} =
\begin{bmatrix}
Cs & -Cs & 0 \\
-Cs + g & \dfrac{1}{R} + Cs & -g - \dfrac{1}{R} \\
-g & -\dfrac{1}{R} & g + \dfrac{1}{R}
\end{bmatrix}
\begin{bmatrix} V_1 \\ V_2 \\ V_3 \end{bmatrix}
$$

that is, $\mathbf{I} = \mathbf{Y}_i\mathbf{V}$. We note that the controlling voltage of the current source is $V_{13} = V_1 - V_3$, and therefore the choice of the reference node is immaterial since only relative voltages between nodes are important. □

---

The indefinite admittance matrix $\mathbf{Y}_i$ possesses several interesting properties, which are listed below.

1. If we add all the equations in Eq. (4-31), we obtain

$$
\begin{aligned}
I_1 + I_2 + \cdots + I_n = 0 = &\ (y_{11} + y_{21} + \cdots + y_{n1})V_1 \\
&+ (y_{12} + y_{22} + \cdots + y_{n2})V_2 \qquad (4\text{-}34) \\
&+ \cdots\cdots + (y_{1n} + y_{2n} + \cdots + y_{nn})V_n
\end{aligned}
$$

This must hold true for any arbitrary $V_1, V_2, \ldots, V_n$. In particular, for $V_1 \neq 0$, $V_2 = V_3 = \cdots = 0$. Next, for $V_2 \neq 0$, $V_1 = V_3 = \cdots = 0$, and so on. Therefore, we conclude that

$$\sum_{j=1}^{n} y_{j1} = \sum_{j=1}^{n} y_{j2} = \cdots = \sum_{j=1}^{n} y_{jn} = 0 \tag{4-35}$$

that is, *the sum of the elements of every column in* $\mathbf{Y}_i$ *is zero.*

2. Now let $V_1 = V_2 = \cdots = V_n = V_0 \neq 0$. Since all the nodes are at the same potential, no currents will flow in the terminals, that is,

$$0 = y_{11}V_0 + y_{12}V_0 + \cdots + y_{1n}V_0$$
$$0 = y_{21}V_0 + y_{22}V_0 + \cdots + y_{2n}V_0$$
$$\vdots \tag{4-36}$$
$$0 = y_{n1}V_0 + y_{n2}V_0 + \cdots + y_{nn}V_0$$

and since $V_0 \neq 0$, we obtain the result

$$\sum_{j=1}^{n} y_{1j} = \sum_{j=1}^{n} y_{2j} = \cdots = \sum_{j=1}^{n} y_{nj} = 0 \tag{4-37}$$

that is, *the sum of the elements of every row in* $\mathbf{Y}_i$ *is zero.*

3. As a result of (1) or (2), $\mathbf{Y}_i$ is singular. Stated otherwise, its determinant is zero and, hence, its rank is less than $n$. It can be shown that its rank is $(n-1)$, which means that only $(n-1)$ of its rows—currents—are independent, and the $n$th row is dependent upon the remaining rows. Indeed, from Eq. (4-31) we obtain the dependence as

$$I_n = -\sum_{k=1}^{n-1} I_k \tag{4-38}$$

4. For a reciprocal N-T network, $\mathbf{Y}_i$ is symmetric

$$y_{jk} = y_{kj} \tag{4-39}$$

5. If we choose the arbitrary reference (datum) as one of the terminals, say the $n$th, then obviously

$$V_n = 0 \tag{4-40}$$

The $n$th column in Eq. (4-30) drops out, and we have $n$ equations with $(n-1)$ variables. However, we also know that

$$I_n = -\sum_{k=1}^{n-1} I_k \tag{4-41}$$

and so the $n$th equation is redundant. Therefore, we have $(n-1)$ equations with $(n-1)$ variables, with the $n$th node as the reference node. But *this is precisely the node analysis described in chapter 3.*[†]

---

[†] We are assuming here that there are no internal nodes in the N-T network.

## Example 14

For the network shown in Fig. 4-28, write the node equations when: (a) Node 1 is grounded, (b) Node 3 is grounded. The indefinite admittance matrix $\mathbf{Y}_i$ of the network is

$$\mathbf{Y}_i = \begin{bmatrix} Y_a + Y_c & -Y_c & -Y_a \\ -Y_c & Y_b + Y_c & -Y_b \\ -Y_a & -Y_b & Y_a + Y_b \end{bmatrix}$$

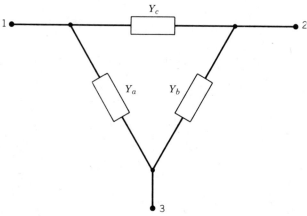

**FIGURE 4-28.** Example 14.

Therefore, with node 1 grounded we delete row 1 and column 1, and the node equations read

$$\begin{bmatrix} Y_b + Y_c & -Y_b \\ -Y_b & Y_a + Y_b \end{bmatrix}\begin{bmatrix} V_2 \\ V_3 \end{bmatrix} = \begin{bmatrix} I_2 \\ I_3 \end{bmatrix}$$

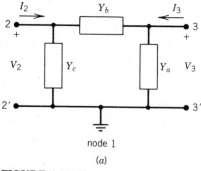

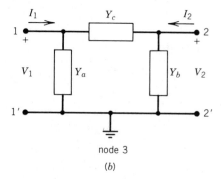

**FIGURE 4-29.** Example 14.

With node 3 as reference, delete row 3 and column 3; the node equations are then

$$\begin{bmatrix} Y_a + Y_c & -Y_c \\ -Y_c & Y_b + Y_c \end{bmatrix}\begin{bmatrix} V_1 \\ V_2 \end{bmatrix} = \begin{bmatrix} I_1 \\ I_2 \end{bmatrix}$$

These two configurations are shown in Fig. 4-29. They are familiar 2-P networks.

☐ **Probs. 4-24**
**4-25**
**4-27**

## PROBLEMS

4-1 Prove in detail the rule for a series connection of impedances

$$Z_{\text{total}}(s) = \sum_{k=1}^{n} Z_k(s)$$

*Hint*: apply the proper source and calculate the resulting zero-state response.

4-2 Prove in detail the rule for a parallel connection of admittances

$$Y_{\text{total}}(s) = \sum_{k=1}^{n} Y_k(s)$$

4-3 Prove the familiar rule for two impedances connected in parallel: "the product over the sum," that is,

$$Z_{\text{total}}(s) = \frac{Z_1 Z_2}{Z_1 + Z_2}$$

(Note: this rule is valid *only for two* impedances).

4-4 The variable $s$ in the Laplace transform is, in general, complex, $s = \sigma + j\omega$. For the special case $s = j\omega$ we have in the Laplace integral

$$e^{-st} = e^{-j\omega t} = \cos \omega t - j \sin \omega t$$

and if we retain either the real part ($\cos \omega t$) or the imaginary part ($-\sin \omega t$) we have the sinusoidal steady-state *phasor analysis*. What are the *sinusoidal impedances* of a resistor ($R$), an inductor ($L$), a capacitor ($C$)?

4-5 Calculate the driving-point impedance of the 1-P network shown. Give it in its final form as a rational function.

4-6 Calculate the driving-point admittance of each network shown.

4-7 Use the informal approach to calculate $\mathbf{Z}_{\text{oc}}$ for each of the two-port networks shown.

4-8 Repeat Problem 4-7 using the formal approach.

4-9 Find $\mathbf{Z}_{\text{oc}}$ for the 2-P shown, using the method of matrix partitioning.

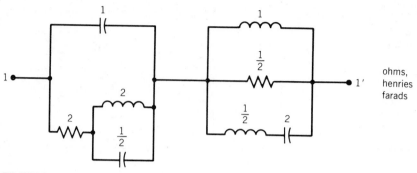

**PROBLEM 4-5.**

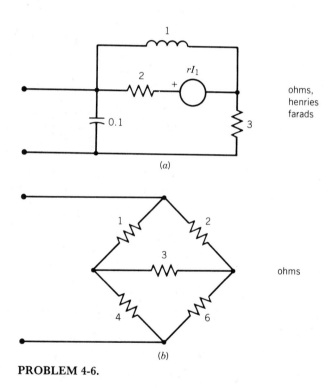

(a)

(b)

**PROBLEM 4-6.**

4-10 The 2-P networks shown below are *equivalent* provided their terminal characterizations are identical, that is, the $v$–$i$ relationships at the ports are the same.
Obtain $\mathbf{Z}_{oc}$ for the network shown in (a) by matrix partitioning, and hence find
the values of $Z_A$, $Z_B$, and $Z_C$ (in terms of $Z_1, Z_2, Z_3$) of the equivalent network
in (b). You have obtained the well-known "$\Delta$-$Y$ transformation." Use this
result to re-work Problem 4-6(b).

4-11 Use an informal approach, with matrix partitioning, to calculate $\mathbf{Y}_{sc}$ for
the 2-P shown in Problem 4-9. Verify your final answer by checking that
$\mathbf{Y}_{sc}\mathbf{Z}_{oc} = \mathbf{U}$, the unit matrix.

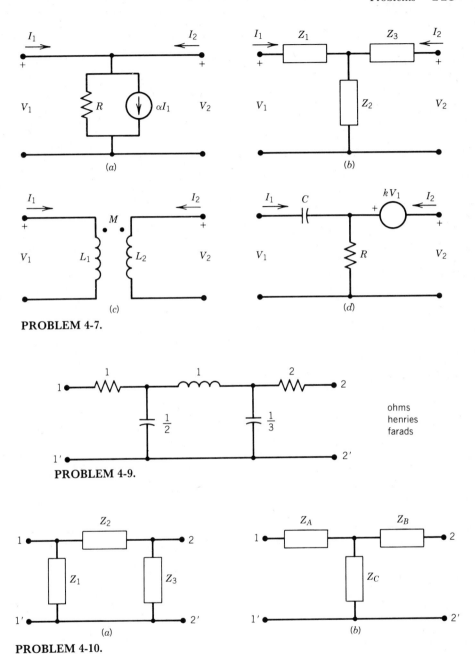

PROBLEM 4-7.

PROBLEM 4-9.

PROBLEM 4-10.

4-12 Using the result of Problem 4-7(c), calculate the sc parameters of a pair of mutually coupled inductors. *Hint*: $\mathbf{Y}_{sc} = \mathbf{Z}_{oc}^{-1}$. What is the condition for the existence of $\mathbf{Y}_{sc}$?

4-13 Show (build) a simple 2-P which has no $\mathbf{Y}_{sc}$ description.

4-14 An *ideal transformer* is a 2-P network, as shown, defined by

$$V_1(s) = \frac{1}{a} V_2(s)$$

$$I_1(s) = -aI_2(s),$$

where $a$ is a given number, positive or negative. (The ideal transformer can be thought of as a limiting case of a pair of perfectly coupled inductors, with $L_1 \to \infty$, $L_2 \to \infty$, and $L_1/L_2$ finite). Investigate the existence of $\mathbf{Z}_{oc}$ and $\mathbf{Y}_{sc}$ for the ideal transformer.

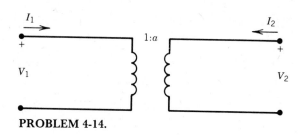

**PROBLEM 4-14.**

4-15 The *gyrator* is a useful 2-P, available commerically. Its ideal model, shown below, is defined by the equations

$$V_1 = -RI_2$$

and

$$V_2 = RI_1,$$

where $R$ is the *gyrating resistance*. Find $\mathbf{Z}_{oc}$ for the gyrator. Is the gyrator reciprocal?

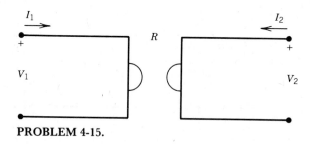

**PROBLEM 4-15.**

4-16 (a) Find the driving-point impedance $Z$ of an overall 1-P network formed by a 1-P network $(Z')$ connected to an ideal transformer as shown. (b) Repeat, to find the overall $\mathbf{Z}_{oc}$ of the 2-P network $(\mathbf{Z}'_{oc})$ connected to two ideal transformers.

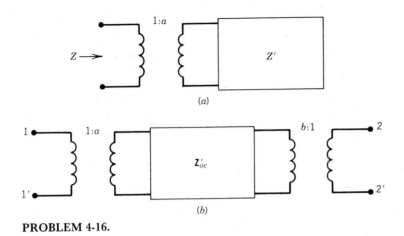

(a)

(b)

**PROBLEM 4-16.**

4-17 Find $\mathbf{Z}_{oc}$ for the 2-P network shown. Notice the ideal transformer.

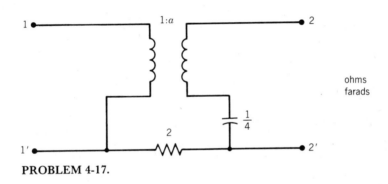

ohms
farads

**PROBLEM 4-17.**

4-18 For the network shown, obtain by matrix partitioning the short-circuit admittance matrix for the two port.

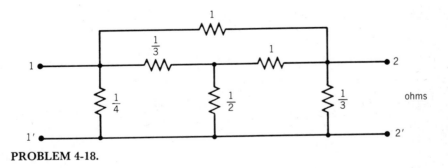

ohms

**PROBLEM 4-18.**

4-19 The 2-P shown is given by its $\mathbf{Z}_{oc}$ parameters. It is loaded at port 2 by $R_L = 1$ ohm. Calculate the zero-state network function $H(s)$ relating $V_L$ to $I_1$.

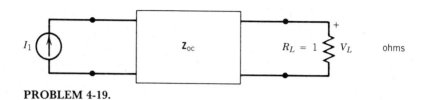

**PROBLEM 4-19.**

4-20 Calculate the inverse hybrid ($\mathbf{g}$) parameters for Problem 4-7(a), (b) and (d) by the informal approach. Next, verify your answers by using $\mathbf{Z}_{oc}$ from Problem 4-7 and Table 4-1.

4-21 Find the chain parameters for an ideal transformer (Problem 4-14).

4-22 Find the chain parameters for the 2-P shown in (a). Repeat for (b). Repeat for (c). Verify that

$$\begin{bmatrix} A & B \\ C & D \end{bmatrix}_c = \begin{bmatrix} A & B \\ C & D \end{bmatrix}_a \begin{bmatrix} A & B \\ C & D \end{bmatrix}_b$$

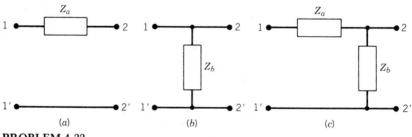

(a)          (b)          (c)

**PROBLEM 4-22.**

4-23 For a certain class of 2-P networks, the following relationship holds for its chain parameters:

$$AD - BC = 1$$

What can be said of the network? (*Hint*: check its $\mathbf{Z}_{oc}$)

4-24 Find $\mathbf{Y}_i$ for a pair of mutually coupled inductors ($L_1, L_2, M^2 < L_1L_2$) as a 4-T network.

4-25 Find $\mathbf{Y}_i$ for a single admittance $Y$ as a 2-T network.

4-26 Calculate the short-circuit parameters of the 2-P shown.

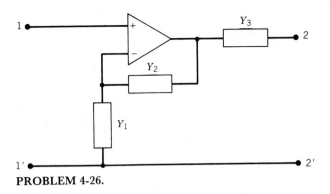

**PROBLEM 4-26.**

4-27 A model of a transistor circuit is shown in the figure as a 2-P network. Obtain its $\mathbf{Y}_{sc}$ and, from it, the sc parameters of the 2-P when terminal 3 is grounded.

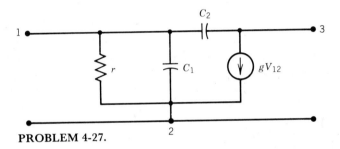

**PROBLEM 4-27.**

# 5

# INTERCONNECTIONS
# OF NETWORKS

In this chapter, we discuss different ways of interconnecting networks, and the resulting characterizations of the composite network. Chapter 4 should be reviewed thoroughly before studying this chapter.

## 5-1 SERIES CONNECTION OF 2-P NETWORKS

The *series connection* of $q$ 2-P networks is shown in Fig. 5-1$a$. These networks are labelled "a," "b,"..., "q," for distinction, and the connection is shown specifically for port 1. The other ports are connected similarly.

For clarity of discussion let us refer to Fig. 5-1$b$, where two 2-P networks are connected in series. The result is a composite 2-P network shown in dotted lines and designated "$t$" (for "total"). By inspecting this connection, we write

$$\begin{bmatrix} I_{1t} \\ I_{2t} \end{bmatrix} = \begin{bmatrix} I_{1a} \\ I_{2a} \end{bmatrix} = \begin{bmatrix} I_{1b} \\ I_{2b} \end{bmatrix} \qquad (5\text{-}1a)$$

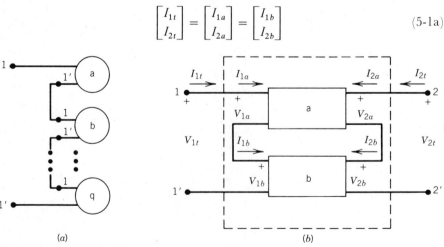

FIGURE 5-1. Series connection of 2-P networks.

that is, as the word "series" implies,

$$\mathbf{I}_t = \mathbf{I}_a = \mathbf{I}_b \qquad (5\text{-}1\text{b})$$

And

$$\begin{bmatrix} V_{1t} \\ V_{2t} \end{bmatrix} = \begin{bmatrix} V_{1a} \\ V_{2a} \end{bmatrix} + \begin{bmatrix} V_{1b} \\ V_{2b} \end{bmatrix} \qquad (5\text{-}2\text{a})$$

that is,

$$\mathbf{V}_t = \mathbf{V}_a + \mathbf{V}_b \qquad (5\text{-}2\text{b})$$

Using open-circuit parameters in Eq. (5-2b), we obtain

$$\mathbf{V}_t = \mathbf{Z}_{oca}\mathbf{I}_a + \mathbf{Z}_{ocb}\mathbf{I}_b \qquad (5\text{-}3)$$

and with Eq. (5-1b) the last result becomes

$$\mathbf{V}_t = [\mathbf{Z}_{oca} + \mathbf{Z}_{ocb}]\mathbf{I}_t \qquad (5\text{-}4)$$

Therefore

$$\mathbf{Z}_{oct} = \mathbf{Z}_{oca} + \mathbf{Z}_{ocb} \qquad (5\text{-}5)$$

In words: for 2-P networks connected in series, the open-circuit impedance matrices add to yield the open-circuit impedance matrix of the composite (total) 2-P network. A restriction to this rule is discussed subsequently.

---

## Example 1

Consider the two 2-P networks shown in Fig. 5-2a and b. To keep the calculations simple (but not less general), all the elements are resistances (in ohms). We have

$$\mathbf{Z}_{oca} = \begin{bmatrix} 4 & 3 \\ 3 & 5 \end{bmatrix} \qquad \mathbf{Z}_{ocb} = \begin{bmatrix} 9 & 5 \\ 5 & 11 \end{bmatrix}$$

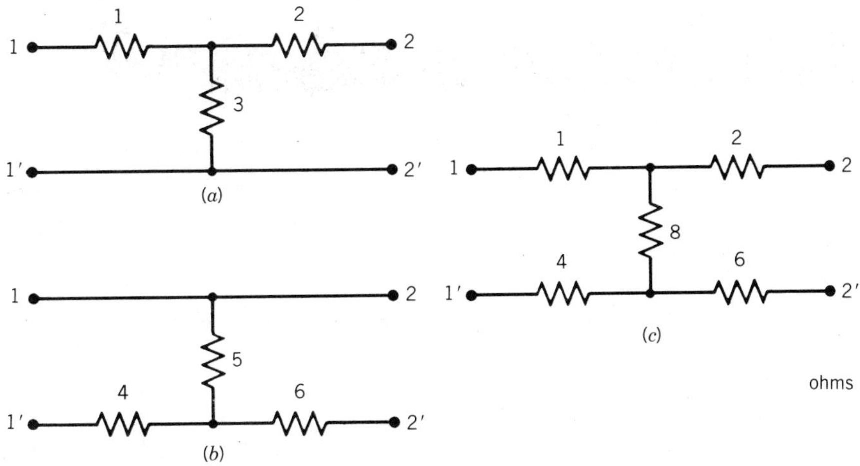

**FIGURE 5-2.** Example 1.

This series connection, shown in Fig. 5-2c, has $\mathbf{Z}_{oct}$ given by

$$\mathbf{Z}_{oct} = \begin{bmatrix} 4 & 3 \\ 3 & 5 \end{bmatrix} + \begin{bmatrix} 9 & 5 \\ 5 & 11 \end{bmatrix} = \begin{bmatrix} 13 & 8 \\ 8 & 16 \end{bmatrix}$$

as can be quickly verified in Fig. 5-2c.                                    □

---

The rule of addition of $\mathbf{Z}_{oc}$ (Eq. (5-5)) is based on the tacit assumption that each of the individual 2-P networks continues to behave like a 2-P network after the connection is made. Before the connection is made, each 2-P network satisfies the relations

$$I_{1a} = I'_{1a}, \qquad I_{2a} = I'_{2a} \tag{5-6a}$$

and

$$I_{1b} = I'_{1b}, \qquad I_{2b} = I'_{2b} \tag{5-6b}$$

respectively, qualifying it as a 2-P network. If these relations are violated after the connection is made, then obviously network "a" or network "b" (or both) cannot be described as a network with ports, and therefore its $\mathbf{Z}_{oc}$ is no longer valid and cannot be used in the addition! See Fig. 5-3a.

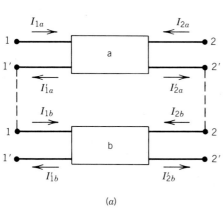
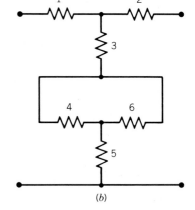

(a)                              (b)

**FIGURE 5-3.** Valid and invalid addition of $\mathbf{Z}_{oc}$.

---

### Example 2

Consider again the two 2-P networks of the previous example. This time, though, we "flip" the second 2-P network upside down—which certainly does not affect either its nature as a port network or its $\mathbf{Z}_{ocb}$. However, when the series connection is made (see Fig. 5-3*b*), the total 2-P network has $\mathbf{Z}_{oct}$ which is *not* $\mathbf{Z}_{oca} + \mathbf{Z}_{ocb}$:

$$\mathbf{Z}_{oct} \neq \begin{bmatrix} 4 & 3 \\ 3 & 5 \end{bmatrix} + \begin{bmatrix} 9 & 5 \\ 5 & 11 \end{bmatrix}$$

Rather, as can be calculated directly from the connection,

$$\mathbf{Z}_{oct} = \begin{bmatrix} 11.4 & 10.4 \\ 10.4 & 12.4 \end{bmatrix}$$

□

What happened in this example? Let us check the validity of Eq. (5-6) for the series connection. First, we connect a current source $I_{1t}$ to terminals 1-1', with terminals 2-2' open, and check Eq. (5-6). See Fig. 5-4*a*. It is easy to see that $I_{1a} = I_{1t} = I'_{1b}$ and $I'_{1a} = I_{1b}$ by virtue of the connection. However, $I'_{1a} = 0.6I_{1t}$ due to the current-divider action of the 4Ω and the 6Ω resistors. Therefore

$$I_{1a} \neq I'_{1a} \qquad I_{1b} \neq I'_{1b} \qquad (5\text{-}7a)$$

Next, connect a current source $I_{2t}$ to terminals 2-2', with terminals 1-1' open. See Fig. 5-4*b*. Here we have $I_{2a} = I_{2t} = I'_{2b}$ and $I'_{2a} = I_{2b}$, but $I'_{2a} = 0.4I_{2t}$ (current divider!) and so

$$I_{2a} \neq I'_{2a} \qquad I_{2b} \neq I'_{2b} \qquad (5\text{-}7b)$$

These results show that networks "a" and "b" lost their port characteristics and Eq. (5-6) is violated.

It is not difficult to recognize that the inequality of currents (after the connection has been made) is caused by a *circulating current* $I_o$, as shown in dotted lines in Fig. 5-4. Let us open, in Fig. 5-4*a*, the connection between terminal 2' of network "a" and terminal 2 of network "b." Now we have $I_o = 0$, and $I_{1t} = I_{1a} = I'_{1a} = I_{1b} = I'_{1b}$. That is, both networks have regained their port characteristics. If, then, $I_o$ would be zero *after* these terminals are reconnected, Eq. (4-6) would hold and matrix

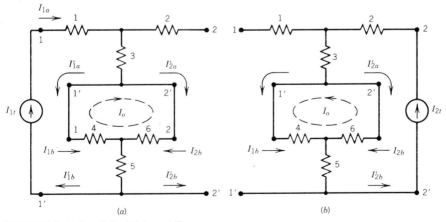

**FIGURE 5-4.** Invalid addition of $\mathbf{Z}_{oc}$.

addition would be valid. But $I_o$ will be zero only if $V_{o2}$, as shown in Fig. 5-5a, is zero. A similar reasoning leads to the requirement that $V_{o1}$ must be zero in Fig. 5-5b.

In summary, to test the validity of adding open-circuit matrices for a series connection, connect $I_{1t}$ to port 1 of the composite network and check if $V_{o2} = 0$ across the open circuit terminals 2' of "a" and 2 of "b". Repeat with $I_{2t}$ across port two of the composite network and check if $V_{o1} = 0$ across the open-circuit terminals 1' of "a" and 1 of "b".

In many cases, this test can be done by inspection. It is also worthwhile to note that the requirements $V_{o2} = 0$ and $V_{o1} = 0$ in Fig. 5-5 are equivalent to the requirement

$$V_{1'2'a} = V_{12b} \tag{5-8}$$

in Fig. 5-6, with proper excitations (currents) applied to the composite network.

When these tests fail, we can use ideal transformers[†] at the input or at the output ports to *make* an entering current equal to a leaving current. In particular, when we choose the turns ratio of the transformers as 1:1, we do not alter any voltages or currents in magnitude. The use of this scheme for example 2 (Fig. 5-4) is illustrated in Fig. 5-7a, and the general scheme is shown in Fig. 5-7b. Notice that since $I_{1t} = I_{1a} = I'_{1a}$ in the composite network, we need at most $(n - 1)$ transformers for the series connection of $n$ 2-P networks.

**Probs. 5-1**
**5-2**

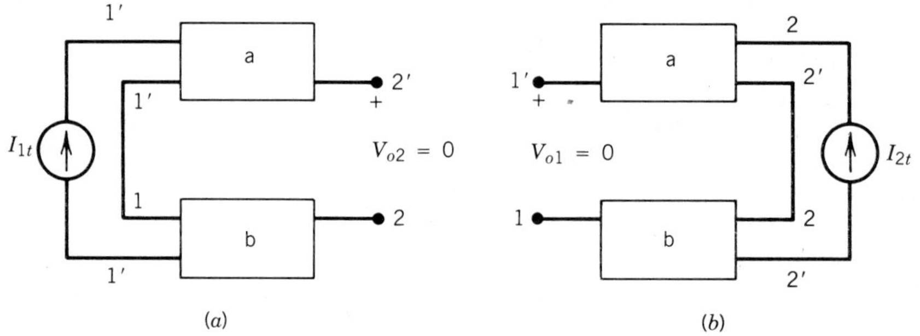

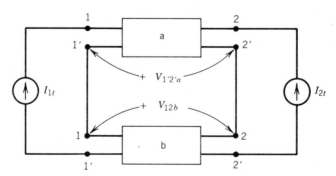

FIGURE 5-5. Tests for valid addition of $\mathbf{Z}_{oc}$.

FIGURE 5-6. Alternate test for valid addition of $\mathbf{Z}_{oc}$.

[†] See problem 4-6.

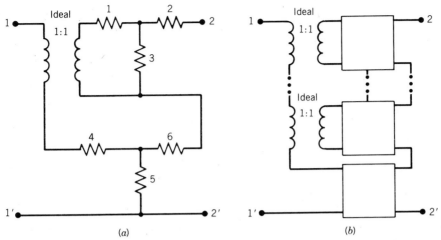

**FIGURE 5-7.** The use of 1:1 ideal transformers for valid addition of $\mathbf{Z}_{oc}$.

## 5-2   PARALLEL CONNECTION OF 2-P NETWORKS

The *parallel connection* of $q$ 2-P networks is shown in Fig. 5-8a. The connection is shown specifically for port 1, and all other ports are similarly connected. Again, for clarity of the subsequent discussion, we consider first two 2-P networks connected in parallel. Here we have, as the word "parallel" implies,

$$\mathbf{V}_t = \mathbf{V}_a = \mathbf{V}_b \tag{5-9}$$

while

$$\mathbf{I}_t = \mathbf{I}_a + \mathbf{I}_b \tag{5-10}$$

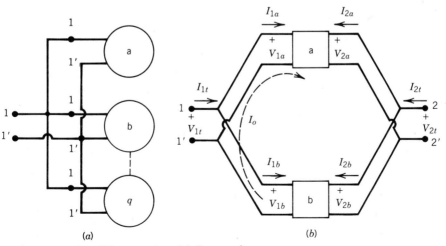

**FIGURE 5-8.** Parallel connection of 2-P networks.

Equation (5-10) suggests the use of short-circuit parameters

$$\mathbf{I}_t = \mathbf{Y}_{sca}\mathbf{V}_a + \mathbf{Y}_{scb}\mathbf{V}_b \tag{5-11}$$

and with Eq. (5-9), the result is

$$\mathbf{I}_t = [\mathbf{Y}_{sca} + \mathbf{Y}_{scb}]\mathbf{V}_t \tag{5-12}$$

Therefore,

$$\mathbf{Y}_{sct} = \mathbf{Y}_{sca} + \mathbf{Y}_{scb} \tag{5-13}$$

That is, for 2-P networks connected in parallel, the short-circuit admittance matrices add to yield the short-circuit admittance matrix of the composite (total) 2-P network. (Here, too, there is a restriction to be discussed later.)

---

### Example 3

Consider the two 2-P shown in Fig. 5-9a. The $\mathbf{Y}_{sc}$ are, respectively,

$$\mathbf{Y}_{sca} = \begin{bmatrix} Y_a & 0 \\ 0 & Y_a \end{bmatrix}, \qquad \mathbf{Y}_{scb} = \begin{bmatrix} Y_c & -Y_c \\ -Y_c & Y_c \end{bmatrix}$$

The parallel connection is shown in Fig. 5-9b, and this 2-P has for its short-circuit parameters

$$\mathbf{Y}_{sct} = \begin{bmatrix} Y_a + Y_c & -Y_c \\ -Y_c & Y_b + Y_c \end{bmatrix} = \mathbf{Y}_{sca} + \mathbf{Y}_{scb}$$

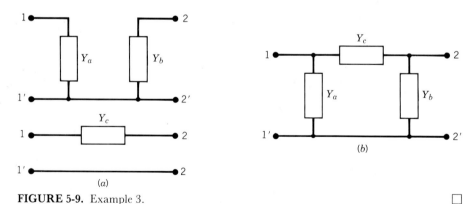

FIGURE 5-9. Example 3.

---

As in the case of $\mathbf{Z}_{oc}$, matrix addition of the various $\mathbf{Y}_{sc}$ is valid only if the *port* characteristics are not destroyed by the parallel connection. If a *circulating current* $I_o$ is present due to the connection, the matrices cannot be added. This circulating

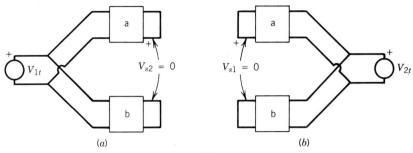

**FIGURE 5-10.** Tests for valid addition of $\mathbf{Y}_{sc}$.

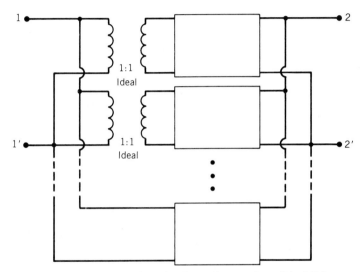

**FIGURE 5-11.** The use of 1:1 ideal transformers for valid addition of $\mathbf{Y}_{sc}$.

current is shown in dotted lines in Fig. 5-8$b$. The test for the valid addition of $\mathbf{Y}_{sc}$ follows in a dual manner to the one for $\mathbf{Z}_{oc}$. Specifically, here we short ports 2 of networks "a" and "b," apply $V_{1t} = V_{1a} = V_{1b}$ and check that $V_{s2} = 0$ between the shorted ports (see Fig. 5-10). Then we repeat the test, with ports one shorted, and $V_{2t} = V_{2a} = V_{2b}$, and check that $V_{s1} = 0$.

If this test fails, we may, again, resort to the use of 1:1 ideal transformers, as shown in Fig. 5-11.

## 5-3  SERIES-PARALLEL CONNECTION OF 2-P NETWORKS

In this configuration, the input ports are connected in series, while the output ports are in parallel. See Fig. 5-12$a$ for the general case, and Fig. 5-12$b$ for two 2-P

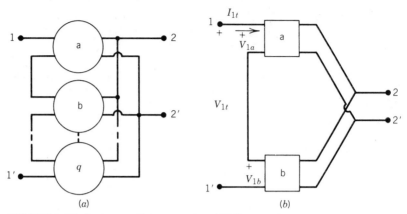

**FIGURE 5-12.** Series-parallel connection of 2-P networks.

networks. For this configuration, we have

$$\begin{bmatrix} V_{1t} \\ I_{2t} \end{bmatrix} = \begin{bmatrix} \sum_{j=1}^{q} V_{1j} \\ \sum_{j=1}^{q} I_{2j} \end{bmatrix} \tag{5-14}$$

and

$$\begin{bmatrix} I_{1t} \\ V_{2t} \end{bmatrix} = \begin{bmatrix} I_{1a} \\ V_{2a} \end{bmatrix} = \cdots = \begin{bmatrix} I_{1q} \\ V_{2q} \end{bmatrix} \tag{5-15}$$

Therefore, the use of the hybrid parameters is most convenient. The details are left as an exercise to the student, and the result is

$$\mathbf{h}_t = \mathbf{h}_a + \mathbf{h}_b \tag{5-16}$$

The test for the validity of this matrix addition makes use of one of the series tests and one of the parallel tests. There are two steps: First, we short the output terminals of each 2-P network and check that $V_{s2} = 0$. See Fig. 5-13a. Next, leave the input

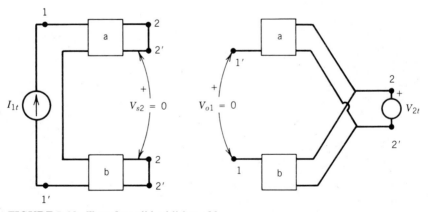

**FIGURE 5-13.** Tests for valid addition of **h**.

terminals open, connect the output terminals in parallel and check that $V_{o1} = 0$. See Fig. 5-13$b$. Ideal transformers may be needed if the tests fail.

## 5-4   PARALLEL-SERIES CONNECTION OF 2-P NETWORKS

In this case, the input terminals are in parallel and the output in series. The derivation here follows that for **h**, except that the words "output" and "input," "1-1'" and "2-2'," "series" and "parallel" are interchanged. The figures shown previously are reversed, and the result is

$$\mathbf{g}_t = \mathbf{g}_a + \mathbf{g}_b \qquad (5\text{-}17)$$

where **g** is the inverse hybrid matrix.

*Probs. 5-3*
*5-4*
*5-5*
*5-6*

## 5-5   THE CHAIN CONNECTION OF 2-P NETWORKS

This connection (known also as the *cascade connection*) is shown in Fig. 5-14, and we can write

$$\begin{bmatrix} V_{1t} \\ I_{1t} \end{bmatrix} = \begin{bmatrix} V_{1a} \\ I_{1a} \end{bmatrix} = \begin{bmatrix} A_a & B_a \\ C_a & D_a \end{bmatrix} \begin{bmatrix} V_{2a} \\ -I_{2a} \end{bmatrix} \qquad (5\text{-}18)$$

$$\begin{bmatrix} V_{2a} \\ -I_{2a} \end{bmatrix} = \begin{bmatrix} V_{1b} \\ I_{1b} \end{bmatrix} = \begin{bmatrix} A_b & B_b \\ C_b & D_b \end{bmatrix} \begin{bmatrix} V_{2b} \\ -I_{2b} \end{bmatrix} \qquad (5\text{-}19)$$

and

$$\begin{bmatrix} V_{2b} \\ -I_{2b} \end{bmatrix} = \begin{bmatrix} V_{2t} \\ -I_{2t} \end{bmatrix} \qquad (5\text{-}20)$$

Substituting Eqs. (5-20) and (5-19) into Eq. (5-18), we obtain

$$\begin{bmatrix} V_{1t} \\ I_{1t} \end{bmatrix} = \begin{bmatrix} A_a & B_a \\ C_a & D_a \end{bmatrix} \begin{bmatrix} A_b & B_b \\ C_b & D_b \end{bmatrix} \begin{bmatrix} V_{2t} \\ -I_{2t} \end{bmatrix} \qquad (5\text{-}21)$$

That is, for 2-P networks connected in chain, the overall chain (transmission) matrix is the product of the individual chain matrices, multiplied *in the same order* as their respective networks.

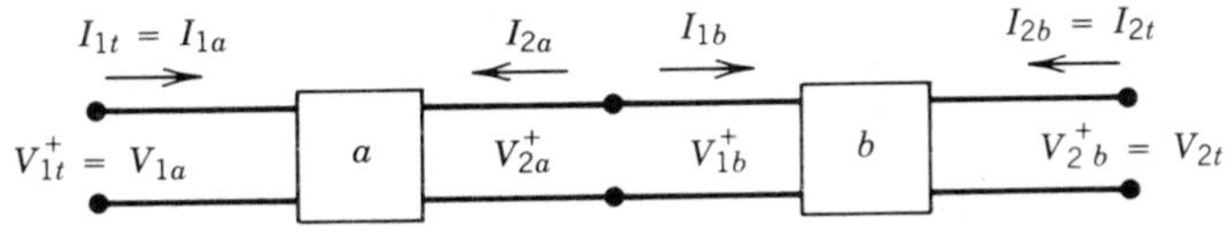

**FIGURE 5-14.**  Chain (cascade) connection of 2-P networks.

### Example 4

Consider the chain connection (shown in dotted lines) in Fig. 5-15a. The individual networks have the following chain matrices (see Prob. 4-22).

$$\begin{bmatrix} A & B \\ C & D \end{bmatrix}_a = \begin{bmatrix} 1 & Z_o \\ 0 & 1 \end{bmatrix} \qquad \begin{bmatrix} A & B \\ C & D \end{bmatrix}_b = \begin{bmatrix} 1 & 0 \\ Y_1 & 1 \end{bmatrix}$$

Hence, the overall 2-P network has the chain parameters

$$\begin{bmatrix} 1 & Z_o \\ 0 & 1 \end{bmatrix}\begin{bmatrix} 1 & 0 \\ Y_1 & 1 \end{bmatrix} = \begin{bmatrix} 1 + Z_o Y_1 & Z_o \\ Y_1 & 1 \end{bmatrix}$$

The 2-P network shown in Fig. 5-14b has the chain parameters

$$\begin{bmatrix} 1 & 0 \\ Y_1 & 1 \end{bmatrix}\begin{bmatrix} 1 & Z_o \\ 0 & 1 \end{bmatrix} = \begin{bmatrix} 1 & Z_o \\ Y_1 & 1 + Z_o Y_1 \end{bmatrix}$$

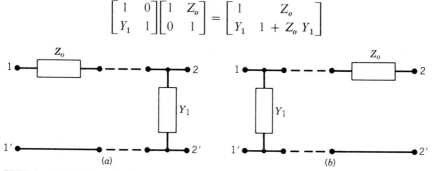

(a)                                                                      (b)

**FIGURE 5-15.** Example 4.

☐ **Probs. 5-7**
**5-8**
**5-9**
**5-10**

## 5-6   PARALLEL CONNECTION OF N-T NETWORKS

The parallel connection of two N-T networks is shown in Fig. 5-16, and the extension to more than two networks is clear: Corresponding terminals are connected together to form the corresponding terminal in the overall network. Due to this connection, we must have

$$V_{ka} = V_{kb} = V_{kt}, \qquad (k = 1, \ldots, N) \tag{5-22}$$

that is,

$$\mathbf{V}_a = \mathbf{V}_b = \mathbf{V}_t \tag{5-23}$$

And

$$I_{ka} + I_{kb} = I_{kt}, \qquad (k = 1, \ldots, N) \tag{5-24a}$$

that is,

$$\mathbf{I}_a + \mathbf{I}_b = \mathbf{I}_t \tag{5-24b}$$

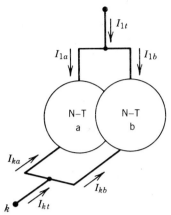

**FIGURE 5-16.** Two N-T networks connected in parallel.

Using the indefinite admittance description, we can write

$$\mathbf{I}_a = \mathbf{Y}_{ia}\mathbf{V}_a \qquad (5\text{-}25)$$

and

$$\mathbf{I}_b = \mathbf{Y}_{ib}\mathbf{V}_b \qquad (5\text{-}26)$$

for the individual N-T networks. Adding Eqs. (5-25) and (5-26), we obtain with Eqs. (5-23) and (5-24b)

$$\mathbf{Y}_{it} = \mathbf{Y}_{ia} + \mathbf{Y}_{ib} \qquad (5\text{-}27)$$

In words, the indefinite admittance matrix of several N-T networks connected in parallel is obtained by adding the individual indefinite admittance matrices.[†]

---

## Example 5

Consider the delta network shown in Fig. 5-17a as a parallel connection of three 3-T networks. These three networks are shown in Fig. 5-17b. The indefinite admittance matrix of the first one is obtained by writing:

$$I_1 = (V_1 - V_2)Y_A$$

$$I_2 = -I_1$$

$$I_3 = 0 \qquad (\text{terminal 3 isolated})$$

that is,

$$\mathbf{Y}_{ia} = \begin{bmatrix} Y_A & -Y_A & 0 \\ -Y_A & Y_A & 0 \\ 0 & 0 & 0 \end{bmatrix}$$

---

[†] The number of terminals is assumed to be the same in both N-T networks. See also problem 5-15.

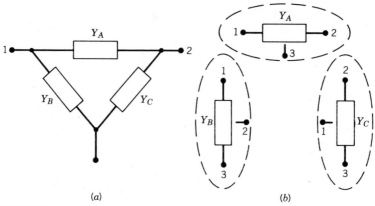

(a)                                      (b)

**FIGURE 5-17.** Example 5.

For the second network we have similarly

$$\mathbf{Y}_{ib} = \begin{bmatrix} Y_B & 0 & -Y_B \\ 0 & 0 & 0 \\ -Y_B & 0 & Y_B \end{bmatrix}$$

and, for the third one

$$\mathbf{Y}_{ic} = \begin{bmatrix} 0 & 0 & 0 \\ 0 & Y_C & -Y_C \\ 0 & -Y_C & Y_C \end{bmatrix}$$

Therefore,

$$\mathbf{Y}_{it} = \mathbf{Y}_{ia} + \mathbf{Y}_{ib} + \mathbf{Y}_{ic} = \begin{bmatrix} Y_A + Y_B & -Y_A & -Y_B \\ -Y_A & Y_A + Y_C & -Y_C \\ Y_B & -Y_C & Y_B + Y_C \end{bmatrix}$$

This example illustrates the idea of introducing isolated terminals in order to enable us to add indefinite admittance matrices for a composite N-T network, when the individual networks have different numbers of terminals. ☐ ***Probs. 5-11 through 5-15***

---

## PROBLEMS

5-1 Find $\mathbf{Z}_{oc}$ for each of the 2-P networks shown; then connect these networks in series, test the validity for adding the matrices and obtain $\mathbf{Z}_{oc}$ for the total network.

5-2 Decompose the "bridged-T" network into simpler 2-P networks connected in series and find its $\mathbf{Z}_{oc}$.

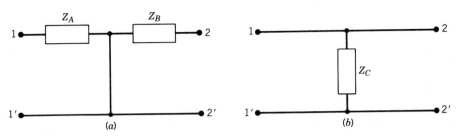

**PROBLEM 5-1.**

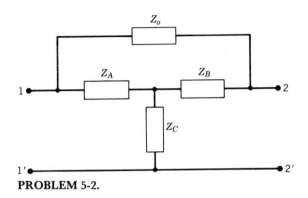

**PROBLEM 5-2.**

5-3 The figure shows a 2-P network with dependent sources. Decompose it into simpler 2-P networks so that matrix addition will be valid for finding its parameters.

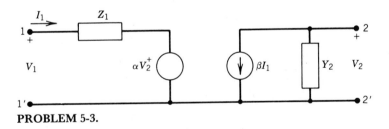

**PROBLEM 5-3.**

5-4 For the symmetric lattice shown, write the $\mathbf{Z}_{oc}$ as the sum of two matrices, one of which is

$$\frac{1}{2}\begin{bmatrix} Z_B & Z_B \\ Z_B & Z_B \end{bmatrix}$$

Find the second matrix. Draw the two resulting simple 2-P networks and check the possibility of connecting them in series.

5-5 Repeat Problem 5-4 with $\mathbf{Y}_{sc}$.

5-6 Repeat Problem 5-3 for the "twin bridged-T" network shown.

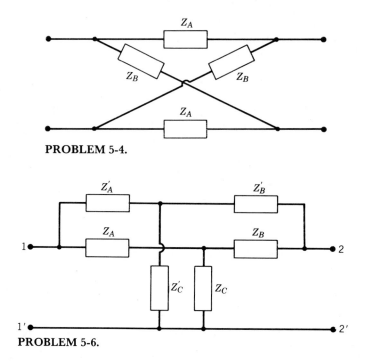

**PROBLEM 5-4.**

**PROBLEM 5-6.**

5-7 A 2-P network, given by any of its parameters, is loaded with $Z_L$ at its output terminals. Find the overall chain parameters, then obtain the overall voltage transfer function $V_2(s)/V_1(s)$.

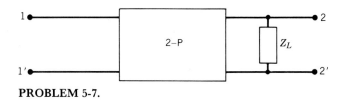

**PROBLEM 5-7.**

5-8 Find the chain parameters of $q$ identical ideal transformers connected in cascade.

5-9 Repeat Problem 5-8, for $q$, gyrators. Try $q$ even; then odd.

5-10 Find the overall chain parameters of a given 2-P network connected in cascade at its input terminals with an ideal transformer. Repeat, when the transformer is connected at the output terminals.

5-11 Find, using Table 4.1 in Chapter 4 (p. 105), $\mathbf{Z_{oc}}$ for the overall networks obtained in Problem 5-10. Are the results familiar?

5-12 Repeat Problem 5-10 with a gyrator (see Problem 4-15) instead of an ideal transformer.

5-13 The three terminals of a transistor circuit are labelled "$b$" (base), "$c$" (collector), and "$e$" (emitter). A 2-P transistor circuit, with the emitter as a common ground, has the sc admittance matrix description:

$$\begin{bmatrix} I_b \\ I_c \end{bmatrix} = \begin{bmatrix} Y_1 & 0 \\ \alpha Y_1 & Y_2 \end{bmatrix} \begin{bmatrix} V_b \\ V_c \end{bmatrix}$$

a) Obtain its indefinite admittance matrix $\mathbf{Y}_i$.
b) Decompose $\mathbf{Y}_i$ into a sum of simpler matrices, then draw the composite 3-T transistor circuit.

5-14 A 3-T network, whose $\mathbf{Y}_i$ is given, is connected as shown. Here $Y_G$ and $Y_L$ are known admittances. Find the overall $\mathbf{Y}_i$, then the node equations when terminal 3 is grounded.

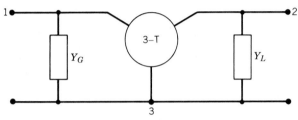

**PROBLEM 5-14.**

5-15 Consider two N-T networks with different numbers of terminals. Their indefinite admittance matrices are consequently of different orders. Develop carefully a procedure for obtaining the indefinite admittance matrix of the composite network obtained by a parallel connection of the two N-T networks.

# 6

---

# NETWORK TOPOLOGY

In this chapter we consider several aspects of the geometrical structure of networks and various problems associated with it. Among these is the concise, *systematic* choice of network variables, and an efficient way of writing *independent* loop or node equations. The intuitively "obvious" choice of loop currents and node voltages—a method used so far—may work for relatively simple networks, but becomes less and less "obvious" for more complicated networks. At any rate, it needs a solid foundation.

---

### Example 1

Consider the cube shown in Fig. 6-1. (For simplicity, all the elements are resistors.) How many node equations are needed? The answer is not too difficult—seven (one node will serve as reference). How many loop equations? The answer is not so "obvious." How to choose the loop currents? On the faces of the cube? Can we write

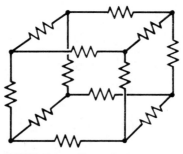

**FIGURE 6-1.** Example 1.

less than seven equations? How do we know that the equations are independent and thus solvable? These and other similar problems will be considered here.   ☐

## 6-1   BASIC PROPERTIES AND DEFINITIONS

By now (if not earlier) the student has undoubtedly recognized two basic facts in network analysis:

1. the $v-i$ relationship for a particular element depend per se on the element itself and *not* on how this element is connected in a network. For example, a linear, constant resistor is defined by $v_R = Ri_R$ irrespective of the network configuration. (Naturally, the specific expression for $i_R$—and hence $v_R$—*will* depend on the network configuration.)

2. Kirchhoff's current and voltage laws, on the other hand, do *not* depend on the types of elements; rather, they depend on the manner in which these elements are connected.

*Example 2*

In Fig. 6-2a, we write Kirchhoff's current law as

$$-i_1 + i_2 - i_3 - i_4 = 0$$

which is true no matter what is connected to node 1. We could have, instead of the diode, another element (say, a nonlinear inductor), instead of the capacitor a diode, etc. (Naturally, the individual expressions for $i_1, i_2, i_3,$ and $i_4$ will be different in each case!) To stress this fact, we redraw Fig. 6-2a as in Fig. 6-2b, showing the geometrical structure—the *topology*—of the network which leads to this equation.

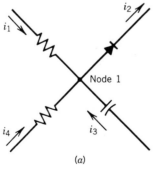
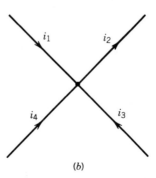

(a)                                              (b)

**FIGURE 6-2.**  Kirchhoff's current law.   ☐

***

### *Example 3*

For the network shown in Fig. 6-3*a*, we write Kirchhoff's voltage law as

$$-v_1 - v_2 + v_3 - v_4 = 0,$$

regardless of the nature of the elements. Rather, the topology of the network (the geometry of their connection) dictates the writing of this equation. Again, the individual expressions for $v_1$, $v_2$, $v_3$, and $v_4$ would indeed depend on the nature of the elements. However, the validity of Kirchhoff's voltage law depends only on the topology of the network as shown in Fig. 6-3*b*.

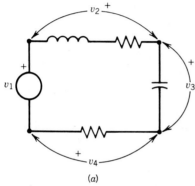 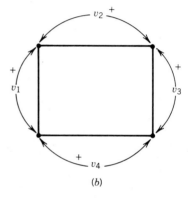

(a)                    (b)

**FIGURE 6-3.** Kirchhoff's voltage law.                    □

***

In general, Kirchhoff's current law at a node is written as

$$\sum i = 0, \tag{6-1}$$

with the following sign convention: a plus sign $(+)$ for a current whose reference direction is away from the node, and a minus sign $(-)$ for a current whose reference direction is towards the node. Equation (6-1) is a restatement of Kirchhoff's current law in the familiar form: "The sum of the currents entering the node is equal to the sum of the currents leaving it." Similarly, Kirchhoff's voltage law around a loop is written as

$$\sum v = 0, \tag{6-2}$$

with a plus sign $(+)$ for a drop and a minus sign $(-)$ for a rise around the loop. Alternatively, Eq. (6-2) can be restated as: "The sum of voltage drops equals the sum of voltage rises around a loop," We have used these conventions and rules in the formulation of loop and node equations in an earlier chapter.

In writing network equations, a combination of (1) and (2) is used; that is, Kirchhoff's current law together with the $v$–$i$ relationships of the elements connected at a node leads to node equations, and Kirchhoff's voltage law together with the $v$–$i$

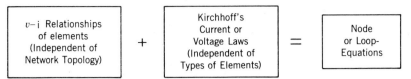

**FIGURE 6-4.** Derivation of network equations.

relationships for the elements that form a loop results in loop equations. These facts are shown schematically in Fig. 6-4.

In this chapter, we explore in some detail several aspects of the geometry of the network. For this purpose, we need a few definitions and terms.

The *graph* of a network shows its geometrical structure, and is drawn quite simply: We redraw the network, replacing each element by a line segment. In Fig. 6-5 we see two networks and their graphs. The length or the curvature of the line segments in a graph is not important; only their interconnections are of interest. Thus, the graph in Fig. 6-5*b* is the same, in this sense, as that in Fig. 6-5*c*. Such identical graphs are called *isomorphic*. Also, what we consider as a line segment depends on our purpose. We can choose to draw each network element as one line segment; or, if convenient, we may want to consider a combination of elements as one line segment (the *RL* combination in Fig. 6-5*a* is represented by one line segment "*a*").

In Fig. 6-5*d* and *e*, we notice that mutual inductance is not drawn as a line segment. Insofar as currents and voltages are concerned, nothing will change if we redraw the graph in Fig. 6-5*e* as shown in Fig. 6-5*f*: This amounts to having a common reference (datum) for all the elements.

A line segment (representing an element), together with its two distinct end points, is called an *edge*. In a graph, some of the individual end points of certain edges

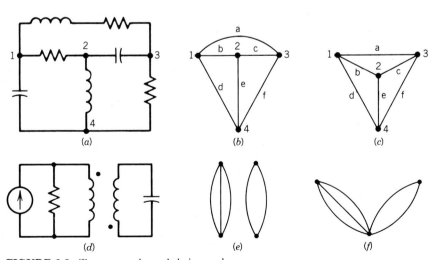

**FIGURE 6-5.** Two networks and their graphs.

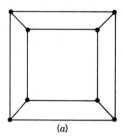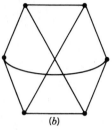

(a) (b)

**FIGURE 6-6.** A planar and nonplanar graph.

may coincide, of course. We shall designate by $e$ the total number of edges in a graph. Thus, in Fig. 6-5$b$, $e = 6$; in Fig. 6-5$f$, $e = 5$.

End points can be joined to form a *node*. The number of nodes in a graph is labelled $n$. For example, $n = 4$ in Fig. 6-5$c$, and $n = 3$ in Fig. 6-5$f$. A graph is, then, a finite collection of edges with no common points other than nodes. A graph is *planar* if we can draw it (or any of its isomorphic graphs) on a plane so that no two edges intersect at a point which is not a node. For example, the graph of the cube (Fig. 6-1) is planar, as shown in Fig. 6-6$a$, but the graph in Fig. 6-6$b$ is *nonplanar*.

An edge and a node are *incident* to each other if the node is one of the two nodes of the edge. The *degree* of a node is the number of edges incident to this node. For example, the degree of node 2 in Fig. 6-5$b$ is 3, since three edges are incident to it. An isolated node (whose degree is zero) is excluded, for the time being, from our discussion.

*Probs. 6-1*
*6-2*
*6-3*

A *subgraph* of a graph is a subset of the graph. The *complement* of a subgraph is the remainder of the graph after the subgraph has been chosen. In Fig. 6-7 we show a chosen subgraph (in full lines) and its complement (in dotted lines). Therefore, a subgraph plus its complement equals the graph.

A *path* between two nodes in a graph is a subgraph where those nodes are each of degree one (in this subgraph) and all other nodes are of degree two (in this subgraph). In Fig. 6-5$b$, a path between node 2 and 3 is $ba$; another one is $eda$; also $c$.

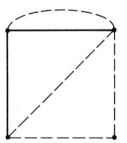

**FIGURE 6-7.** A subgraph
and its complement.

**FIGURE 6-8.** A single-edge loop

A graph is *connected* if there is a path between every two nodes. The graphs in Fig. 6-5$b$, $c$ and $f$ are connected, while that in Fig. 6-5$e$ is not. The graph in Fig. 6-5$e$ has two *separate parts* ($p = 2$); in a connected graph, $p = 1$.

A *loop* (*circuit*) is a closed path, i.e., it is a subgraph in which every node is of degree two. Some loops in Fig. 6-5$b$ are *1-b-2-e-4-d-1*; *1-a-3-f-4-d-1*; *3-c-2-e-4-f-3*. Excluded from the present discussion is a single-edge loop, as shown by a full line in Fig. 6-8. This exclusion is obvious, electrically.

The next definition is of primary importance: A *tree* of a connected graph is a connected subgraph containing all the nodes but no loops. Notice that it must be *connected*, contain *all* the nodes and *no* loops. In Fig. 6-9 are shown several trees of the graph in Fig. 6-5$b$. The construction of a tree follows the definition: First, draw all the nodes, then fill in enough edges to connect all the nodes without creating any loops. The edges of a tree are called (naturally!) *branches* and their number is $b$. We prove heuristically the following relation between $n$, the number of nodes of a connected graph, and $b$:

$$b = n - 1 \tag{6-3}$$

As before, we draw, first of all, the $n$ nodes. Next, draw the first branch of the tree: It requires two nodes. Every subsequent branch will require only *one* new node. (It cannot have *two* new nodes because the tree must be connected; on the other hand, if it needs *no* new nodes it will form a loop.) Hence, the total number of branches is $n - 1$.

The complement of a tree is called, briefly, a *co-tree*. Its edges are *links* and their number is $l$. Obviously, then

$$b + l = e \tag{6-4}$$

since a tree and its complement form the original graph. Using Eq. (6-3), we obtain the important result

$$l = e - (n - 1) = e - n + 1. \tag{6-5}$$

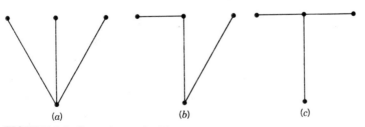

(a)          (b)          (c)

**FIGURE 6-9.** Several trees for Fig. 6-5$b$.

Notice that every link forms a *unique* loop with some branches of the tree. We shall return to this topic later.

## 6-2   THE INCIDENCE MATRIX. CUT-SETS

### The Incidence Matrix

The first step towards a systematic application of Kirchhoff's current law requires that we assign an orientation to every edge in a graph. This is done by placing an arrowhead on the edge. In Fig. 6-10, we show an *oriented (directed)* graph; the one in Fig. 6-5$b$ is not oriented.

For this oriented graph, let us write Kirchhoff's current law at each node, using the orientation of an edge as the positive reference for the current in that edge. With the convention adopted for Eq. (6-1), we obtain for nodes 1, 2, 3, and 4:

$$-i_a + i_b + i_d = 0 \tag{6-6a}$$

$$-i_b - i_c - i_e = 0 \tag{6-6b}$$

$$i_a + i_c - i_f = 0 \tag{6-6c}$$

and

$$-i_d + i_e + i_f = 0 \tag{6-6d}$$

respectively. These equations can be rewritten in matrix form as follows:

$$\begin{bmatrix} -1 & 1 & 0 & 1 & 0 & 0 \\ 0 & -1 & -1 & 0 & -1 & 0 \\ 1 & 0 & 1 & 0 & 0 & -1 \\ 0 & 0 & 0 & -1 & 1 & 1 \end{bmatrix} \begin{bmatrix} i_a \\ i_b \\ i_c \\ i_d \\ i_e \\ i_f \end{bmatrix} = 0 \tag{6-6e}$$

or

$$\mathbf{A}_a\mathbf{I}_e = \mathbf{0} \tag{6-6f}$$

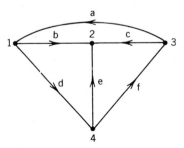

**FIGURE 6-10.** An oriented graph.

where $\mathbf{I}_e$ is the *edge current matrix* (or *edge current vector*), and $\mathbf{A}_a$ is the *incidence matrix* of the oriented graph. In general, $\mathbf{A}_a$ is defined as follows:

$\mathbf{A}_a = [a_{ij}]$ is of order $(n \times e)$, where each row corresponds to a node of the graph, and each column corresponds to an oriented edge. An element $a_{ij}$ of $\mathbf{A}_a$ is given by

$$a_{ij} = \begin{cases} 1 \text{ if the edge } j \text{ is incident to node } i \text{ and oriented } away \text{ from it} \\ -1 \text{ if the edge } j \text{ is incident to node } i \text{ and oriented } towards \text{ it} \\ 0 \text{ if the edge } j \text{ is not incident to node } i. \end{cases}$$

We see, therefore, that the incidence matrix $\mathbf{A}_a$ completely identifies its associated graph, showing which nodes and edges are incident to each other, and the orientations of the edges. There is a one-to-one correspondence between a graph and its incidence matrix: given one, we can obtain the other. In addition, from the standpoint of network analysis, the incidence matrix serves to formulate *Kirchhoff's current law in matrix form for the entire network*, Eq. (6-6f). This law depends only on the graph of the network.

The incidence matrix has several interesting properties.

1. Every column of $\mathbf{A}_a$ of a connected graph has *exactly* one $(-1)$ and one 1, and all other elements 0. Why?

2. The rank of $\mathbf{A}_a$ of a connected graph is, at most, $n - 1$. The rank is not $n$, for, if we add to the first row all the other rows, we obtain a row full of zeros [due to property (1)]. Such an addition is an elementary matrix transformation which does not change the rank of $\mathbf{A}_a$.[†] Every $(n \times n)$ determinant in $\mathbf{A}_a$ vanishes. Hence, the rank of $\mathbf{A}_a$ is at most $n - 1$. Later in this chapter, we shall prove that it is *exactly* $n - 1$, and therefore only $n - 1$ Kirchhoff's current law equations are independent (a result which we knew intuitively!).

3. Any one row can be deleted from $\mathbf{A}_a$ without any loss of information [due to property (1)]. From the point of view of networks, the resulting matrix is the one obtained in node analysis when the node corresponding to the eliminated row is grounded. The subscript "$a$" in $\mathbf{A}_a$ indicates "*all* nodes," and by deleting one row from $\mathbf{A}_a$ we get the *reduced incidence matrix* $\mathbf{A}$. ***Probs. 6-4***
***6-5***
***6-6***

The student will undoubtedly recognize that the relationship between $\mathbf{A}_a$ and $\mathbf{A}$ is similar to that between $\mathbf{Y}_i$ and $\mathbf{Y}$ (the indefinite and node admittance matrices) discussed in chapter 4. Further relationships between $\mathbf{A}$ and some of its submatrices will be developed subsequently.

## The Cut Set

Consider again the graph in Fig. 6-10 and its current equations, Eq. (6-6a) through

[†] See Appendix A.

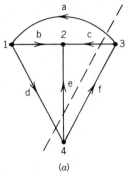

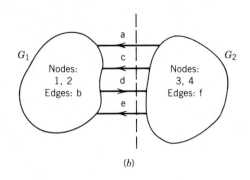

**FIGURE 6-11.**  A cut set.

(d). Let us add Eq. (6-6c) to Eq. (6-6d). The result is

$$i_a + i_c - i_d + i_e = 0. \tag{6-7}$$

What does this equation mean? If we introduce an imaginary plane that "cuts" edges $a, c, d$ and $e$, as shown in dotted lines in Fig. 6-11a, the total current crossing this plane is zero. Seen from a slightly different point of view, if we cut edges $a, c, d$ and $e$, as indicated by the dotted lines in Fig. 6-11b, the connected graph is separated into two parts, $G_1$ and $G_2$. The edges $a, c, d$ and $e$ are called a *cut set*. Another cut set is illustrated in Fig. 6-12a and b, consisting of the edges $a, b,$ and $d$. This set divides the connected graph into two separate parts, one of which is an isolated node.

We are ready now for the formal definition: a *cut set* (of a connected graph) is a *minimal* set of edges which, when cut, will separate the graph into two separate parts. Notice the emphasis on the word "minimal": no smaller number of edges must have this property. For example, edges $a, c,$ and $d$ are *not* a cut set (Fig. 6-11), nor are edges $a$ and $b$ (Fig. 6-12), since in both cases the graph remains connected after they are cut.

As we can see, the cut set equations are a generalization of Kirchhoff's current law. In fact, a cut set equation is a linear combination (by addition or subtraction) of Kirchhoff's law equations at two or more nodes. A systematic approach to cut sets is based on the following property: Each branch (of a tree of the graph) together with some links (of the co-tree) gives a *unique* cut set. Such a cut set is a *fundamental cut set*.

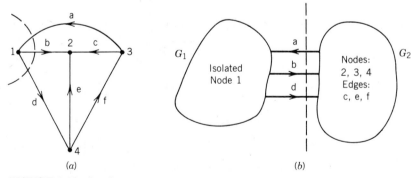

**FIGURE 6-12.**  Another cut set.

## Example 4

If we choose a tree of the graph in Fig. 6-10 as shown in Fig. 6-13 the following are the fundamental cut sets:

$$abd \qquad \text{(for branch } a)$$
$$cbdf \qquad \text{(for branch } c)$$
$$edf \qquad \text{(for branch } e).$$

Again, a fundamental cut set contains only one branch. The edges *acde* are certainly a cut set, but it is not a fundamental one in this example. In general, there are $n - 1$ fundamental cut sets, one for each branch.

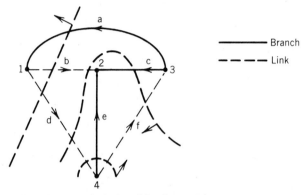

**FIGURE 6-13.** A choice of fundamental cut sets.

Let us continue with this example and write the equations for the fundamental cut sets. Also, let us assign an orientation to the cutting plane, chosen as that of the branch for each fundamental cut set (see Fig. 6-13). For the cut set *abd*, we have

$$i_a - i_b - i_d = 0 \qquad (6\text{-}8a)$$

Similarly, for the cut set *cfbd*,

$$i_c + i_b + i_d - i_f = 0 \qquad (6\text{-}8b)$$

and for the cut set *edf*,

$$i_e - i_d + i_f = 0 \qquad (6\text{-}8c)$$

In matrix form, these equations can be written as follows:

$$
\begin{bmatrix}
-1 & -1 & 0 & \vdots & 1 & 0 & 0 \\
1 & 1 & -1 & \vdots & 0 & 1 & 0 \\
0 & -1 & 1 & \vdots & 0 & 0 & 1
\end{bmatrix}
\begin{bmatrix}
i_b \\
i_d \\
i_f \\
\hdashline
i_a \\
i_c \\
i_e
\end{bmatrix}
= 0 \qquad (6\text{-}8d)
$$

Here, the edge current vector (column matrix) is written as

$$\mathbf{I}_e = \begin{bmatrix} \mathbf{I}_{link} \\ \hline \mathbf{I}_{branch} \end{bmatrix} \qquad (6-9)$$

As a result of this partitioning of $\mathbf{I}_e$ and of the way fundamental cut sets are formed, the matrix premultiplying $\mathbf{I}_e$ contains $\mathbf{U}$, the unit submatrix whose order is $(n - 1) \times (n - 1)$, and whose determinant is nonzero. Furthermore, it is the largest nonzero determinant in this matrix; and hence, the rank of this matrix is $(n - 1)$. Stated otherwise, it means that *the $(n - 1)$ fundamental cut set equations are linearly independent*. This conclusion is not surprising. Every fundamental cut set contains one branch which does not appear in any of the other fundamental cut sets. Therefore, the equation for one fundamental cut set cannot possibly be obtained from a linear combination of the other equations!   □   **Probs. 6-7**

**6-8**

**6-9**

One additional conclusion can be drawn: Since there are $n - 1$ independent fundamental cut sets, there must be no less than $n - 1$ independent Kirchhoff's current law equations. However, we found previously that there are, at most, $n - 1$ such independent equations. Therefore, there are *precisely* $n - 1$ independent Kirchhoff's current law equations. This means also that the rank of $\mathbf{A}_a$ (or of $\mathbf{A}$) is $n - 1$.

Next, we generalize the previous results. In particular, we wish to discuss the nature of the matrix that premultiplies $\mathbf{I}_e$ in Eq. (6-8d). For this purpose, let us define the *fundamental cut set* matrix $\mathbf{Q}_f$ as follows:

$\mathbf{Q}_f = [q_{ij}]$ is of order $(n - 1) \times e$, where each row corresponds to an oriented fundamental cut set, and each column corresponds to an oriented edge. An element $q_{ij}$ of $\mathbf{Q}_f$ is given by

$$q_{ij} = \begin{cases} 1 \text{ if edge } j \text{ is in cut set } i \text{ and their orientations agree} \\ -1 \text{ if edge } j \text{ is in cut set } i \text{ and their orientations disagree} \\ 0 \text{ if edge } j \text{ is not in cut set } i. \end{cases}$$

Also, let us arrange the columns of $\mathbf{Q}_f$ so that the first $e - n + 1$ are the links and the remaining $n - 1$ are the branches for the chosen tree. Let the first fundamental cut set (the first row of $\mathbf{Q}_f$) be the one formed by the branch listed first, the second fundamental cut set—by the branch listed second, etc. In this fashion, $\mathbf{Q}_f$ will have the partitioned form

$$\mathbf{Q}_f = [\mathbf{Q}_{f11} \vdots \mathbf{U}], \qquad (6-10)$$

exhibiting the unit submatrix $\mathbf{U}$ for the branches. The edge current matrix is partitioned accordingly, as before

$$\mathbf{I}_e = \begin{bmatrix} \mathbf{I}_{link} \\ \hline \mathbf{I}_{branch} \end{bmatrix} \qquad (6-9)$$

The fundamental cut set equations are

$$\mathbf{Q}_f \mathbf{I}_e = 0 \qquad (6-11a)$$

that is,

$$[\mathbf{Q}_{f11} \vdots \mathbf{U}]\begin{bmatrix} \mathbf{I}_{\text{link}} \\ \hdashline \mathbf{I}_{\text{branch}} \end{bmatrix} = \mathbf{0} \tag{6-11b}$$

Carrying out this product, we obtain

$$\mathbf{I}_{\text{branch}} = -\mathbf{Q}_{f11}\mathbf{I}_{\text{link}} \tag{6-12}$$

This result suggests that, if we know all the link currents, the branch currents can be computed. For example, in Fig. 6-13, the knowledge of $i_b$, $i_d$ and $i_f$ will determine $i_a$, $i_c$ and $i_e$ by using Eq. (6-12). However, what is $\mathbf{Q}_{f11}$? To answer this question, let us go back to the reduced incidence matrix $\mathbf{A}$ for the graph. If $\mathbf{I}_e$ is retained as in Eq. (6-9), we can partition $\mathbf{A}$ into

$$\mathbf{A} = [\mathbf{A}_{11} \vdots \mathbf{A}_{12}] \tag{6-13}$$

where $\mathbf{A}_{11}$ is of order $(n - 1) \times (e - n + 1)$ and corresponds to the links, and where $\mathbf{A}_{12}$ is of order $(n - 1) \times (n - 1)$ and corresponds to the branches. Then

$$\mathbf{AI}_e = [\mathbf{A}_{11} \vdots \mathbf{A}_{12}]\begin{bmatrix} \mathbf{I}_{\text{link}} \\ \hdashline \mathbf{I}_{\text{branch}} \end{bmatrix} = \mathbf{0} \tag{6-14}$$

Multiplying, we get

$$\mathbf{A}_{11}\mathbf{I}_{\text{link}} + \mathbf{A}_{12}\mathbf{I}_{\text{branch}} = \mathbf{0} \tag{6-15}$$

It can be shown (although we shall not do it) that $\mathbf{A}_{12}$, corresponding to a tree, is always nonsingular. Therefore, we obtain from Eq. (6-15)

$$\mathbf{I}_{\text{branch}} = -\mathbf{A}_{12}^{-1}\mathbf{A}_{11}\mathbf{I}_{\text{link}} \tag{6-16}$$

Comparison of Eqs. (6-12) and (6-16) yields the answer

$$\mathbf{Q}_{f11} = \mathbf{A}_{12}^{-1}\mathbf{A}_{11} \tag{6-17}$$

In summary, we have shown the relationships between link currents and branch currents, Eq. (6-12), as well as the derivation of the fundamental cut set matrix from the reduced incidence matrix, Eqs. (6-10) and (6-17). To conclude this discussion, let us illustrate with the previous example, continued and concluded here.

---

## Example 5

For the tree chosen in Fig. 6-13, the fundamental cut set equation $\mathbf{Q}_f\mathbf{I}_e = \mathbf{0}$ is given in Eq. (6-8d). Let us write the reduced incidence matrix $\mathbf{A}$,

$$\mathbf{A} = [\mathbf{A}_{11} \vdots \mathbf{A}_{12}] = \begin{matrix} & \begin{matrix} b & d & f & a & c & e \end{matrix} \\ \begin{matrix} 1 \\ 2 \\ 3 \end{matrix} & \begin{bmatrix} 1 & 1 & 0 & -1 & 0 & 0 \\ -1 & 0 & 0 & 0 & -1 & -1 \\ 0 & 0 & -1 & 1 & 1 & 0 \end{bmatrix} \end{matrix}$$

We have labelled the rows $(1, 2, 3)$ and the columns $(b, d, f, a, c, e)$ to help identify the nodes, the links, and the branches. Here,

$$\mathbf{A}_{12}^{-1} = \begin{bmatrix} -1 & 0 & 0 \\ 1 & 0 & 1 \\ -1 & -1 & -1 \end{bmatrix}$$

$$\therefore \mathbf{Q}_{f11} = \begin{bmatrix} -1 & 0 & 0 \\ 1 & 0 & 1 \\ -1 & -1 & -1 \end{bmatrix} \begin{bmatrix} 1 & 1 & 0 \\ -1 & 0 & 0 \\ 0 & 0 & -1 \end{bmatrix} = \begin{bmatrix} -1 & -1 & 0 \\ 1 & 1 & -1 \\ 0 & -1 & 1 \end{bmatrix}$$

which agrees with Eq. (6-8d).

$$\therefore \mathbf{I}_{\text{branch}} = -\mathbf{Q}_{f11}\mathbf{I}_{\text{link}} = \begin{bmatrix} 1 & 1 & 0 \\ -1 & -1 & 1 \\ 0 & 1 & -1 \end{bmatrix} \mathbf{I}_{\text{link}}$$

$$\therefore \begin{bmatrix} i_a \\ i_c \\ i_e \end{bmatrix} = \begin{bmatrix} 1 & 1 & 0 \\ -1 & -1 & 1 \\ 0 & 1 & -1 \end{bmatrix} \begin{bmatrix} i_b \\ i_d \\ i_f \end{bmatrix}$$

which reads $i_a = i_b + i_d$; $i_c = -i_b - i_d + i_f$; $i_e = i_d - i_f$ as expected.   □

---

As a final note, let us compare the fundamental cut set equations with those using the reduced incidence matrix. Once again,

$$\mathbf{AI}_e = [\mathbf{A}_{11} \vdots \mathbf{A}_{12}]\mathbf{I}_e = 0 \tag{6-14}$$

and

$$\mathbf{Q}_f\mathbf{I}_e = [\mathbf{Q}_{f11} \vdots \mathbf{U}]\mathbf{I}_e = 0 \tag{6-11b}$$

From this comparison, as well as from our previous discussion, it is apparent that the matrix $\mathbf{Q}_f$ can be obtained from matrix $\mathbf{A}$ by elementary matrix transformations.[†] In other words, the rows of $\mathbf{Q}_f$ are linear combinations of the rows of $\mathbf{A}$—a conclusion that is also in agreement from the electrical point of view. So, we can perform elementary matrix operations on $\mathbf{A}$ until we obtain the unit submatrix $\mathbf{U}$ in the right position. The left submatrix is then $\mathbf{Q}_f$. This result, needless to say, is identical to the one obtained by using Eq. (6-17).

## 6-3   THE CIRCUIT MATRIX

At the end of section 6-1, we observed that every link (of a co-tree) forms a unique circuit (loop) with some branches of the chosen tree. Such a circuit is a *fundamental circuit*.

[†] See Appendix A.

## Example 6

For the tree shown in Fig. 6-14a, the following are the fundamental circuits:

$$bca \qquad \text{(for link } b)$$
$$deca \qquad \text{(for link } d)$$
$$fce \qquad \text{(for link } f)$$

To repeat, a fundamental circuit contains only *one* link. The edges $adf$ are certainly a circuit, but it is not a fundamental circuit in this example. In general, there are (in a connected graph) as many fundamental circuits as links, that is, $l = e - n + 1$.

Let us assign an orientation to each fundamental circuit, chosen as that of its link (see Fig. 6-14). Also, let the orientation of each edge be the positive reference for the voltage *drop* across the edge. Kirchhoff's voltage law, written for each fundamental circuit, is

$$v_b - v_c + v_a = 0 \tag{6-18a}$$

for the fundamental circuit $bca$, and

$$v_d + v_e - v_c + v_a = 0 \tag{6-18b}$$

and

$$v_f + v_c - v_e = 0 \tag{6-18c}$$

for $deca$ and $fce$. Rewriting these in matrix form, we obtain

$$\begin{bmatrix} 1 & 0 & 0 & 1 & -1 & 0 \\ 0 & 1 & 0 & 1 & -1 & 1 \\ 0 & 0 & 1 & 0 & 1 & -1 \end{bmatrix} \begin{bmatrix} v_b \\ v_d \\ v_f \\ v_a \\ v_c \\ v_e \end{bmatrix} = 0 \tag{6-18d}$$

Here, the edge voltage vector (column matrix) is written as

$$\mathbf{V}_e = \begin{bmatrix} \mathbf{V}_{link} \\ \hline \mathbf{V}_{branch} \end{bmatrix} \tag{6-19}$$

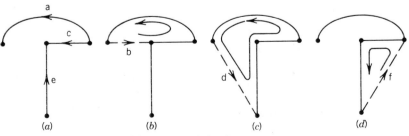

**FIGURE 6-14.** A choice of fundamental circuits.

and, as a result, the matrix premultiplying $\mathbf{V}_e$ in Eq. (6-18d) contains a unit submatrix $\mathbf{U}$ of order $(e - n + 1) \times (e - n + 1)$. Therefore, the rank of this matrix is $e - n + 1$. In other words, *the $(e - n + 1)$ fundamental circuit equations are linearly independent*. This conclusion should be also obvious from the way we construct fundamental circuits: Each one contains one *new* link, and therefore the equation for one fundamental circuit cannot be obtained from a linear combination of the other equations.

☐   ***Probs. 6-10***<br>***6-11***

---

We have, therefore, reached the conclusion that there are *at least* $l = e - n + 1$ independent Kirchhoff's voltage law equations. To establish their exact number, let us reason as follows: We need, to begin with, $2e$ independent equations for solving the network, since there are $e$ edges and for each edge we want to know the voltage and the current. (If the edge happens to be a voltage or a current source, this is a degenerate case where the voltage or the current of the edge is already known.) The requirement of $2e$ equations can be reduced to $e$, because of the given $v$–$i$ relationship for every edge. Next, we use the topology of the network: Kirchhoff's current law, as we saw, provides $n - 1$ independent equations. Hence, Kirchhoff's voltage law must account for the remaining $e - (n - 1) = e - n + 1$ independent equations. In brief, there are *exactly* $l = e - n + 1$ independent Kirchhoff's voltage law equations.

Let us generalize the discussion that lead to Eq. (6-18d). We define the *fundamental circuit matrix* $\mathbf{B}_f$ as follows:

$\mathbf{B}_f = [b_{ij}]$ is of order $(e - n + 1) \times e$, where each row corresponds to an oriented fundamental circuit, and each column corresponds to an oriented edge. An element $b_{ij}$ of $\mathbf{B}_f$ is given by

$$
b_{ij} = \begin{cases} 1 \text{ if edge } j \text{ is in circuit } i \text{ and their orientations agree} \\ -1 \text{ if edge } j \text{ is in circuit } i \text{ and their orientations disagree} \\ 0 \text{ if edge } j \text{ is not in circuit } i. \end{cases}
$$

Also, let us arrange the columns of $\mathbf{B}_f$ so that the first $e - n + 1$ are the links and the remaining $n - 1$ are the branches of the chosen tree. The first fundamental circuit (the first row in $\mathbf{B}_f$) will be the one formed by the link listed first, the second fundamental circuit, by the link listed second, etc. Thus, $\mathbf{B}_f$ will have the form

$$
\mathbf{B}_f = [\mathbf{U} \vdots \mathbf{B}_{f12}] \tag{6-20}
$$

and Eq. (6-18d) can be generalized into

$$
\mathbf{B}_f \mathbf{V}_e = [\mathbf{U} \vdots \mathbf{B}_{f12}] \begin{bmatrix} \mathbf{V}_{link} \\ \hline \mathbf{V}_{branch} \end{bmatrix} = 0 \tag{6-21}
$$

Upon carrying out the multiplication, we obtain

$$
\mathbf{V}_{link} + \mathbf{B}_{f12} \mathbf{V}_{branch} = 0 \tag{6-22}
$$

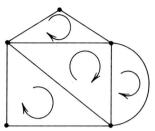

**FIGURE 6-15.** Loops traced around windowpanes of a planar graph.

or

$$\mathbf{V}_{\text{link}} = -\mathbf{B}_{f12}\mathbf{V}_{\text{branch}} \tag{6-23}$$

The last result relates the link voltages to the branch voltages. In effect, it means that any link voltage is a linear combination, by addition or subtraction, of branch voltages, since $\mathbf{B}_{f12}$ consists only of 0, 1, or $-1$ as elements. This conclusion is also obvious from the electrical network point of view.

At this point it is proper to consider a well-known artifice, the "obvious" choice of loops around the windowpanes, as shown in Fig. 6-15.[†] Are the resulting equations independent? To answer this question, we must stress, first of all, that such window-panes can be meaningful only for *planar* graphs. For a planar graph, it is easy to verify that there are $e - n + 1$ such windowpanes: Draw a tree and fill in, one at a time, the links. Whenever a windowpane is formed it may, or may not, be also a fundamental circuit. To verify whether the resulting equations around these windowpanes are independent, assume that they are linearly dependent; if so, we must have[‡]

$$x_1 K_1 + x_2 K_2 + \cdots + x_l K_l = 0 \qquad l = e - v + 1 \tag{6-24}$$

where some $x$'s are non-zero, and $K_i (1 \le i \le l)$ is the left-hand side of the $i$th mesh equation. Equation 6-24 states our assumption that a mesh equation is a linear combination of some of the other equations. Let $x_1, x_2, x_3, \ldots, x_q (q \le l)$ be nonzero, i.e., let the first mesh equation depend linearly on the second, the third, ..., the $q$th. Thus

$$x_1 K_1 + x_2 K_2 + \cdots + x_q K_q = 0 \tag{6-25}$$

Remove from the planar graph the edges not contained in those $q$ meshes. The remaining graph (these $q$ meshes) will certainly have one or more edges belonging only to one mesh—a "border" edge. Therefore, one of the $K$'s contains a term corresponding to this "border" edge. This term does not appear in the other $K$'s. Therefore, this mesh equation cannot be a linear combination of the other equations. Hence our assumption was false and the $e - n + 1$ mesh equations are linearly independent. ***Prob. 6-12***

[†] Such loops are also called "meshes."
[‡] See appendix A.

In summary, there are exactly $e - n + 1$ independent Kirchhoff's voltage law equations. A most general and systematic way to obtain these is by fundamental circuits. For planar graphs, the windowpane (mesh) method is also satisfactory.

The discussion of the fundamental circuit matrix has proceeded, so far, along lines dual to those of the fundamental cut set matrix. Before obtaining an expression for $\mathbf{B}_{f12}$, Eq. (6-23), we shall prove a result that is both significant and elegant: Let the columns (edges) of the matrices $\mathbf{A}$ and $\mathbf{B}_f$ be arranged in the same order—links first, then branches. Then,

$$\mathbf{AB}_f^T = \mathbf{0} \tag{6-26a}$$

and

$$\mathbf{B}_f\mathbf{A}^T = \mathbf{0} \tag{6-26b}$$

where the superscript "$T$" indicates the transpose. As an example, the reader should check these results in Eq. (6-18d) and in example 5 of the previous section.

To prove this result in general, we must realize first that the $(p, q)$ element of the product $\mathbf{AB}_f^T$ is obtained from the $p$th row of $\mathbf{A}$—corresponding to the $p$th node—and the $q$th *row* of $\mathbf{B}_f$—corresponding to the $q$th circuit. The multiplication is the usual matrix product of corresponding terms, then their addition. There are now two possibilities to consider: node $p$ is in circuit $q$, and node $p$ is not in circuit $q$. In the first case, the various orientations of the two edges incident to node $p$ (there are *precisely* two such edges) and the orientation of circuit $q$ are shown in Fig. 6-16. These are the only nonzero contributions in the product, and, in all four cases, they are $+1$ and $-1$, adding to zero. In the second case, when node $p$ is not in circuit $q$, the edges incident to node $p$ are not in circuit $q$; therefore, the product of the corresponding terms will be zero.

The result in Eq. (6-26b) is obtained by merely transposing Eq. (6-26a).

It is recognized that the previous argument will hold for $\mathbf{A}_a$ and $\mathbf{B}_a$, or for the reduced matrices $\mathbf{A}$ and $\mathbf{B}$. Due to their importance, these results are listed here

$$\mathbf{A}_a\mathbf{B}_a^T = \mathbf{0} \tag{6-26c}$$

$$\mathbf{B}_a\mathbf{A}_a^T = \mathbf{0} \tag{6-26d}$$

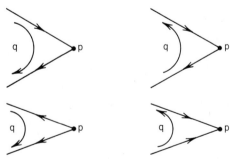

**FIGURE 6-16.** Relative orientations of a circuit and edges.

$$\mathbf{AB}^T = 0 \tag{6-26e}$$

$$\mathbf{BA}^T = 0 \tag{6-26f}$$

In each case, the columns in $\mathbf{A}_a$ and $\mathbf{B}_a$, or in $\mathbf{A}$ and $\mathbf{B}$, must be arranged in the same edge order. Similar results can be obtained with the cut set matrix instead of the incidence matrix.

Let us now find the expression for $\mathbf{B}_{f12}$ Eq. (6-22) in terms of submatrices of $\mathbf{A}$. For this purpose, use Eq. (6-26a), together with Eqs. (6-13) and (6-20),

$$[\mathbf{A}_{11}\!\vdots\!\mathbf{A}_{12}]\left[\begin{array}{c}\mathbf{U}\\ \text{------} \\ \mathbf{B}_{f12}^T\end{array}\right] = 0 \tag{6-27}$$

that is,

$$\mathbf{A}_{11} + \mathbf{A}_{12}\mathbf{B}_{f12}^T = 0 \tag{6-28}$$

And since $\mathbf{A}_{12}^{-1}$ exists,

$$\mathbf{B}_{f12}^T = -\mathbf{A}_{12}^{-1}\mathbf{A}_{11} = -\mathbf{Q}_{f11} \tag{6-29}$$

when compared with Eq. (6-17). Then Eqs. (6-10) and (6-20) become

$$\mathbf{Q}_f = [\mathbf{Q}_{f11}\!\vdots\!\mathbf{U}] \tag{6-10a}$$

$$\mathbf{B}_f = [\mathbf{U}\!\vdots\!-\mathbf{Q}_{f11}^T] \tag{6-20a}$$

In summary, we have developed the relationships between link voltages and branch voltages, Eq. (6-23), and derived the fundamental circuit matrix from the reduced incidence matrix, Eqs. (6-20) and (6-29). Let us illustrate with the choice of circuits shown in Fig. 6-14.

---

## Example 7

From example 6 of the previous section, pertaining to the same graph, we have

$$\mathbf{Q}_{f11} = \begin{bmatrix} -1 & -1 & 0 \\ 1 & 1 & -1 \\ 0 & -1 & 1 \end{bmatrix}$$

and so

$$\mathbf{B}_{f12} = -\mathbf{Q}_{f11}^T = \begin{bmatrix} 1 & -1 & 0 \\ 1 & -1 & 1 \\ 0 & 1 & -1 \end{bmatrix}$$

in agreement with Eq. (6-18d). Furthermore, to illustrate Eq. (6-23), we write

$$\begin{bmatrix} v_b \\ v_d \\ v_f \end{bmatrix} = \begin{bmatrix} -1 & 1 & 0 \\ -1 & 1 & -1 \\ 0 & -1 & 1 \end{bmatrix} \begin{bmatrix} v_a \\ v_c \\ v_e \end{bmatrix}$$

which reads $v_b = -v_a + v_c$; $v_d = -v_a + v_c - v_e$; and $v_f = -v_c + v_e$ as expected.

□ **Probs. 6-13**
**6-14**
**6-15**

## 6-4   TRANSFORMATION OF VARIABLES: LOOP AND NODE EQUATIONS

So far in our discussion, Kirchhoff's current and voltage laws were expressed in terms of *edge* variables,

$$\mathbf{AI}_e = 0 \tag{6-30}$$

and

$$\mathbf{BV}_e = 0 \tag{6-31}$$

These variables are the currents and voltages of each element in the network.

We should like to relate these edge variables to *loop* currents and *node* voltages, so commonly used in network analysis. For this purpose, consider the following transformations of variables:

$$\mathbf{I}_e = \mathbf{B}^T \mathbf{I}_m \tag{6-32}$$

$$\mathbf{V}_e = \mathbf{A}^T \mathbf{V}_n \tag{6-33}$$

where $\mathbf{I}_m$ is a vector (column matrix) of the $(e - n + 1)$ *loop* (or *mesh*) *currents*

$$\mathbf{I}_m = \begin{bmatrix} i_{m1} \\ i_{m2} \\ \vdots \\ i_{m,e-n+1} \end{bmatrix}$$

and $\mathbf{V}_n$ is a column matrix of $(n - 1)$ *node-to-datum* (or briefly, *node*) *voltages*,

$$\mathbf{V}_n = \begin{bmatrix} v_{n1} \\ v_{n2} \\ \vdots \\ v_{n,n-1} \end{bmatrix}$$

Consider first the current transformation, Eq. (6-32). The justification for substituting these variables rests on Eq. (6-26e), $\mathbf{AB}^T = 0$; for, if we substitute Eq. (6-32) into Eq. (6-30), we obtain

$$\mathbf{AI}_e = \mathbf{A}(\mathbf{B}^T \mathbf{I}_m) = (\mathbf{AB}^T)\mathbf{I}_m = 0 \tag{6-36}$$

as required. Let us illustrate with an example.

## Example 8

For the graph in Fig. 6-17, a reduced circuit matrix is

$$
B = \begin{array}{c} \\ 1 \\ 2 \\ 3 \\ 4 \end{array}
\begin{array}{cccccccc}
a & b & c & d & e & f & g & h \\
\left[\begin{array}{cccccccc}
1 & 1 & 0 & 0 & -1 & 0 & 0 & 0 \\
0 & -1 & 1 & 1 & 0 & 0 & 0 & 0 \\
0 & 0 & 0 & -1 & 0 & 0 & 1 & 1 \\
1 & 0 & 1 & 0 & 0 & 1 & 0 & 1
\end{array}\right]
\end{array}
$$

Here, again, we have labelled rows (circuits) and columns (edges) for ease of identification. The four circuits are: *1-a-2-b-3-e-1*, *2-c-4-d-3-b-2*, *3-d-4-h-5-g-3*, and *1-a-2-c-4-h-5-f-1*. The loop transformation, Eq. (6-32), reads here

$$
\begin{bmatrix} i_a \\ i_b \\ i_c \\ i_d \\ i_e \\ i_f \\ i_g \\ i_h \end{bmatrix}
=
\begin{bmatrix}
1 & 0 & 0 & 1 \\
1 & -1 & 0 & 0 \\
0 & 1 & 0 & 1 \\
0 & 1 & -1 & 0 \\
-1 & 0 & 0 & 0 \\
0 & 0 & 0 & 1 \\
0 & 0 & 1 & 0 \\
0 & 0 & 1 & 1
\end{bmatrix}
\begin{bmatrix} i_{m1} \\ i_{m2} \\ i_{m3} \\ i_{m4} \end{bmatrix}
$$

that is

$$
i_a = i_{m1} + i_{m4}; \; i_b = i_{m1} - i_{m2}; \; i_c = i_{m2} + i_{m4};
$$

$$
i_d = i_{m2} - i_{m3}; \; i_e = -i_{m1}; \; i_f = i_{m4};
$$

$$
i_g = i_{m3}; \; i_h = i_{m3} + i_{m4};
$$

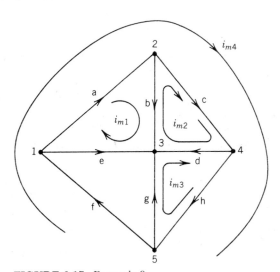

**FIGURE 6-17.** Example 8.

giving the edge currents in terms of the loop currents. ☐ ***Probs. 6-16***
***6-17***

In all cases, link currents are always a proper choice for a set of independent loop currents, in planar and in nonplanar graphs.

*Example 9*

In Fig. 6-17, let the tree be *abdf*, then:

a) Loop current $i_{m1}$ is chosen as $i_g$, and the first loop is *5-g-3-b-2-a-1-f-5*
b) Loop current $i_{m2}$ is chosen as $i_e$, and the second loop is *1-e-3-b-2-a-1*
c) Loop current $i_{m3}$ is chosen as $i_c$, and the third loop is *2-c-4-d-3-b-2*
d) Loop current $i_{m4}$ is chosen as $i_h$, and the fourth loop is *4-h-5-f-1-a-2-b-3-d-4* ☐

*Example 10*

In Fig. 6-17, let the tree be *bdeg*. Then the loop currents, chosen as the link currents $i_a$, $i_c$, $i_h$, and $i_f$, are the familiar window pane (mesh) currents. As stressed earlier, window pane loops are restricted to planar graphs. Link currents, i.e., fundamental circuits, are valid for all graphs and constitute a systematic method for obtaining independent Kirchhoff's voltage law equations. ☐

Next, consider the voltage transformation, $V_e = A^T V_n$. This substitution is justified, since Kirchhoff's voltage law is preserved

$$BV_e = B(A^T V_n) = (BA^T)V_n = 0 \qquad (6\text{-}37)$$

because of Eq. (6-26f).

*Example 11*

For the graph in Fig. 6-17, the reduced incidence matrix **A** is given by

$$
A = \begin{array}{c} \\ 1 \\ 2 \\ 3 \\ 4 \end{array}
\begin{array}{c} a \quad\;\; b \quad\;\; c \quad\;\; d \quad\;\; e \quad\;\; f \quad\;\; g \quad\;\; h \end{array}
\left[ \begin{array}{cccccccc}
1 & 0 & 0 & 0 & 1 & -1 & 0 & 0 \\
-1 & 1 & 1 & 0 & 0 & 0 & 0 & 0 \\
0 & -1 & 0 & -1 & -1 & 0 & -1 & 0 \\
0 & 0 & -1 & 1 & 0 & 0 & 0 & 1
\end{array} \right]
$$

where row 5 is omitted (note 5 is grounded). The node transformation, Eq. (6-33), becomes here

$$
\begin{bmatrix} v_a \\ v_b \\ v_c \\ v_d \\ v_e \\ v_f \\ v_g \\ v_h \end{bmatrix}
=
\begin{bmatrix}
1 & -1 & 0 & 0 \\
0 & 1 & -1 & 0 \\
0 & 1 & 0 & -1 \\
0 & 0 & -1 & 1 \\
1 & 0 & -1 & 0 \\
-1 & 0 & 0 & 0 \\
0 & 0 & -1 & 0 \\
0 & 0 & 0 & 1
\end{bmatrix}
\begin{bmatrix} v_{n1} \\ v_{n2} \\ v_{n3} \\ v_{n4} \end{bmatrix}
$$

where $v_{n1}$ is the voltage of node 1 with respect to ground, etc. These relations read

$$ v_a = v_{n1} - v_{n2}; \; v_b = v_{n2} - v_{n3}; \; v_c = v_{n1} - v_{n4}; $$

$$ v_d = -v_{n3} + v_{n4}; \; v_e = v_{n1} - v_{n3}; \; v_f = -v_{n1}; $$

$$ v_g = -v_{n3}; \; v_h = v_{n4}; $$

giving all the edge voltages in terms of the new variables.    □

---

Finally, we turn our attention to the derivation of loop and node equations; these are the familiar equations used in chapter 3. In the present derivation, initial conditions are assumed to be zero; this assumption is not too severe. On the other hand, the following derivation is made easier with this assumption. The scheme of this derivation is shown earlier, in Fig. 6-4. The $v$–$i$ relationship for an element is of the general form

$$
V_e(s) = \left( R_e + L_e s + \frac{1}{C_e s} \right) I_e(s) + E_e(s)
$$

$$
= Z_e(s) I_e(s) + E_e(s) \tag{6-38}
$$

or

$$
I_e(s) = \left( G_e + \frac{1}{L_e s} + C_e s \right) V_e(s) + J_e(s)
$$

$$
= Y_e(s) V_e(s) + J_e(s) \tag{6-39}
$$

Both equations are Laplace transformed. The subscript $e$ indicates "element" (or "edge"). An element may have a voltage source in it and/or a current source across it. These are accounted for by $E_e(s)$ and $J_e(s)$, respectively. See Fig. 6-18. These $v$–$i$ relations are written for all the elements of the network in matrix form, dropping for convenience the letter $s$

$$
\mathbf{V}_e = \mathbf{Z}_e \mathbf{I}_e + \mathbf{E}_e \tag{6-40}
$$

or

$$
\mathbf{I}_e = \mathbf{Y}_e \mathbf{V}_e + \mathbf{J}_e \tag{6-41}
$$

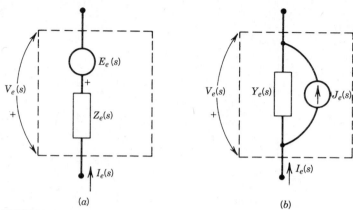

*(a)* *(b)*

**FIGURE 6-18.** An element in a network.

Now combine Eq. (6-40) with Kirchhoff's voltage law, Eq. (6-31):

$$\mathbf{BV}_e = \mathbf{B}(\mathbf{Z}_e\mathbf{I}_e + \mathbf{E}_e) = 0 \tag{6-42}$$

$$\therefore \ \mathbf{BZ}_e\mathbf{I}_e = -\mathbf{BE}_e \tag{6-43}$$

Introduce the new variables, the loop currents of Eq. (6-32):

$$\mathbf{BZ}_e\mathbf{B}^T\mathbf{I}_m = -\mathbf{BE}_e \tag{6-44}$$

Equation (6-44) is the set of loop equations discussed in Chapter 3 ($\mathbf{ZI} = \mathbf{E}$). We therefore identify

$$\mathbf{BZ}_e\mathbf{B}^T = \mathbf{Z} \tag{6-45}$$

as the square loop impedance matrix, $\mathbf{I}_m$ as the column matrix (vector) of the unknown transform loop currents, $\mathbf{I}$, and

$$-\mathbf{BE}_e = \mathbf{E} \tag{6-46}$$

as the column matrix of the voltage sources around each loop.

---

### Example 12

Let us use the example shown in Fig. 3-3 of Chapter 3, with all initial conditions zero. The network (with initial conditions zero), its oriented graph, and a tree are shown in Fig. 6-19. The fundamental loops are: *3-a-1-d-4-e-2-c-3*, *1-b-2-e-4-d-1*, and *4-f-3-c-2-e-4*. Accordingly, $\mathbf{B}_f$ is given by

$$\mathbf{B}_f = \begin{array}{c} \\ 1 \\ 2 \\ 3 \end{array} \begin{array}{cccccc} a & b & f & c & d & e \\ \left[\begin{array}{ccc|ccc} 1 & 0 & 0 & -1 & 1 & 1 \\ 0 & 1 & 0 & 0 & -1 & -1 \\ 0 & 0 & 1 & 1 & 0 & -1 \end{array}\right] \end{array}$$

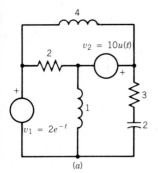
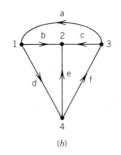
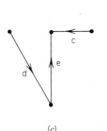

ohms
henries
farads

**FIGURE 6-19.** Example 12.

The $v$–$i$ relations for the edges Eq. (6-40) are written by inspection

$$
\begin{bmatrix} V_a \\ V_b \\ V_f \\ V_c \\ V_d \\ V_e \end{bmatrix} =
\begin{bmatrix}
4s & 0 & 0 & 0 & 0 & 0 \\
0 & 2 & 0 & 0 & 0 & 0 \\
0 & 0 & \left(3 + \dfrac{2}{s}\right) & 0 & 0 & 0 \\
0 & 0 & 0 & 0 & 0 & 0 \\
0 & 0 & 0 & 0 & 0 & 0 \\
0 & 0 & 0 & 0 & 0 & s
\end{bmatrix}
\begin{bmatrix} I_a \\ I_b \\ I_f \\ I_c \\ I_d \\ I_e \end{bmatrix} +
\begin{bmatrix} 0 \\ 0 \\ 0 \\ \dfrac{10}{s} \\ \dfrac{2}{s+1} \\ 0 \end{bmatrix}
$$

The fundamental loop equations, according to Eq. (6-44), are then

$$
\begin{bmatrix}
1 & 0 & 0 & -1 & 1 & 1 \\
0 & 1 & 0 & 0 & -1 & -1 \\
0 & 0 & 1 & 1 & 0 & -1
\end{bmatrix}
\begin{bmatrix}
4s & 0 & 0 & 0 & 0 & 0 \\
0 & 2 & 0 & 0 & 0 & 0 \\
0 & 0 & \left(3 + \dfrac{2}{s}\right) & 0 & 0 & 0 \\
0 & 0 & 0 & 0 & 0 & 0 \\
0 & 0 & 0 & 0 & 0 & 0 \\
0 & 0 & 0 & 0 & 0 & s
\end{bmatrix}
$$

$$
\times \begin{bmatrix}
1 & 0 & 0 \\
0 & 1 & 0 \\
0 & 0 & 1 \\
-1 & 0 & 1 \\
1 & -1 & 0 \\
1 & -1 & -1
\end{bmatrix} \mathbf{I}_m = -
\begin{bmatrix}
1 & 0 & 0 & -1 & 1 & 1 \\
0 & 1 & 0 & 0 & -1 & -1 \\
0 & 0 & 1 & 1 & 0 & -1
\end{bmatrix}
\begin{bmatrix} 0 \\ 0 \\ 0 \\ \dfrac{10}{s} \\ \dfrac{2}{(s+1)} \\ 0 \end{bmatrix}
$$

and, upon multiplying out, we obtain

$$
\begin{bmatrix}
5s & -s & -s \\
-s & 2+s & s \\
-s & s & 3+\dfrac{2}{s}+s
\end{bmatrix}
\begin{bmatrix}
I_{m1} \\
I_{m2} \\
I_{m3}
\end{bmatrix}
=
\begin{bmatrix}
\dfrac{10}{s}-\dfrac{2}{s+1} \\
\dfrac{2}{s+1} \\
\dfrac{10}{s}
\end{bmatrix}
$$

the three fundamental loop equations.

☐ **Probs. 6-18**
**6-19**

On a dual basis, we derive the node equations. Starting with the $v$–$i$ relations of the edges, Eq. (6-41], and Kirchhoff's current law, Eq. (6-30), we get

$$\mathbf{AI}_e = \mathbf{A}(\mathbf{Y}_e\mathbf{V}_e + \mathbf{J}_e) = 0 \tag{6-47}$$

$$\therefore\ \mathbf{AY}_e\mathbf{V}_e = -\mathbf{AJ}_e \tag{6-48}$$

Now introduce the new variables, the node voltages of Eq. (6-33),

$$\mathbf{AY}_e\mathbf{A}^T\mathbf{V}_n = -\mathbf{AJ}_e \tag{6-49}$$

Equation (6-49) is the set of node equations (see Chapter 3, $\mathbf{YV} = \mathbf{J}$). Thus,

$$\mathbf{AY}_e\mathbf{A}^T = \mathbf{Y} \tag{6-50}$$

is the (square) node admittance matrix, $\mathbf{V}_n$ is the column matrix of the unknown transform node voltages, $\mathbf{V}$, and

$$-\mathbf{AJ}_e = \mathbf{J} \tag{6-51}$$

is the column matrix of the current sources at each node.

### Example 13

The example shown in Fig. 3-5, Chapter 3, is reviewed here, with all initial conditions zero. The network, its oriented graph, and a tree are shown in Fig. 6-20. (By the way, notice that, for convenience, edge $a$ represents here two elements; similarly, edge $f$ represented two elements in the previous example.) With node 3 grounded, the reduced incidence matrix is

$$
\mathbf{A} =
\begin{array}{c}
\\
1 \\
2
\end{array}
\begin{array}{cccccc}
a & b & e & c & d \\
\begin{bmatrix}
1 & -1 & 0 & 1 & 0 \\
-1 & 0 & 1 & 0 & 1
\end{bmatrix}
\end{array}
$$

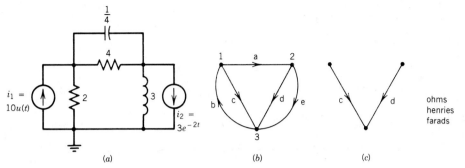

**FIGURE 6-20.** Example 13.

The $v$–$i$ relations for the edges are written by inspection

$$
\begin{bmatrix} I_a \\ I_b \\ I_e \\ I_c \\ I_d \end{bmatrix} = \begin{bmatrix} \left(\dfrac{s}{4}+\dfrac{1}{4}\right) & 0 & 0 & 0 & 0 \\ 0 & 0 & 0 & 0 & 0 \\ 0 & 0 & 0 & 0 & 0 \\ 0 & 0 & 0 & \dfrac{1}{2} & 0 \\ 0 & 0 & 0 & 0 & \dfrac{1}{3s} \end{bmatrix} \begin{bmatrix} V_a \\ V_b \\ V_e \\ V_c \\ V_d \end{bmatrix} + \begin{bmatrix} 0 \\ \dfrac{10}{s} \\ \dfrac{3}{s+2} \\ 0 \\ 0 \end{bmatrix}
$$

The node equations are, according to Eq. (6-49)

$$
\begin{bmatrix} 1 & -1 & 0 & 1 & 0 \\ -1 & 0 & 1 & 0 & 1 \end{bmatrix} \begin{bmatrix} \left(\dfrac{s}{4}+\dfrac{1}{4}\right) & 0 & 0 & 0 & 0 \\ 0 & 0 & 0 & 0 & 0 \\ 0 & 0 & 0 & 0 & 0 \\ 0 & 0 & 0 & \dfrac{1}{2} & 0 \\ 0 & 0 & 0 & 0 & \dfrac{1}{3s} \end{bmatrix} \begin{bmatrix} 1 & -1 \\ -1 & 0 \\ 0 & 1 \\ 1 & 0 \\ 0 & 1 \end{bmatrix} \mathbf{V}_n
$$

$$
= -\begin{bmatrix} 1 & -1 & 0 & 1 & 0 \\ -1 & 0 & 1 & 0 & 1 \end{bmatrix} \begin{bmatrix} 0 \\ \dfrac{10}{s} \\ \dfrac{3}{s+2} \\ 0 \\ 0 \end{bmatrix}
$$

After multiplication, we get

$$
\begin{bmatrix}
\left(\dfrac{s}{4}+\dfrac{3}{4}\right) & -\left(\dfrac{s}{4}+\dfrac{1}{4}\right) \\
-\left(\dfrac{s}{4}+\dfrac{1}{4}\right) & \left(\dfrac{s}{4}+\dfrac{1}{4}+\dfrac{1}{3s}\right)
\end{bmatrix}
\begin{bmatrix}
V_{n1} \\
V_{n2}
\end{bmatrix}
=
\begin{bmatrix}
\dfrac{10}{s} \\
-\dfrac{3}{s+2}
\end{bmatrix}
$$

the two node equations. □ *Prob. 6-20*

## 6-5 POWER AND ENERGY: PASSIVE AND ACTIVE ELEMENTS

Consider an element, represented by an oriented edge. This orientation gives the reference directions for the edge current and edge voltage, as explained earlier: The positive reference direction of $i(t)$ is that of the arrow, and the positive reference direction of $v(t)$ is at the tail of the arrow. The instantaneous power $p_r(t)$ delivered at time $t$ to element $r$ is then

$$p_r(t) = v_r(t)i_r(t) \tag{6-52}$$

The total instantaneous power is the sum of the instantaneous powers in all the elements,

$$p(t) = \sum_{r=1}^{e} v_r(t)i_r(t) \tag{6-53}$$

We shall prove Tellegen's theorem: *The total power is zero, provided the elements in the network obey Kirchhoff's laws.* Notice the generality of this statement: Nothing is said about the *nature* of the elements. They can be linear or nonlinear, passive or active, constant or time-varying—as long as they satisfy Kirchhoff's current and voltage law. And recall that the nature of the elements, per se, does not affect Kirchhoff's laws; only the topology does.

To prove this result we note that the sum in Eq. (6-53) can be expressed as the scalar product of the column vectors $\mathbf{i}(t)$ and $\mathbf{v}(t)$, that is,

$$\sum_{r=1}^{e} v_r(t)i_r(t) = v_1 i_1 + v_2 i_2 + \cdots + v_n i_n = \mathbf{i}^T(t)\mathbf{v}(t) \tag{6-54}$$

where $\mathbf{i}(t)$ is the column matrix of the edge currents and $\mathbf{v}(t)$ is the column matrix of the edge voltages. Now introduce the loop and node transformations, $\mathbf{i}(t) = \mathbf{B}^T \mathbf{i}_m(t)$ and $\mathbf{v}(t) = \mathbf{A}^T \mathbf{v}_n(t)$. Then we have $\mathbf{i}^T(t) = \mathbf{i}_m^T(t)\mathbf{B}$ and Eq. (6-54) becomes

$$\mathbf{i}^T(t)\mathbf{v}(t) = \mathbf{i}_m^T(t)(\mathbf{B}\mathbf{A}^T)\mathbf{v}_n(t) = 0 \tag{6-55}$$

because $\mathbf{B}\mathbf{A}^T = \mathbf{0}$. This proves Tellegen's theorem.

*Example 14*

Consider the network shown in Fig. 6-19 and choose currents and voltages, at a particular time $t$, to satisfy Kirchhoff's laws. For example,

$$i_a = 1 \quad i_b = 2 \qquad \therefore i_d = -1 \quad \text{(at node 1)}$$

$$i_c = 3 \quad i_b = 2 \qquad \therefore i_e = -5 \quad \text{(at node 2)}$$

$$i_a = 1 \quad i_c = 3 \qquad \therefore i_f = 4 \qquad \text{(at node 3)}$$

and since edges $d$, $e$, and $f$ constitute a cut set, $-i_d + i_e + i_f = 0$, as it should. Likewise, let

$$v_a = 1 \quad v_c = 10 \qquad \therefore v_b = 9 \qquad \text{(loop } abc)$$

$$v_b = 9 \quad v_d = 1 \qquad \therefore v_e = 8 \qquad \text{(loop } bde)$$

$$v_c = 10 \quad v_e = 8 \qquad \therefore v_f = -2 \quad \text{(loop } cef)$$

Therefore,

$$p(t) = (1)(1) + (9)(2) + (10)(3) + (1)(-1) + (8)(-5) + (-2)(4) = 0$$

(Question: what does negative instantaneous power, as $-40$ watts or $-8$ watts, mean? *Hint*: review the definition of $p_r(t)$, Eq. (6-52).)   □

Since the instantaneous power $p_r(t)$ is, by definition, the rate of change of the energy $w_r(t)$ at time $t$,

$$p_r(t) = \frac{dw_r(t)}{dt} \tag{6-56}$$

then Tellegen's theorem implies the principle of *conservation of energy*

$$\sum_{r=1}^{e} p_r(t) = 0 \tag{6-57}$$

$$\therefore \sum_{r=1}^{e} w_r(t) = \text{constant} \tag{6-58}$$

This is an amazing conclusion! In other words, the principle of conservation of energy *is included already in Kirchhoff's laws*, and therefore need not be stipulated separately. Or, writing Kirchhoff's laws for a network is essentially a formulation of the principle of conservation of energy for that network, in terms of currents and voltages.

We conclude with a brief discussion of passive and active elements. In particular, consider a one-port (1-P) network, as shown in Fig. 6-21a. We have access only at the terminals of this port, as explained in Chapter 4. Let $v_1$ and $i_1$ be the voltage and current at the port, and at a particular time $t = t_0$, let the energy stored

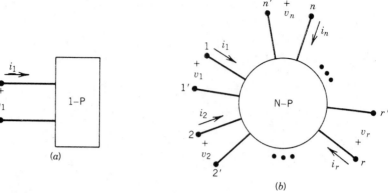

**FIGURE 6-21.** 1-P and N-P networks.

inside this 1-P be $\omega_0$. Let an *arbitrary* current or voltage source excite the 1-P from $t_0$ on. Then the energy delivered by this source to the 1-P network at $t \geq t_0$ is, from Eq. (6-56),

$$w(t) = \int_{t_0}^{t} v_1(\tau)i_1(\tau) \, d\tau \qquad (6\text{-}59)$$

The 1-P is *passive* if

$$w(t) + w_0 \geq 0 \qquad (6\text{-}60)$$

for all $t_0$ and for all $t \geq t_0$. Otherwise it is *active*.

The extension of these concepts to an N-port (N-P) network, Fig. 6-21b, is straightforward: The N-P network is passive, if, for arbitrary excitations at the ports, and with initial energy $w_0$ stored inside this network, Eq. (6-60) holds. Here, $w(t)$ is the energy delivered by *all* the sources to the network, between $t_0$ and $t$.

$$w(t) = \sum_{r=1}^{n} \int_{t_0}^{t} v_r(\tau)i_r(\tau) \, d\tau \qquad (6\text{-}61)$$

---

### Example 15

Consider a nonlinear, constant resistor. Its $v$–$i$ relationship is given by the curve shown in Fig. 6-22. Since the resistor does not store energy, $w_0 = 0$. Also, we see that the product $vi$ (the instantaneous power) supplied to this resistor, is never negative. Hence $w(t) \geq 0$ always, and Eq. (6-60) holds. This resistor is passive. If the $v$–$i$ curve were in the second or fourth quadrant, the product $vi$ would be negative and Eq. (6-60) would *not* hold for all $t_0$ and $t$: The resistor would then be an *active* element.

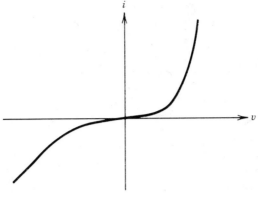

**FIGURE 6-22.** Example 15.

☐ *Probs. 6-21*
*through*
*6-25*

## Example 16

Consider a gyrator (see Prob. 4-15) defined by the relationships $v_1 = -Ri_2$, $v_2 = Ri_1$ $(R = \text{constant})$, and let $w_0 = 0$. Then $v_1i_1 + v_2i_2 = -Ri_1i_2 + Ri_2i_1 = 0$. Thus $w(t) = 0$, and the gyrator is passive. ☐

## PROBLEMS

6-1 Devise a systematic, foolproof method to find the number of edges, $e$ in a complicated graph. (Note: marking off each one while counting is *not* foolproof!) Use the graph shown below to illustrate your method. *Hint*: every edge is incident to precisely two nodes.

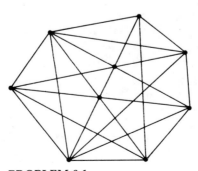

**PROBLEM 6-1.**

6-2 Prove that in any graph the number of nodes of odd degree is always even. *Hint*: problem 6-1.

6-3  Prove that the number of edges $e$ and nodes $n$ in a path are related by

$$e = n - 1$$

*Hint*: do it by mathematical induction.

6-4  Label properly the nodes and the edges of the graph shown in Fig. 6-5$f$, orient this graph and write its incidence matrix $\mathbf{A}_a$.

6-5  Repeat problem 6-4 for the nonplanar graph of Fig. 6-6$b$.

6-6  Consider the following incidence matrix

$$\begin{bmatrix} -1 & 1 & 0 & 1 & 0 & 0 \\ 0 & 0 & -1 & -1 & -1 & 0 \\ 1 & 0 & 1 & 0 & 0 & -1 \\ 0 & -1 & 0 & 0 & 1 & 1 \end{bmatrix}$$

Obtain the graph associated with this matrix. How is this graph related to that in Fig. 6-10?

6-7  For the graph shown in Fig. 6-6$b$: label its edges and nodes, assign orientations, then write Kirchhoff's current law at each node. Next, form linear combinations of these equations and identify the cut sets obtained.

6-8  Enumerate the fundamental cut sets for the graph in Fig. 6-10 for

a)  the tree in Fig. 6-9$a$

b)  the tree in Fig. 6-9$b$

c)  the tree in Fig. 6-9$c$

6-9  Just as we have defined *the incidence matrix* $\mathbf{A}_a$ for *all* vertices, we define $\mathbf{Q}_a$ for *all possible* cut sets. The subscript $a$ is for "*all* cut sets," and $\mathbf{Q}_a$ contains one row for every possible cut set, and one column for each edge. The entries are $+1$, $-1$ or $0$—by the same criterion as for $\mathbf{Q}_f$. The rank of $\mathbf{Q}_a$ is $n-1$.

a)  Obtain $\mathbf{Q}_a$ for the graph of Fig. 6-10.

b)  Verify that $\mathbf{Q}_f$ is a submatrix of $\mathbf{Q}_a$.

c)  Why is it preferable to work with $\mathbf{Q}_f$ rather than with $\mathbf{Q}_a$?

6-10  For the graph of Fig. 6-10 and each of the trees shown in Fig. 6-9, obtain the fundamental circuits and write Kirchhoff's voltage law in matrix form, $\mathbf{B}_f \mathbf{V}_e = \mathbf{0}$.

6-11  For the nonplanar graph shown in Fig. 6-6, write Kirchhoff's voltage law for its fundamental circuits. (First, the graph has to be oriented, a tree chosen, etc.)

6-12  How many independent Kirchhoff's voltage law equations are needed for the cube shown in Fig. 6-1? Assume each resistor to be one ohm and obtain a set of fundamental circuit equations in matrix form. Next, verify the relations among the link voltages and the branch voltages, Eq. (6-23), for this cube.

6-13 Just as we have defined $\mathbf{A}_a$ and $\mathbf{Q}_a$ (for *all* the nodes and *all* the cut sets), we define the *circuit matrix* $\mathbf{B}_a$ for *all possible circuits*. $\mathbf{B}_a$ contains one row for every possible circuit, and one column for each edge. The entries are $+1$, $-1$ or $0$— by the same criterion as for $\mathbf{B}_f$. The rank of $\mathbf{B}_a$, as discussed in the text, is $e - n + 1$.

a)  Obtain $\mathbf{B}_a$ for the graph of Fig. 6-10.

b)  Verify that $\mathbf{B}_f$ (the *fundamental* circuit matrix) is a submatrix of $\mathbf{B}_a$.

c)  Use $\mathbf{Q}_a$ from problem 6-9 to evaluate the matrix product $\mathbf{Q}_a\mathbf{B}_a^T$, where the columns of $\mathbf{Q}_a$ and $\mathbf{B}_a$ are arranged in the same element order.

6-14 Read problem 6-13. Let $\mathbf{B}$ be a submatrix of $\mathbf{B}_a$, of order $(e - n + 1) \times e$ and of rank $e - n + 1$. This is a *reduced* circuit matrix, and since its rank is $e - n + 1$, the equations obtained are linearly independent. ($\mathbf{B}_f$ is a special case of $\mathbf{B}$.) From $\mathbf{B}_a$ obtained in problem 6-13, find a *reduced* circuit matrix by trial-and-error: Pick any $e - n + 1$ rows, then check the rank. After finding such a $\mathbf{B}$, sketch the independent circuits. Finally, verify (with a couple of examples) that the columns of any nonsingular square submatrix of $\mathbf{B}$, of order $e - n + 1$, correspond to a set of links.

6-15 In Fig. 6-17, a student chose the following four loops

$$1\text{-}a\text{-}2\text{-}c\text{-}4\text{-}d\text{-}3\text{-}e\text{-}1 \qquad 1\text{-}a\text{-}2\text{-}b\text{-}3\text{-}g\text{-}5\text{-}f\text{-}1$$
$$1\text{-}e\text{-}3\text{-}d\text{-}4\text{-}h\text{-}5\text{-}f\text{-}1 \qquad 2\text{-}c\text{-}4\text{-}h\text{-}5\text{-}g\text{-}3\text{-}b\text{-}2$$

hoping that the equations will be independent. Check this choice and explain.

6-16 A (somewhat inexperienced) student chose the following *three* loops in Fig. 6-17:

$$1\text{-}a\text{-}2\text{-}c\text{-}4\text{-}d\text{-}3\text{-}e\text{-}1$$
$$2\text{-}c\text{-}4\text{-}h\text{-}5\text{-}g\text{-}3\text{-}b\text{-}2$$
$$1\text{-}e\text{-}3\text{-}d\text{-}4\text{-}h\text{-}5\text{-}f\text{-}1$$

At this stage the student said, "No edge has been left out, and every loop contains at least one new edge. That should be sufficient, then." Criticize fully.

6-17 Develop the relationship between the loop currents $\mathbf{I}_m$, and link and branch currents for non-fundamental circuits, that is, when $\mathbf{B} \neq \mathbf{B}_f$.

6-18 Consider the network shown in problem 3-8 (Chapter 3), with all initial conditions zero.

a)  Draw its oriented graph.

b)  Choose a tree so that all the voltage sources are branches.

c)  Write $\mathbf{B}_f$.

d)  Write the $v$–$i$ relations for the edges.

()  From (c) and (d), obtain the fundamental loop equations.

6-19 Repeat problem 6-18 for the network shown in problem 3-9 (Chapter 3). Observe what happens to $\mathbf{Z}_e$, as compared with $\mathbf{Z}_e$ in problem 6-18.

6-20 Consider the network in problem 3-10 (Chapter 3) with all initial conditions zero.

   a) Draw its oriented graph.

   b) Choose a tree so that all the current sources are links.

   c) Write $\mathbf{A}$.

   d) Write the $v$–$i$ relations for the edges.

   e) From (c) and (d), obtain the node equations.

6-21 Solve the network shown in problem 3-9 (Chapter 3) and find the total instantaneous power for it at $t = 1$ sec. and verify Tellegen's theorem. (Assume all initial conditions to be zero.)

6-22 Repeat problem 6-21 for the network shown in problem 3-10 (Chapter 3) at $t = 0.2$ sec.

6-23 Is the ideal transformer (Chapter 4, problem 4-14) passive?

6-24 Repeat problem 6-23 for a regular transformer, i.e., two inductances, $L_1 > 0$, $L_2 > 0$, with mutual coupling $M^2 < L_1 L_2$.

6-25 Repeat problem 6-23 for a perfect transformer, $M^2 = L_1 L_2$.

# 7

---

# NETWORK FUNCTIONS
# AND RESPONSES

---

In previous chapters (Chapters 3, 4, and 5), the Laplace transform was used as a convenient tool to describe the behavior of a network. Specifically, the analysis of lumped, linear, and constant (time-invariant) networks was handled by Laplace transform methods. In this chapter, we explore some of the mathematical functions associated with this analysis.

## 7-1  NETWORK FUNCTIONS:
## DEFINITIONS AND DERIVATIONS

---

### Example 1

Consider the network shown in Fig. 7-1. It consists of a 2-P network, excited by a voltage source, $v_{in}(t)$. The capacitor has an initial voltage $v_c(0_-)$ across it. The desired output is $v_{out}(t)$ across $R$. By methods studied in Chapter 3, we obtain, with little

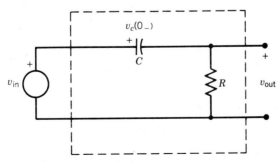

**FIGURE 7-1.** Example 1.

difficulty, the following result

$$V_{out}(s) = \frac{RCs}{RCs + 1} V_{in}(s) - \frac{RC}{RCs + 1} v_c(0_-)$$

Let us concentrate on the *zero-state response*, that is, when $v_c(0_-)$ is zero. It is given by the first term, and we observe that this response is related to the excitation $V_{in}(s)$ by a function that depends only on the parameters of the network:

$$V_{out}(s) = \frac{RCs}{RCs + 1} V_{in}(s) \qquad \square$$

---

We define now, in general, a *network function $H(s)$* as

$$R(s) = H(s)E(s) \tag{7-1}$$

where $R(s)$ is the zero-state transform response, $E(s)$ is the transform excitation. The network function, being characteristic of the network, contains only those parameters that are inherent to the network. Therefore, for the definition of $H(s)$, we assume that *the network has no independent sources* (and this includes initial conditions). In Chapter 4, we have used some special cases of network functions such as open-circuit impedances, short-circuit admittances, etc. In fact, we wrote many network functions in matrix form.

Depending on the type of the excitation and the response, network functions fall into four broad classifications, as follows:

| Excitation | Response | Network function | Symbol |
|---|---|---|---|
| Current | Voltage | Impedance | $Z(s)$ |
| Voltage | Current | Admittance | $Y(s)$ |
| Current | Current | Current transfer | $\alpha(s)$ |
| Voltage | Voltage | Voltage transfer | $G(s)$ |

The symbols $Z(s)$, $Y(s)$, $\alpha(s)$ and $G(s)$ are used for specific network functions. The designation $H(s)$ will be used as a generic symbol.

One additional distinction should be made for the first two classes listed previously:

a)  The excitation and the response are at the same port; in this case we speak of the *driving-point (dp) impedance* $Z_{dp}(s)$ or the *driving-point (dp) admittance* $Y_{dp}(s)$.

b)  The excitation is at one port and the response is at a different port; here, we are concerned with a *transfer impedance* $Z_{tr}(s)$ or a *transfer admittance* $Y_{tr}(s)$. Quite obviously, $\alpha(s)$ and $G(s)$ are transfer functions by nature.

Figure 7-2 shows these relations in a graphical manner. A different subscript notation is sometimes convenient for transfer functions: When the different ports are designated by numbers, the subscripts are listed in the order "output port,

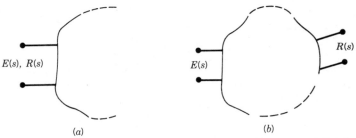

**FIGURE 7-2.** Excitation and response for network functions: (*a*) Driving-point; (*b*) Transfer.

input port." Thus,

$$H_{21}(s) = \frac{R_2(s)}{E_1(s)} \tag{7-3}$$

means a transfer function relating the response at port 2 (or terminals 2–2′) to the excitation at port 1 (terminals 1–1′).

The most general approach to finding a network function is quite straightforward, and can be summarized in five steps:

1. Identify the excitation and its port.
2. Identify the response and its port.
3. Verify that there are no independent sources (other than the excitation, of course), including all initial conditions.
4. By methods of analysis studied previously, find the transform response $R(s)$.
5. The ratio $R(s)/E(s)$ yields the desired $H(s)$.

These steps were essentially taken in the solutions of problems 3-11, 3-12 and 3-14, and the student is urged to review them carefully.

## 7-2  DRIVING-POINT FUNCTIONS

The computation of $Z_{dp}(s)$ or $Y_{dp}(s)$ implies that two terminals are accessible, and the remainder of the network—containing passive elements, and possibly dependent sources and other terminals—may be considered to be enclosed in the proverbial "black box." These driving-point terminals can be excited by a current source $i_{dp}$, the response being $v_{dp}$. The pertinent network function is then

$$Z_{dp}(s) = \frac{V_{dp}(s)}{I_{dp}(s)} \tag{7-4}$$

See Fig. 7-3a. Alternately, if $v_{dp}$ is the excitation and $i_{dp}$ the response, we have

$$Y_{dp}(s) = \frac{I_{dp}(s)}{V_{dp}(s)} \tag{7-5}$$

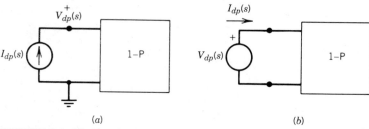

(a)                                              (b)

**FIGURE 7-3.** Finding driving-point functions.

These two driving-point functions are reciprocal of each other, that is,

$$Y_{dp}(s) = \frac{1}{Z_{dp}(s)} \qquad (7\text{-}6)$$

This is an interesting feature of dp functions: The reciprocal of one network function (*dp*) is also a network function. *In general, the reciprocal of H(s) is NOT a network function* for the same network. This is important to realize, and we shall return to it later.

Let us illustrate with several examples.

---

## Example 2

Find $Z_{dp}(s)$ for the network shown in Fig. 7-4. To stress the importance of the basic approach, let us solve this problem step by step, as outlined previously. (Later, we shall verify it by a short-cut method.) The excitation is $I_{dp}(s)$, according to Eq. (7-4), and is shown in Fig. 7-3 in dotted lines. There are no independent sources, including initial conditions. To solve for the response $V_{dp}(s)$, use node analysis and label the only other node with $V_2(s)$. The node equations are written by inspection

$$\begin{bmatrix} \dfrac{1}{3s} + 1 & -1 \\ -1 & 2s + 1 \end{bmatrix} \begin{bmatrix} V_{dp} \\ V_2 \end{bmatrix} = \begin{bmatrix} I_{dp} \\ 0 \end{bmatrix}$$

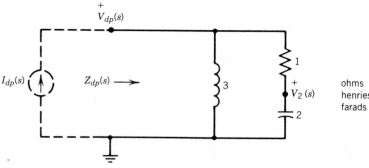

**FIGURE 7-4.** Example 2.

Therefore,

$$V_{dp} = \frac{3s(2s + 1)}{6s^2 + 2s + 1} I_{dp}$$

and

$$Z_{dp}(s) = \frac{V_{dp}(s)}{I_{dp}(s)} = \frac{3s(2s + 1)}{6s^2 + 2s + 1}$$

As a quick check, we note that the network consists of series and parallel combinations of passive elements. Therefore,

$$\frac{1}{2s} + 1 = \frac{2s + 1}{2s} \qquad \text{(series } RC\text{)}$$

$$Y_{dp}(s) = \frac{2s}{2s + 1} + \frac{1}{3s} = \frac{6s^2 + 2s + 1}{6s^2 + 3s} \qquad \text{(total parallel)}$$

and

$$Z_{dp}(s) = \frac{1}{Y_{dp}(s)} = \frac{6s^2 + 3s}{6s^2 + 2s + 1}$$

as before. The student is encouraged to use such shortcuts (provided they are valid!). However, the importance of using the basic approach cannot be overstressed.   ☐

## Example 3

In Fig. 7-5 we have the equivalent circuit of a transistor network known as an emitter follower. The letters "$b$", "$c$", and "$e$" stand for the base, the collector and the emitter.

The output (driving-point) impedance $Z_o$ seen between the emitter and the collector is found by connecting a current source $i_o$ as shown and calculating $v_o$ with

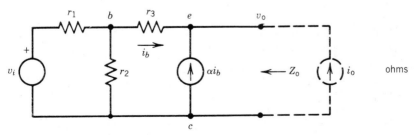

**FIGURE 7-5.**  An emitter-follower circuit.

$v_i = 0$ (the independent source set to zero). *KCL* at the emitter reads then

$$\frac{v_e}{r_3 + (r_1 \| r_2)} - \alpha i_b - i_o = 0$$

where the symbol "$\|$" stands for "parallel connection." Also

$$\alpha i_b = \alpha \left( \frac{-v_e}{r_3 + (r_1 \| r_2)} \right)$$

Therefore

$$\frac{v_e}{r_3 + (r_1 \| r_2)} (1 + \alpha) = i_o$$

and

$$v_o = v_e = \frac{r_3 + (r_1 \| r_2)}{1 + \alpha} i_o$$

yielding

$$Z_o = \frac{r_3 + (r_1 \| r_2)}{1 + \alpha}$$

Note that in this purely resistive network we did not have to use the Laplace transform: the time domain equations are the same (algebraic), with no initial conditions.     $\square$

## Example 4

Find the driving-point admittance of the 1-P network shown in Fig. 7-6. Although this network contains no dependent sources, there are no series or parallel combinations for shortcut calculations. Therefore, let up apply the general method.

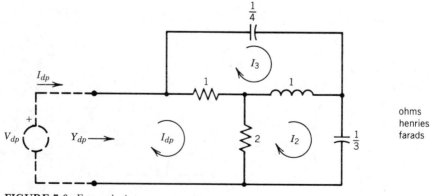

**FIGURE 7-6.** Example 4.

Again, $Y_{dp}(s) = I_{dp}(s)/V_{dp}(s)$, so the excitation is $V_{dp}(s)$, as shown in dotted lines. To solve for the response $I_{dp}(s)$, use loop currents and choose the first loop as the only one with $I_{dp}(s)$ in it. The other two loops are labelled 2 and 3, as shown. The loop equations are written by inspection in the form $\mathbf{ZI} = \mathbf{E}$ as in Chapter 3:

$$
\begin{bmatrix}
3 & -2 & -1 \\
-2 & s + 2 + \dfrac{3}{s} & -s \\
-1 & -s & s + 1 + \dfrac{4}{s}
\end{bmatrix}
\begin{bmatrix}
I_{dp} \\
I_2 \\
I_3
\end{bmatrix}
=
\begin{bmatrix}
V_{dp} \\
0 \\
0
\end{bmatrix}
$$

Solve for $I_{dp}(s)$:

$$
I_{dp}(s) = \frac{
\begin{vmatrix}
s + 2 + \dfrac{3}{s} & -s \\
-s & s + 1 + \dfrac{4}{s}
\end{vmatrix}
}{
\begin{vmatrix}
3 & -2 & -1 \\
-2 & s + 2 + \dfrac{3}{s} & -s \\
-1 & -s & s + 1 + \dfrac{4}{s}
\end{vmatrix}
} V_{dp}(s)
$$

The denominator is recognized as the determinant of the loop impedance matrix $\mathbf{Z}$. Denote it by $\det \mathbf{Z}$. In the numerator we see the cofactor of the $(1, 1)$ element in $\det \mathbf{Z}$. Call this cofactor $\det_{11} \mathbf{Z}$. Therefore,

$$
I_{dp}(s) = \frac{\det_{11} \mathbf{Z}}{\det \mathbf{Z}} V_{dp}(s)
$$

and

$$
Y_{dp}(s) = \frac{\det_{11} \mathbf{Z}}{\det \mathbf{Z}} \tag{7-7}
$$

In our example, the evaluation of these two determinants, after clearing all fractions, yields the answer

$$
Y_{dp}(s) = \frac{3s^3 + 9s^2 + 11s + 12}{21s^2 + 14s + 36}
$$

Reviewing the steps of this example, we recognize that Eq. (7-7) is valid in general for 1-P networks with no dependent sources (and, of course, no independent sources!). In such a case we can write by inspection $\mathbf{ZI} = \mathbf{E}$, and since $\mathbf{E}$ contains only one (the first) nonzero entry, we arrive at the result in Eq. (7-7).

On a dual basis, we can derive the general expression for $Z_{dp}(s)$, under the same conditions. To illustrate, consider the same network as before. It is redrawn in Fig. 7-7. Since $Z_{dp}(s) = V_{dp}(s)/I_{dp}(s)$, the excitation is $I_{dp}(s)$, as shown, and we

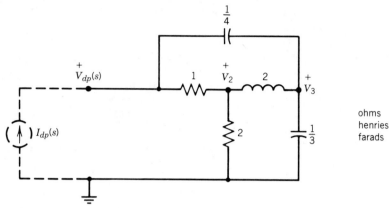

**FIGURE 7-7.** Example 4.

write the node equations in the form $\mathbf{YV} = \mathbf{J}$ as in Chapter 3:

$$
\begin{bmatrix}
1 + \dfrac{s}{4} & -1 & -\dfrac{s}{4} \\[2mm]
-1 & \dfrac{3}{2} + \dfrac{1}{s} & -\dfrac{1}{s} \\[2mm]
-\dfrac{s}{4} & -\dfrac{1}{s} & \dfrac{7s}{12} + \dfrac{1}{s}
\end{bmatrix}
\begin{bmatrix}
V_{dp} \\[2mm]
V_2 \\[2mm]
V_3
\end{bmatrix}
=
\begin{bmatrix}
I_{dp} \\[2mm]
0 \\[2mm]
0
\end{bmatrix}
$$

Solve for $V_{dp}(s)$:

$$
V_{dp}(s) = \frac{\det_{11} \mathbf{Y}}{\det \mathbf{Y}} I_{dp}(s)
$$

where $\det \mathbf{Y}$ is the determinant of the node admittance matrix and $\det_{11} \mathbf{Y}$ is the cofactor of the $(1, 1)$ element. Finally,

$$
Z_{dp}(s) = \frac{\det_{11} \mathbf{Y}}{\det \mathbf{Y}}
$$

In our example, this evaluation becomes

$$
Z_{dp}(s) = \frac{21s^2 + 14s + 36}{3s^3 + 9s^2 + 11s + 12}
$$

the reciprocal of $Y_{dp}(s)$, as expected.

□ *Probs.* **7-1**
*through*
**7-7**

In addition to the series-parallel combination, there is one structure that deserves mentioning because of its frequent use in analysis and in design: the *ladder network*, as shown in Fig. 7-8. The series (horizontal) arms of the ladder are given

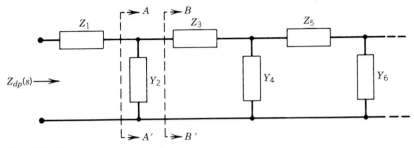

**FIGURE 7-8.** A ladder network.

by their individual dp impedances $Z_1$, $Z_3$, etc., while the shunt (vertical) arms are given by their dp admittances. The driving-point impedance $Z_{dp}(s)$ may be computed as follows: It is the sum of $Z_1$ and of $Z_{AA'}$, where $Z_{AA'}$ is the driving-point impedance of the ladder network to the right of $AA'$. This is true because $Z_1$ is in series with $Z_{AA'}$. Therefore, $Z_{dp}(s) = Z_1 + Z_{AA'}$. Now $Y_{AA'} = 1/Z_{AA'} = Y_2 + Y_{BB'}$ because the admittance $Y_2$ is in parallel with the remaining network to the right of $BB'$. Continuing in this fashion, we write

$$Z_{dp}(s) = Z_1(s) + \cfrac{1}{Y_2(s) + \cfrac{1}{Z_3(s) + \cfrac{1}{Y_4(s) + \ddots}}} \tag{7-8}$$

The right-hand side of Eq. (7-8) is known as a *continued fraction*.

*Probs. 7-8 through 7-12*

Let us discuss, at this point, another method suitable for computing $H(s)$ in ladder networks. Its outline, in brief, is: Start at the output and find the corresponding input. The following example illustrates this.

---

*Example 5*

In the ladder network in Fig. 7-9, it is required to find $G_{21}(s) = V_2(s)/V_1(s)$. For the purpose of this problem, we have labeled several currents, $I_1$, $I_2$, $I_3$, and $I_4$, as well as

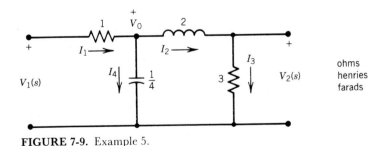

**FIGURE 7-9.** Example 5.

$V_0$. Working from the output "backwards," we write successively:

$$I_3 = \frac{V_2}{3}$$

$$I_2 = I_3 = \frac{V_2}{3}$$

$$V_0 = V_2 + 2sI_2 = V_2 + 2s\frac{V_2}{3} = V_2\left(\frac{2s}{3} + 1\right)$$

$$I_4 = \frac{s}{4}V_0 = V_2\left(\frac{s^2}{6} + \frac{s}{4}\right)$$

$$I_1 = I_2 + I_4 = V_2\left(\frac{s^2}{6} + \frac{s}{4} + \frac{1}{3}\right)$$

$$V_1 = V_0 + I_1 = V_2\left(\frac{s^2}{6} + \frac{11}{12}s + \frac{4}{3}\right)$$

And therefore,

$$G_{21}(s) = \frac{V_2(s)}{V_1(s)} = \frac{12}{2s^2 + 11s + 16}$$

This method can be applied sometimes to non-ladder networks (see Prob. 7-15).

□ **Probs.  7-13**
**7-14**
**7-15**

## 7-3   TRANSFER FUNCTIONS

Whereas there is little room for misunderstanding when a dp function is required, we must be very careful in stating the precise nature of a desired transfer function. Also, in practice, there is no difference whether we excite with $i_{dp}$ or $v_{dp}$ in the computation of dp functions. In the case of transfer functions the careful identification of the excitation and the response is crucial. Previous examples of transfer functions can be found in Chapter 4 in the descriptions of N-P and N-T networks.

*Example 6*

For the 2-P network shown in Fig. 7-10, find the open-circuit voltage transfer function $G_{21}(s) = V_2(s)/V_1(s)$. The specification "open circuit" indicates that the output port (2-2′) is unloaded. The proper excitation is $V_1(s)$, as shown in dotted lines. We use loop analysis and the two loop currents are shown. (Note: Terminals 2-2′ are open, hence only *two* loops!) In terms of these loop currents, the response

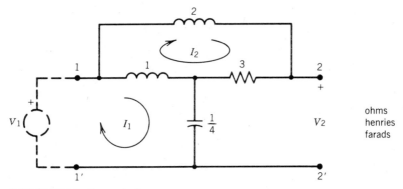

**FIGURE 7-10.** Example 6.

$V_2(s)$ is given by $V_2 = 3I_2 + (4/s)I_1$, and so we must obtain both $I_1$ and $I_2$. The loop equations are

$$
\begin{bmatrix} s + \dfrac{4}{s} & -s \\ -s & 3s + 3 \end{bmatrix} \begin{bmatrix} I_1 \\ I_2 \end{bmatrix} = \begin{bmatrix} V_1 \\ 0 \end{bmatrix}
$$

and therefore,

$$
I_1 = \frac{\det_{11} \mathbf{Z}}{\det \mathbf{Z}} V_1, \qquad I_2 = \frac{\det_{12} \mathbf{Z}}{\det \mathbf{Z}} V_1
$$

Here we have expressed the currents in terms of $V_1$, of the determinant of the loop impedance matrix and of the appropriate cofactors. Then

$$
\begin{aligned}
V_2 &= 3 \frac{\det_{12} \mathbf{Z}}{\det \mathbf{Z}} V_1 + \frac{4}{s} \frac{\det_{11} \mathbf{Z}}{\det \mathbf{Z}} V_1 \\
&= \left( 3 \frac{\det_{12} \mathbf{Z}}{\det \mathbf{Z}} + \frac{4}{s} \frac{\det_{11} \mathbf{Z}}{\det \mathbf{Z}} \right) V_1
\end{aligned}
$$

exhibiting

$$
G_{21}(s) = \frac{3s^2 + 12s + 12}{2s^3 + 3s^2 + 12s + 12}
$$

where the last result has been obtained after evaluating all the determinants and clearing of fractions. ☐

---

*Example 7*

In the network shown in Fig. 7-7, find the transfer impedance $Z_{tr}(s) = V_L(s)/I_{dp}(s)$, where $V_L = V_2 - V_3$, the voltage across the inductor. The excitation is $I_{dp}(s)$ and the response is $V_2(s) - V_3(s)$. The appropriate equations are found there, and we solve for $V_2$ and $V_3$ as follows:

$$V_2 = \frac{\det_{12} \mathbf{Y}}{\det \mathbf{Y}} I_{dp}, \qquad V_3 = \frac{\det_{13} \mathbf{Y}}{\det \mathbf{Y}} I_{dp}$$

Therefore,

$$Z_{tr}(s) = \frac{\det_{12} \mathbf{Y} - \det_{13} \mathbf{Y}}{\det \mathbf{Y}} = \frac{\dfrac{5}{24} s}{3s^3 + 9s^2 + 11s + 12} \qquad \square$$

## Example 8

For the network shown in Fig. 7-11, find $\alpha(s) = I_L(s)/I_1(s)$. The excitation is $i_1$ and the desired response is $i_L$. Note the dependent source. We apply node analysis, labeling the nodes 1, 2 and 3.

$$\begin{bmatrix} \dfrac{3}{s} + s & -\dfrac{3}{s} & 0 \\ -\dfrac{3}{s}(1+k) & 3 + \dfrac{3}{s}(1+k) & -1 \\ \dfrac{3k}{s} & -\left(1 + \dfrac{3k}{s}\right) & (2s+1) \end{bmatrix} \begin{bmatrix} V_1 \\ V_2 \\ V_3 \end{bmatrix} = \begin{bmatrix} I_1 \\ 0 \\ 0 \end{bmatrix}$$

which, again, is of the general form $\mathbf{YV} = \mathbf{J}$. By the way, notice that $\mathbf{Y}$ is not symmetric, $y_{jk} \neq y_{kj}$, because of the dependent source. We solve for $V_1$ and $V_2$:

$$V_1 = \frac{\det_{11} \mathbf{Y}}{\det \mathbf{Y}} I_1, \qquad V_2 = \frac{\det_{12} \mathbf{Y}}{\det \mathbf{Y}} I_1$$

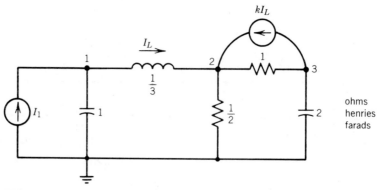

**FIGURE 7-11.** Example 8.

Then,

$$I_L = \frac{3}{s}(V_1 - V_2) = \frac{3I_1(\det{}_{11}\mathbf{Y} - \det{}_{12}\mathbf{Y})}{s\det\mathbf{Y}}$$

and finally,

$$\alpha(s) = \frac{I_L(s)}{I_1(s)} = \frac{\dfrac{3}{s}(\det{}_{11}\mathbf{Y} - \det{}_{12}\mathbf{Y})}{\det\mathbf{Y}}$$

$$= \frac{18s + 6}{6s^3 + (8 + 6k)s^2 + 21s + 30 + 18k} \qquad \square$$

In summary, the basic approach, requiring no memorization, should be followed always. Specialized results such as Eq. (7-7), although nice, can be often misused. The five steps of the fundamental approach, on the other hand, are rather natural: All that we need is the basic definition of $H(s)$, Eq. (7-1).

For a different approach to network functions using a signal-flow graph technique, see Appendix D.

## 7-4    PROPERTIES OF NETWORK FUNCTIONS: POLES AND ZEROS AND PARTS OF $H(s)$

From the previous discussion, several conclusions may be drawn about the network functions. These are developed in the remainder of this chapter as well as in Chapter 9.

For lumped, constant, and linear networks under consideration, $H(s)$ is a ratio of two polynomials. Such a ratio is called a *rational function*.

$$\begin{aligned}
H(s) &= \frac{c_0 s^m + c_1 s^{m-1} + \cdots + c_{m-1}s + c_m}{d_0 s^n + d_1 s^{n-1} + \cdots + d_{n-1}s + d_n} \\
&= K\frac{s^m + a_1 s^{m-1} + \cdots + a_{m-1}s + a_m}{s^n + b_1 s^{n-1} + \cdots + b_{n-1}s + b_n} = K\frac{P(s)}{Q(s)} \qquad (7\text{-}9)
\end{aligned}$$

where $K = c_0/d_0$, $a_1 = c_1/c_0$, $b_1 = d_1/d_0$, etc. The second form of the expression is written with unity coefficients for the highest powers of $s$. The numerator and the denominator polynomials are $P(s)$ and $Q(s)$, respectively.

Note that all the coefficients in $H(s)$, i.e., all the $c$'s and $d$'s and, therefore, all the $a$'s and $b$'s, are real. This is so because they are related to the values of the network elements. As a result, if we choose $s$ to be a real number, $H(s)$ must be a real number. Briefly then, we have,

**Property 1**    $H(s)$ is real when $s$ is real.

This is an important result, although it seems "obviously" simple.

By the fundamental theorem of algebra, the numerator in Eq. (7-9) has $m$ factors and the denominator has $n$.

$$K\frac{P(s)}{Q(s)} = K\frac{(s - z_1)(s - z_2) \cdots (s - z_m)}{(s - p_1)(s - p_2) \cdots (s - p_n)} \tag{7-10}$$

where $z_i$, $(1 \leq i \leq m)$, are the roots of the equation

$$P(s) = 0 \tag{7-11a}$$

and $p_j$, $(1 \leq j \leq n)$, are the roots of the equation

$$Q(s) = 0 \tag{7-11b}$$

The rational function $H(s)$ has a *pole of order* (or *multiplicity*) $k$ at $s = s_0$ if

$$\lim_{s \to s_0} H(s) \to \infty \tag{7-12a}$$

but

$$\lim_{s \to s_0} H(s)(s - s_0)^k \tag{7-12b}$$

is finite, nonzero. If $k = 1$, the pole is *simple*. From the expression of $H(s)$ in Eq. (7-10), we note that $H(s)$ has finite poles at $s = p_1$, $s = p_2, \ldots, s = p_n$. These are simple poles if $p_1 \neq p_2 \neq \cdots \neq p_n$. If $k$ of these are repeated, say $p_1 = p_2 = \cdots = p_k \neq p_{k+1} \neq \cdots$, then this is a pole of order $k$. In addition, we see that if $m > n$,

$$\lim_{s \to \infty} H(s) = \lim_{s \to \infty} Ks^{m-n} \tag{7-13}$$

by applying l'Hôpital's rule to Eq. (7-9). Therefore, $H(s)$ has a pole of order $(m - n)$ at $s = \infty$.

The rational function $H(s)$ has a *zero of order* (or *multiplicity*) $k$ at $s = s_0$ if $1/H(s)$ has a pole of order $k$ there. Again, if $k = 1$, the zero is called *simple*. Noting Eq. (7-10), $H(s)$ is seen to have zeros at $s = z_1$, $s = z_2, \ldots, z_m$. These zeros are simple if they are all distinct; if $k$ of them are repeated, the zero is of order $k$. Furthermore, if $m < n$

$$\lim_{s \to \infty} \frac{1}{H(s)} = \lim_{s \to \infty} Ks^{n-m} \tag{7-14}$$

and so $H(s)$ has a zero of order $(n - m)$ at $s = \infty$.

---

## Example 9

Find the poles and zeros of

$$H(s) = \frac{(s + 1)^4}{s^3 + 4s^2 + 20s}$$

The roots of $Q(s) = 0$ are $s = 0$, $s = -2 + j4$ and $s = -2 - j4$, and these are the finite (and simple) poles. In addition, $H(s)$ has a simple pole at $s = \infty$ because

$s^{m-n} = s^{4-3} = s$. There is a zero of order 4 at $s = -1$, the repeated root of $P(s) = 0$. □

---

## Example 10

Find the poles and zeros of

$$H(s) = 2 \frac{s^2 - s - 6}{s^2 + 4}$$

Here, there are two simple poles at $s = j2$ and $s = -j2$, the roots of $Q(s) = 0$; there are two simple zeros at $s = 3$ and $s = -2$, the roots of $P(s) = 0$. The point $s = \infty$ is a regular point since $\lim_{s \to \infty} H(s) = 2$. □

---

If we recall now the first property—$H(s)$ is real when $s$ is real—we can draw the following conclusion,

**Property 2** Poles and zeros are either real or complex conjugate.

A complex pole (or zero) cannot occur without its conjugate; if it did, $H(s)$ would not be real for real values of $s$.

If we count a multiple pole or zero $k$ times, then the following is true for the rational function $H(s)$,

**Property 3** The number of poles equals the number of zeros. This number is $m$ or $n$, whichever is greater.

To prove it, we consider the three possibilities: $m > n$, $m = n$, and $m < n$. If $m > n$, there are $m$ finite zeros, roots of $P(s) = 0$; there are $n$ finite poles, roots of $Q(s) = 0$, plus a pole of order $(m - n)$ at $s = \infty$. Altogether, then, there are $n + m - n = m$ poles also. In the second case, $m = n$, there are $m$ finite zeros and $n$ finite poles. At $s = \infty$, $H(s)$ has a finite, nonzero value. Finally, if $m < n$, there are $m$ finite zeros, $n$ finite poles, and in addition, a zero of order $(n - m)$ at $s = \infty$. All told, there are $m + n - m = n$ zeros. It is, therefore, easy to decide by inspection how many poles (and zeros) $H(s)$ has: This number is the highest power of $s$ in $H(s)$.

Graphically, we show the location of poles and zeros in the complex $s$-plane, using **o** for a zero and **x** for a pole. The order is shown in parentheses next to the pole or the zero. With no parentheses, the pole or the zero is assumed to be simple.

---

## Example 11

Show the pole-zero configuration of

$$H(s) = K \frac{s(s^2 + 2s + 10)}{(s^2 + 4)(s + 4)^2}$$

on the $s$-plane. Since the highest power is 4, there are four zeros and four poles.

*Zeros:* At $s = 0$, $s = -1 + j3$, $s = -1 - j3$, and at $s = \infty$.

*Poles:* At $s = j2$, $s = -j2$, $s = -4$ (double).

The pole-zero configuration is shown in Fig. 7-12. Note that the constant multiplier $K$ does not affect the poles and zeros.

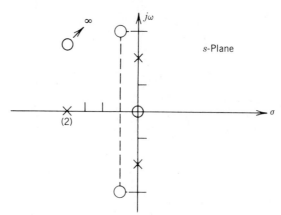

**FIGURE 7-12.** Example 11.

☐ *Probs. 7-16*
*7-17*

Let us digress here briefly to explain the zero at $s = \infty$, shown in Fig. 7-12 as a zero **o** with an arrow pointing to infinity. We may as well start by asking, "where is $s = \infty$ on the complex plane?" A complex number $\sigma + j\omega$ tends to infinity when its magnitude, $(\sigma^2 + \omega^2)^{1/2}$, becomes infinitely large. Therefore, either $\sigma$, or $\omega$, or both can be a (positive or negative) large number. This makes the concept $s = \infty$ rather vague. To overcome this ambiguity, mathematicians have devised the following scheme: Consider placing a sphere on the $s$-plane touching it at $s = 0$. See Fig. 7-13. Let this point on the sphere be its south pole $S$. Then its north pole $N$ is directly above $S$. Next, connect any point $s_1$ in the $s$-plane to $N$ with a straight line. This line cuts the

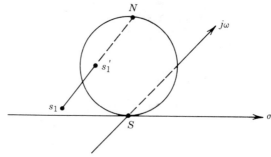

**FIGURE 7-13.** Mapping of the $s$-plane onto a sphere.

sphere at *one* particular point, $s_1'$, which can be considered the *mapping* of point $s_1$. Thus, every point in the $s$-plane maps *uniquely* onto the sphere. The ambiguous point $s = \infty$ in the $s$-plane maps onto *one* point—the north pole $N$! Hence, with this mapping in mind, we shall speak of *the* point $s = \infty$ without any misunderstanding.

It is clear that $H(s)$ is completely known if we are given: 1) all its finite poles and zeros, and 2) one extra piece of information, to determine the constant multiplier $K$ in Eq. (7-9); this information can be, for example, the value of $H(s)$ at $s = s_0$, where $s_0$ is neither a zero nor a pole.

---

## Example 12

Find $H(s)$, given the following data: finite zeros at $s = 0$, $s = -2$; finite poles at $s = -1$ (double). Also, at $s = 1$, $H(s) = 3$. From the pole-zero information we write

$$H(s) = K \frac{s(s + 2)}{(s + 1)^2}$$

then

$$H(1) = K \frac{3}{4} = 3$$

$$\therefore \quad K = 4$$

$$\therefore H(s) = 4 \frac{s(s + 2)}{(s + 1)^2} \qquad \square \quad \textbf{Probs. \ 7-18}$$
$$\textbf{7-19}$$

---

Returning to the discussion of general properties of $H(s)$, we consider the evaluation of $H(s)$ for purely imaginary $s$, $s = j\omega$. In general, $H(j\omega)$ will be a complex number; written in rectangular form, it reads

$$H(j\omega) = R(\omega) + jX(\omega) \tag{7-15}$$

with its real part $R(\omega)$ and its imaginary part $X(\omega)$ both depending on $\omega$. The real part of any complex number is one-half the sum of the complex number and its conjugate. So,

$$R(\omega) = \tfrac{1}{2}[H(j\omega) + H^*(j\omega)] \tag{7-16}$$

However, because $H(s)$ has only real coefficients (Property 1) we can write

$$H^*(j\omega) = H(-j\omega) \tag{7-17}$$

that is, for $s = j\omega$, the conjugate of $H(j\omega)$ is simply obtained by substituting $-j\omega$ for $j\omega$. From Eqs. (7-16) and (7-17) we get

$$R(\omega) = R(-\omega) \tag{7-18a}$$

which means that $R(\omega)$ is an *even* function. Similarly, we have

$$X(\omega) = \frac{1}{2j} [H(j\omega) - H^*(j\omega)]$$

$$= \frac{1}{2j} [H(j\omega) - H(-j\omega)]$$

$$= -X(-\omega) \qquad\qquad (7\text{-}18b)$$

and so $X(\omega)$ is an *odd* function.

**Property 4**   For $s = j\omega$, the real part of $H(j\omega)$ is an even function, $R(\omega) = R(-\omega)$, and the imaginary part of $H(j\omega)$ is an odd function, $X(\omega) = -X(-\omega)$.

---

**Example 13**

Let

$$H(s) = \frac{1}{s + 1},$$

Then

$$H(j\omega) = \frac{1}{j\omega + 1} = \frac{-j\omega + 1}{-j\omega + 1} \cdot \frac{1}{j\omega + 1}$$

$$= \frac{1}{1 + \omega^2} + j \frac{-\omega}{1 + \omega^2}$$

$$= R(\omega) + jX(\omega)$$

Also, as a quick check

$$H^*(j\omega) = \frac{1}{-j\omega + 1} = H(-j\omega)$$

and

$$R(\omega) = \frac{1}{2} [H(j\omega) + H(-j\omega)] = \frac{1}{1 + \omega^2}$$

$$X(\omega) = \frac{1}{2j} [H(j\omega) - H(-j\omega)] = \frac{-\omega}{1 + \omega^2}$$

Finally,

$$R(-\omega) = \frac{1}{1 + (-\omega)^2} = \frac{1}{1 + \omega^2} = R(\omega)$$

and

$$X(-\omega) = -\frac{(-\omega)}{1 + (-\omega)^2} = \frac{\omega}{1 + \omega^2} = -X(\omega)$$

The plots of these two functions are shown in Fig. 7-14, displaying their even and odd nature. □

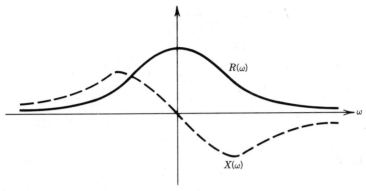

**FIGURE 7-14.** Plots of $R(\omega)$ and $X(\omega)$, Example 13.

Alternately, $H(j\omega)$ can be written in its polar form as

$$H(j\omega) = |H(j\omega)|e^{j\theta(\omega)} \qquad (7\text{-}19)$$

where $|H(j\omega)|$ is the magnitude of $H(j\omega)$ and $\theta(\omega)$ is the angle. The relations among the rectangular form and polar form are well-known:

$$|H(j\omega)|^2 = [R(\omega)]^2 + [X(\omega)]^2 = H(j\omega) \cdot H(-j\omega) \qquad (7\text{-}20)$$

$$\tan \theta(\omega) = \frac{X(\omega)}{R(\omega)} \qquad (7\text{-}21)$$

from which we can conclude

**Property 5**  For $s = j\omega$, the magnitude of $H(j\omega)$, $|H(j\omega)|$, is an even function. The angle $\theta(\omega)$ is an odd function if $\theta(0) = 0$.

## Example 14

From example 13 we have

$$|H(j\omega)| = \left| \frac{1}{j\omega + 1} \right| = \sqrt{\frac{1}{\omega^2 + 1}} = \sqrt{R^2 + X^2}$$

$$= \left[ \left( \frac{1}{j\omega + 1} \frac{1}{-j\omega + 1} \right) \right]^{1/2}$$

and $\theta(\omega) = \tan^{-1}(-\omega)$ with $\theta(0) = 0$. The plots of $|H(j\omega)|$ and $\theta(\omega)$ are shown in Fig. 7-15.

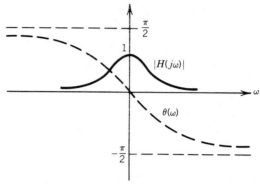

**FIGURE 7-15.**  Plots of $|H(j\omega)|$ and $\theta(\omega)$, Example 14.

□  ***Probs.*** ***7-20***
***7-21***

## 7-5  NATURAL FREQUENCIES AND THE CHARACTERISTIC EQUATION

Initial conditions can be thought of as the *state* of the network at $t = 0$. As discussed in an earlier chapter, the *state of a network* is the minimal amount of information necessary at any time to characterize completely the future behavior of the network. Thus, initial conditions—the state of the network at $t = 0$—together with a given input $e(t)$, $t \geq 0$, will determine the output $r(t)$ for $t > 0$.

Consider again the examples in Chapter 3 illustrating loop or node analysis. In the resulting equations, $\mathbf{ZI} = \mathbf{E}$ or $\mathbf{YV} = \mathbf{J}$, the right-hand side, $\mathbf{E}$ or $\mathbf{J}$, contains two types of sources: 1) inputs, and 2) initial conditions. Let us concentrate our attention on the *zero-input response*, that is, the response (output) of the network with no inputs. Naturally, this response will depend on the initial state of the network and on its topology and element values.[†]

*Example 15*

For the network shown in Fig. 3-3 of Chapter 3, the zero-input response may be found

---

[†] From the viewpoint of differential equations, the zero input response is the *complementary function*, i.e., the solution of the *homogeneous* differential equations, with the inputs—the terms on the right-hand side—set to zero.

from the equations

$$
\begin{bmatrix}
s+2 & -s & -2 \\
-s & s+3+\dfrac{1}{2s} & 0 \\
-2 & 0 & 4s+2
\end{bmatrix}
\begin{bmatrix} I_1 \\ I_2 \\ I_3 \end{bmatrix}
=
\begin{bmatrix} -2 \\ \dfrac{2}{s}+2 \\ 2 \end{bmatrix}
$$

where the inputs $v_1(t)$ and $v_2(t)$ have been set to zero. We obtain

$$
\mathbf{I} = \begin{bmatrix} I_1 \\ I_2 \\ I_3 \end{bmatrix}
= \mathbf{Z}^{-1}
\begin{bmatrix} -2 \\ \dfrac{2}{s}+2 \\ 2 \end{bmatrix}
= \frac{1}{\det \mathbf{Z}} \operatorname{Adj} \mathbf{Z}
\begin{bmatrix} -2 \\ \dfrac{2}{s}+2 \\ 2 \end{bmatrix}
$$

where the inverse of $\mathbf{Z}$ is formed according to Eq. (A-36) in Appendix A.   $\square$

Any one of the loop currents $I_1$, $I_2$, and $I_3$ will be—after evaluating det $\mathbf{Z}$ and clearing fractions—a rational function. The inverse Laplace of $I_1(s)$, for example, will be of the form

$$
i_1(t) = K_1 e^{s_1 t} + K_2 e^{s_2 t} + K_3 e^{s_3 t}
$$

where $s_1$, $s_2$, and $s_3$ are the roots (assumed to be simple) of the *characteristic equation*

$$
\det \mathbf{Z} = 0 \tag{7-22}
$$

since det $\mathbf{Z}$ is the denominator (here of degree 3) of $I_1(s)$. Similar results hold for $i_2(t)$ and $i_3(t)$, with different $K$'s but with the same exponentials. These exponentials are the *natural frequencies* (or *natural modes*) of $i_1(t)$, $i_2(t)$ or $i_3(t)$. These will also be natural frequencies of the branch currents, since the latter are linear combinations of the loop currents. Furthermore, they are also natural frequencies of branch voltages, since a branch voltage is related to a branch current by $V = RI$, $V = LsI$, or $V = (1/Cs)I$, or by linear combinations of such terms.

Collectively, all these natural frequencies are called the *natural frequencies* of the network. They certainly do not depend on choice of the method of analysis, and therefore can be found by solving the characteristic equation

$$
\det \mathbf{Z} = 0 \tag{7-23}
$$

or

$$
\det \mathbf{Y} = 0 \tag{7-24}
$$

The adjectives "natural" and "characteristic" are well chosen, since we are talking about entities that are inherent to the network: The matrix $\mathbf{Z}$, and hence the roots of det $\mathbf{Z} = 0$, depend only on the topology of the network and its element values.

## Example 16

In the network shown in Fig. 7-16, find the natural frequencies of $i_R(t)$, the current through the two-ohm resistor. The initial state is: $i_L(0_-)$ as shown, $v_c(0_-) = 0$. The transform equations are

$$\begin{bmatrix} s + 3 & -2 \\ -2 & \dfrac{2}{s} + 2 \end{bmatrix} \begin{bmatrix} I_1 \\ I_2 \end{bmatrix} = \begin{bmatrix} i_L(0_-) \\ 0 \end{bmatrix}$$

yielding

$$I_1 = \frac{i_L(0_-)(s + 1)}{s^2 + 2s + 3} \qquad I_2 = \frac{2i_L(0_-)s}{s^2 + 2s + 3}$$

The natural frequencies of $i_1(t)$ and of $i_2(t)$ are the roots of det $\mathbf{Z} = s^2 + 2s + 3 = 0$, that is, $s_1 = -1 + j\sqrt{2}$ and $s_2 = -1 - j\sqrt{2}$. The current $I_R(s)$ is given by

$$I_R(s) = I_1(s) - I_2(s) = \frac{-i_L(0_-)(s - 1)}{s^2 + 2s + 3}$$

and so,

$$i_R(t) = Ke^{-t}\cos{(\sqrt{2}t + \theta)}$$

Alternately, the natural frequencies can be found from det $\mathbf{Y} = 0$; with the nodes labeled 1 and 2, as shown, we write the characteristic equation as

$$\det \mathbf{Y} = \begin{vmatrix} 1 + \dfrac{1}{s} & -\dfrac{1}{s} \\ -\dfrac{1}{s} & \dfrac{1}{2} + \dfrac{s}{2} + \dfrac{1}{s} \end{vmatrix} = 0$$

which yields $s^2 + 2s + 3 = 0$ as before.

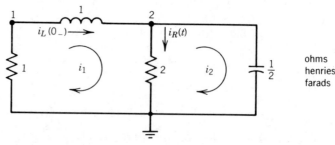

**FIGURE 7-16.** Example 16.

## Example 17

In the network shown in Fig. 7-17, the initial state is given as $i_{L_1}(0_-)$ and $i_{L_2}(0_-)$. The natural frequencies of the network are found from the characteristic equation

$$\det \mathbf{Z} = \begin{vmatrix} s + 2 & -1 \\ -1 & s + 2 \end{vmatrix} = 0$$

$$\therefore s_1 = -1 \qquad s_2 = -3$$

Solving for $I_1(s)$, we get

$$I_1(s) \frac{i_{L_1}(0_-) + i_{L_2}(0_-)}{2(s + 1)} - \frac{i_{L_2}(0_0) - i_{L_1}(0_-)}{2(s + 3)}$$

Note: In general, both natural frequencies are present in $i_1$ and $i_2$. However, with a special choice of the initial state, $i_{L_1}(0_-) = i_{L_2}(0_-)$, the natural frequency $s = -3$ is absent from $i_1(t)$. Similarly, $s = -1$ is suppressed in $i_1(t)$ if $i_{L_1}(0_-) = -i_{L_2}(0_-)$ in this example.

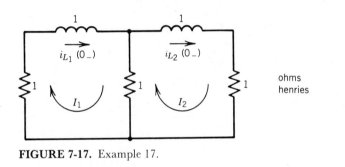

**FIGURE 7-17.** Example 17. ☐

The previous example illustrates a general statement: All the natural frequencies of a network need not be present in a particular current or voltage. Furthermore, if $s = 0$ is a natural frequency of an edge voltage (or current), it need not be present in the edge current (or voltage), as illustrated below.

## Example 18

Consider a single capacitor, with a constant voltage $V_0$ across it. In this (somewhat trivial) network, the voltage $v_c(t) = V_0$ has a natural frequency at $s = 0$, $V_c(s) = V_0/s$, but the current does not have it, since $i_c(t) = 0$. Similarly, a constant inductor current has a natural frequency at $s = 0$, but not the voltage. ☐

In summary, the natural frequencies of a network are found by solving its characteristic equation det $\mathbf{M} = 0$, where $\mathbf{M}$ is the loop impedance matrix, or the node admittance matrix, or the matrix of *any* consistent set of equations describing the network.

**Prob. 7-22**

## 7-6   NETWORK FUNCTIONS AND THE IMPULSE RESPONSE

From the definition of a network function, Eq. (7-1), we can write

$$R(s) = H(s)\, E(s) \qquad (7\text{-}25)$$

yielding the (transform) *zero-state response*, i.e., the response of the network to an excitation applied at $t = 0$, provided the state of the network at $t = 0_-$, the initial conditions, is zero.

Let us consider the zero-state response to a very special excitation,

$$e(t) = \delta(t) \qquad (7\text{-}26)$$

the unit impulse. This response will be called *the impulse response*. Since

$$E(s) = \mathcal{L}\,\delta(t) = 1 \qquad (7\text{-}27)$$

we obtain from Eq. (7-25)

$$R(s)]_{E(s)=1} = H(s) \qquad (7\text{-}28)$$

This important result gives a physical significance to the network function which, until now, was just a mathematical definition. Specifically, $H(s)$ *is the transform of the impulse response* of the network. This impulse response will be denoted by $h(t)$,

$$\mathcal{L}^{-1}R(s)]_{E(s)=1} = \mathcal{L}^{-1}H(s) = h(t) \qquad (7\text{-}29)$$

See Fig. 7-18.

**FIGURE 7-18.** The impulse response.

---

*Example 19*

Find the response to a current $i(t) = \delta(t)$ of the 1-P shown in Fig. 7-19. The pertinent network function is $Z_{dp}(s)$, and we write by inspection

$$Z_{dp}(s) = \frac{1}{s} + \frac{1}{s + 1}$$

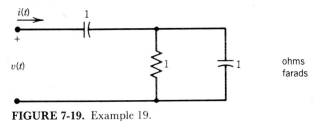

**FIGURE 7-19.** Example 19.

Therefore $h(t) = v(t)$ here and is given by

$$h(t) = \mathcal{L}^{-1}Z_{dp}(s) = (1 + e^{-t})u(t)$$

The presence of the unit step voltage in this response is also evident from a physical point of view. $\qquad\square$

## Example 20

Find the response to a voltage $v(t) = \delta(t)$ of the 1-P network shown in Fig. 7-4. Here the network function is $Y_{dp}(s)$,

$$Y_{dp}(s) = \frac{6s^2 + 2s + 1}{6s^2 + 3s}$$

and

$$h(t) = i(t) = \mathcal{L}^{-1}Y_{dp}(s) = \delta(t) + \tfrac{1}{3}u(t) - \tfrac{1}{2}e^{-t/2}u(t) \qquad\square$$

## Example 21

Find the impulse response of the network shown in Fig. 7-9. The network function is

$$G_{21}(s) = \frac{V_2(s)}{V_1(s)} = \frac{12}{2s^2 + 11s + 16}$$

and with $v_1(t) = \delta(t)$ the output will be

$$h(t) = v_2(t) = \mathcal{L}^{-1}G_{21}(s) = \mathcal{L}^{-1}\frac{6}{(s + \tfrac{11}{4})^2 + \tfrac{7}{16}}$$

$$= \frac{24}{\sqrt{7}}e^{-(11t/4)}\sin\frac{\sqrt{7}}{4}tu(t) \qquad\square$$

In general, assume that we want to find the impulse response to $v_k(t) = \delta(t)$ located in the $k$th link. Writing the fundamental loop equations, we get

$$\mathbf{ZI} = \begin{bmatrix} 0 \\ 0 \\ \vdots \\ 1 \\ 0 \\ \vdots \\ 0 \end{bmatrix}$$

and the impulse response current vector is

$$\mathbf{I} = \mathbf{Z}^{-1} \begin{bmatrix} 0 \\ 0 \\ \vdots \\ 1 \\ 0 \\ \vdots \\ 0 \end{bmatrix} = \frac{1}{\det \mathbf{Z}} \mathrm{Adj}\, \mathbf{Z} \begin{bmatrix} 0 \\ 0 \\ \vdots \\ 1 \\ 0 \\ \vdots \\ 0 \end{bmatrix}$$

As discussed earlier, we note that it has the nonzero natural frequencies of the network, the solutions of the characteristic equation $\det \mathbf{Z} = 0$. Needless to say, a particular branch current, say $i_p$, will have these natural frequencies too. On the other hand, the network function

$$Z_{pk}(s) = \frac{I_p(s)}{V_k(s)} \tag{7-30}$$

relates the excitation and the response; its inverse Laplace yields the same impulse response. Therefore, *the poles of the network function, $H(s)$, are the natural frequencies of the pertinent response.* In this connection, review the derivations of the various network functions, such as Eqs. (7-6), (7-7), and the examples in section 7-3, where the denominator of $H(s)$ is the determinant $\det \mathbf{Z}$ or $\det \mathbf{Y}$.

It is clear now why the impulse response is so important: it is characteristic of the network itself. Also, the present discussion illustrates again the earlier statement concerning $H(s)$; namely, its reciprocal *is not*, in general, a network function.

**Prob. 7-23**

## 7-7   COMPLETE RESPONSE

When we have both a nonzero initial state and an input, *the complete response is the sum of the zero-state response plus the zero-input response.* This is true for all linear systems, for then we can use superposition. First, we compute the zero-input response due to all the initial conditions without any inputs (section 7-5), then we compute the zero-state response due to the inputs, with all initial conditions zero (section 7-6). We add the two to obtain the complete response.

## Example 22

Consider again the network shown in Fig. 7-10. Let the input be $v_1(t)$, and let the initial state be given as follows: a current in the top inductor $i_2(0_-)$ and a voltage $v_c(0_-)$ across the capacitor (top plate positive). Using transform diagrams, as studied in Chapter 3, we write the following matrix equation $\mathbf{ZI} = \mathbf{E}$:

$$\begin{bmatrix} s + \dfrac{4}{s} & -s \\ -s & 3s + 3 \end{bmatrix} \begin{bmatrix} I_1 \\ I_2 \end{bmatrix} = \begin{bmatrix} V_1 - \dfrac{v_c(0_-)}{s} \\ 2i_2(0_-) \end{bmatrix}$$

Note that $\mathbf{E}$ can be written as follows:

$$\mathbf{E} = \begin{bmatrix} V_1 - \dfrac{v_c(0_-)}{s} \\ 2i_2(0) \end{bmatrix} = \begin{bmatrix} V_1 \\ 0 \end{bmatrix} + \begin{bmatrix} \dfrac{-v_c(0_-)}{s} \\ 2i_2(0_-) \end{bmatrix} = \mathbf{E}_0 + \mathbf{E}_i$$

and therefore $\mathbf{I}$ can be obtained as the sum $\mathbf{I} = \mathbf{I}_0 + \mathbf{I}_i$, where $\mathbf{I}_0$ is the zero-state response, the solution of

$$\begin{bmatrix} s + \dfrac{4}{s} & -s \\ -s & 3s + 3 \end{bmatrix} \mathbf{I}_0 = \mathbf{E}_0 = \begin{bmatrix} V_1 \\ 0 \end{bmatrix}$$

and $\mathbf{I}_i$ is the zero-input response, the solution of

$$\begin{bmatrix} s + \dfrac{4}{s} & -s \\ -s & 3s + 3 \end{bmatrix} \mathbf{I}_i = \mathbf{E}_i = \begin{bmatrix} \dfrac{-v_c(0_-)}{s} \\ 2i_2(0_-) \end{bmatrix}$$

Specifically,

$$\mathbf{I}_0 = \mathbf{Z}^{-1}\mathbf{E}_0 = \frac{V_1(s)}{\det \mathbf{Z}} \begin{bmatrix} 3s + 3 \\ s \end{bmatrix}$$

and

$$\mathbf{I}_i = \frac{1}{\det \mathbf{Z}} \begin{bmatrix} \dfrac{v_c(0_-)}{s}(3s + 3) + 2si_2(0_-) \\ 2\left(s + \dfrac{4}{s}\right)i_2(0_-) - v_c(0_-) \end{bmatrix}$$

exhibiting the dependence of $\mathbf{I}_0$ only on the inputs and of $\mathbf{I}_i$ only on the initial state.  □ *Probs. 7-24*
*7-25*

---

## Example 23

Consider a simple $RL$ series circuit, Fig. 7-20, with a given initial state $i(0_-)$. Suppose the input for $t \geq 0$ is $v(t) = V$, a constant. It is required to find $i(t)$, the total response. We note immediately the natural frequency of the network to be $s = -1$, being the root of det $\mathbf{Z} = s + 1 = 0$ or the pole of the network function $Y_{dp} = 1/(s + 1)$. The loop equation for the transform network reads

$$(s + 1)I = \frac{V}{s} + i(0_-)$$

and therefore,

$$I(s) = \frac{V}{s(s + 1)} + \frac{i(0_-)}{s + 1}$$

yielding the complete response:

$$i(t) = Vu(t) - Ve^{-t}u(t) + i(0_-)e^{-t}u(t) = i_0(t) + i_i(t)$$

with $i_0(t) = Vu(t) - Ve^{-t}$ the zero-state response and $i_i(t) = i(0_-)e^{-t}$ the zero-input response, as discussed before. We can, however, look at $i(t)$, the complete response, as the sum

$$i(t) = Vu(t) + [-V + i(0_-)]e^{-t}u(t)$$

where the first term, $Vu(t)$, remains constant for large $t$ ($t \to \infty$) and thus constitutes the *steady-state* part of the complete response. The second term, $[-V + i(0_-)]e^{-t}$, decays exponentially to zero for large $t$, and therefore is the *transient* part.

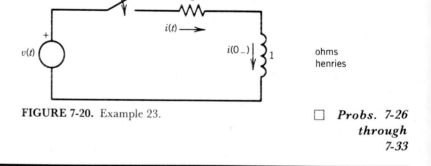

**FIGURE 7-20.** Example 23.

□ **Probs. 7-26 through 7-33**

---

We make the following important observations:

1. The steady-state response is due to the zero-state response alone. This is reasonable, since the initial state (at $t = 0$) must not affect the total response for $t \to \infty$.

2. The zero-state response "resembles" the input: a constant if the input is constant, a sine wave if the input is a sine wave (of the same frequency, with a possible phase shift), an exponential if the input is exponential (with the same exponent).

3. The zero-input response "resembles" the impulse response, within arbitrary multiplying constants. This is another important feature of the impulse response.

4. Sometimes, as in our example, the transient part can be eliminated by a proper choice of the initial state (here, with $i(0_-) = V$).

5. Sometimes the complete response may consist entirely of a transient part, decaying to zero for large $t$, as illustrated in the following example.

---

**Example 24**

Consider the previous network, with the input $v(t) = Ve^{\alpha t}$, for $t \geq 0$, where $\alpha < 0$. This time we write

$$(s + 1)I = \frac{V}{s - \alpha} + i(0_-)$$

and therefore,

$$I(s) = \frac{V}{(s + 1)(s - \alpha)} + \frac{i(0_-)}{s + 1}$$

yielding the complete response:

$$i(t) = \left[\frac{-V}{\alpha + 1} + i(0_-)\right]e^{-t} + \frac{V}{\alpha + 1}e^{\alpha t} \qquad \alpha \neq -1 \qquad t \geq 0$$

two decaying exponentials—a transient with no steady-state part.

It is interesting to see what happens in the previous example when $\alpha = -1$. In this case we get

$$I(s) = \frac{V}{(s + 1)^2} + \frac{i(0_-)}{s + 1}$$

and

$$i(t) = Vte^{-t}u(t) + i(0_-)e^{-t}u(t)$$

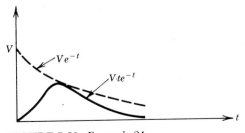

**FIGURE 7-21.** Example 24.

What happened? The value $s = -1$ is both a natural frequency of the network *and* the exponent of the input. We have excited a natural frequency of the network and "resonance" occurred,[†] resulting in the *critically damped* term $Vte^{-t}$ (notice the multiplication by $t$). See Fig. 7-21.    □

## PROBLEMS

7-1  Prove, by using the basic definition of network functions, the familiar rule: "Driving-point impedances in series add up" for $n$ passive elements connected in series.

7-2  Repeat problem 7-1 for driving-point admittances connected in parallel.

7-3  Find $Y_{dp}(s)$ for the 1-P network shown.

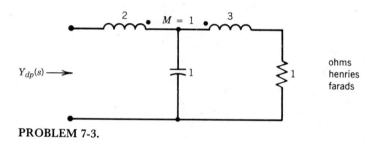

**PROBLEM 7-3.**

7-4  Repeat problem 7-3 for the network shown. Notice the dependent source. What happens if $k = 2$? In this case the 2-P network, shown in dotted lines, is known as a *negative impedance converter* (*NIC*).

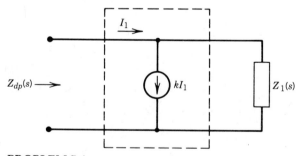

**PROBLEM 7-4.**

[†] Quotation marks are used in the word "resonance," since the natural frequency and the exponent of the excitation are real. If both were purely imaginary, $s = j\omega_0$, we would have resonance in the familiar sense of the word. See problems 7-30 and 7-31.

7-5 Find $Z_{dp}(s)$ for the network shown. Every resistor is $1\Omega$.

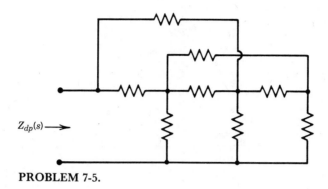

**PROBLEM 7-5.**

7-6 Find $Y_{dp}(s)$ for the network shown.

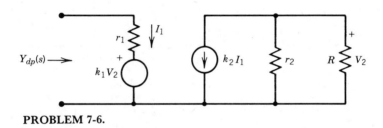

**PROBLEM 7-6.**

7-7 Find $Z_{dp}(s)$ for the *lattice* network shown.

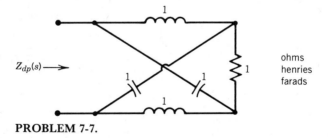

ohms
henries
farads

**PROBLEM 7-7.**

7-8 Find $Y_{dp}(s)$ for the network shown. The ideal transformer is $1:1$.

7-9 A general 2-P network is given by its open-circuit parameters ($z_{11}$, $z_{12}$, $z_{21}$, and $z_{22}$). A load impedance $Z_L$ is connected to port 2. Find the driving-point impedance at port 1.

7-10 Find $Z_{dp}(s)$ of each ladder by writing it as a continued fraction. Then clear the fractions and obtain the final answer as a ratio of two polynomials.

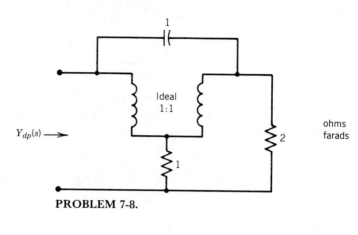

**PROBLEM 7-8.**

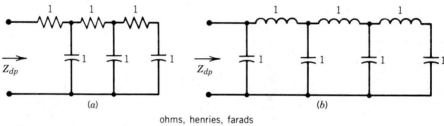

ohms, henries, farads

**PROBLEM 7-10.**

7-11  Repeat problem 7-10, but in each ladder interchange the series elements with the shunt elements.

7-12  Find $Z_{dp}(s)$ for the network shown. The gyrating resistance is 1.

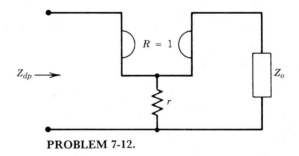

**PROBLEM 7-12.**

7-13  Find the input impedance, i.e., $Z_{dp}(s)$ at port 1, of the circuit shown in Fig. 7-10 in the text. Port 2 remains open-circuited.

7-14  An equivalent circuit of a transistor amplifier is shown below, using the hybrid parameters. The load impedance $Z_L$ is connected to port 2. Find (a) the input impedance at port 1; (b) the current transfer function $\alpha(s) = I_L/I_1$.

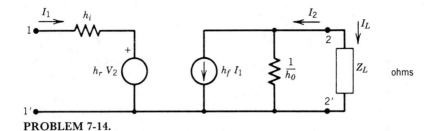

**PROBLEM 7-14.**

7-15 a) In the network shown, find $G_{21}(s) = V_2(s)/V_1(s)$ by working "backwards," from the output to the input. Assume $V_2(s)$, then, by calculating all the necessary currents and voltages, arrive at $V_1(s)$. See example 5 in section 7-2.

b) Repeat part (a), with the input this time at the right and the output at the left.

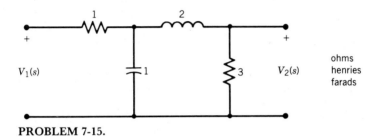

**PROBLEM 7-15.**

7-16 For each of the following $H(s)$, find the poles and zeros, and plot the pole-zero configuration:

a) $H(s) = \dfrac{s^2 - s + 2}{s^3 + 19s^2 + 10s + 100}$

b) $H(s) = 3.7\,\dfrac{(s + 1)^2(s - 2)}{s^5 + s^4 + 3s^3 + 2s^2 + 2s + 1}$

c) $H(s) = \dfrac{(6s^2 - s + 4)(s + 1)}{(s^2 + s + 2)(s + 1)^3}$

d) $H(s) = \dfrac{s^2 - 1}{s^3 + 2s^2 + 3s + 2}$

7-17 Plot the pole-zero configuration of:

a) $Y_{dp}(s)$ of problem 7-3.
b) $Y_{dp}(s)$ of problem 7-6.
c) $Z_{dp}(s)$ of problem 7-7.
d) $Y_{dp}(s)$ of problem 7-8.

e) each $Z_{dp}(s)$ of problem 7-10.

f) $G_{21}(s)$ of problem 7-15.

7-18  Find completely $H(s)$, given:
zeros at $s = -2$, $s = 2$.
poles at $s = 0$, $s = -1$ (double).
$dH/ds$ at $s = 1$ is 0.6.

7-19  A network function has finite zeros at $s = 0$, $s = +j3$, $s = -j3$; and finite poles at $s = +j1$, $s = -j1$, $s = +j2$, and $s = -j2$. Also $dH/ds$ at $s = j3$ is 0.24. Find $H(s)$.

7-20  For each of the following network functions, find and sketch: $R(\omega)$, $X(\omega)$, $|H(j\omega)|$, and $\theta(\omega)$.

a) $\dfrac{s - 1}{s + 1}$

b) $\dfrac{1}{s^2 + s + 1}$

c) $\dfrac{s}{s + 1}$

d) $\dfrac{s}{s^2 + s + 1}$

7-21  a) If we write

$$H(j\omega) = \frac{P(j\omega)}{Q(j\omega)} = \frac{P_1(\omega) + jP_2(\omega)}{Q_1(\omega) + jQ_2(\omega)}$$

find the expressions for $R(\omega)$, $(X\omega)$, $|H(j\omega)|$ and $\theta(\omega)$ in terms of $P_1(\omega)$, $P_2(\omega)$, $Q_1(\omega)$ and $Q_2(\omega)$.

b) If we know that $R(\omega) \equiv 0$, draw conclusions about $H(s)$.

7-22  Find the characteristic equations and the natural frequencies of:

a) the network shown in problem 7-3.

b) the network in problem 7-7.

c) the network in problem 7-8.

d) the network in problem 7-10.

e) the network in problem 7-15(a).

7-23  Find the impulse response of:

a) the network shown in problem 7-3.

b) the network shown in problem 7-7.

c) the network shown in problem 7-8.

d) the network shown in problem 7-10.

e) the network shown in problem 7-15(a).

Compare with your answers to problem 7-22.

7-24 a) In the network of problem 7-3, the capacitor has an initial charge $q_c(0)$. Find the zero-input current $i_{dpi}(t)$ in the (shorted) driving-point terminals. Compare with problem 7-23(a).

b) Find the zero-state response $i_{dpo}(t)$ of this network to the input $v(t) = 10e^{-t}$.

c) Find the complete response under the given state and input.

7-25 a) In the network of problem 7-15(a), the inductor has an initial current $i_L(0)$. Find the zero-input ($v_1 \equiv 0$) response, $v_{2i}(t)$. Compare with problem 7-23(e).

b) Find the zero-state response, $v_{20}(t)$, to the input $v_1(t) = 10u(t)$.

c) Find the complete response under the given state and input.

7-26 Show with the help of a simple example (such as the one in Fig. 7-20 in the text) that the zero-state response is linearly dependent on the input, obeying additivity and homogeneity (see Chapter 1).

7-27 Show with the help of a simple example that the complete response is *not* linearly dependent on the input when the initial state is nonzero.

7-28 A network function $H(s)$ has finite poles at $s = -1$, $s = +j2$, and $s = -j2$. It has a zero of order 2 at $s = 2$. In addition, we know that

$$\frac{dH}{ds}\bigg]_{s=1} = -0.087$$

a) Find $H(s)$

b) This network is excited, for $t > 0$, by a d.c. source, the initial state being zero. With a minimal effort, write the form of the complete response.

c) In this complete response, indicate the transient and the steady-state parts.

d) Repeat parts (b) and (c) for an input of the form $A \sin 2t$.

7-29 A network function $H(s)$ has finite zeros at $s = 0$ (simple) and $s = 1$ (of order 2). It has finite simple poles at $s = -\frac{1}{2} + j\frac{3}{2}$, $s = -\frac{1}{2} - j\frac{3}{2}$, and $s = -1$. Also $H(2) = 21$.

a) Find $H(s)$.

b) With the initial state being zero, the network is excited by a d.c. input. Write, within arbitrary multiplying constants, the expression for the complete response. What happens to the natural frequency at $s = 0$?

7-30 For the 1-P network shown, the input is of the form $v(t) = Ve^{j\omega_0 t}$. Determine the value (values) of $\omega_0$ for resonance.

7-31 Repeat problem 7-30 for the 1-P network shown.

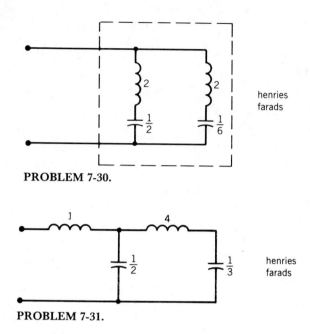

**PROBLEM 7-30.**

**PROBLEM 7-31.**

**7-32** For each of the following networks, give the form (or forms) of input that will excite one or more of the natural frequencies:

a) the network shown in problem 7-3.

b) the network shown in problem 7-7.

c) the network shown in problem 7-8.

d) the network shown in problem 7-10.

e) the network shown in problem 7-15(a).

See your answers to problem 7-22.

**7-33** Since we don't have (normally) a handy impulse source, we wish to obtain the impulse response of a network using a step input (e.g., a dc source applied at $t = 0$), which is a commonly available source. Using the Laplace transform, calculate the expression for $h(t)$, the impulse response, from an experiment in which you obtain the *step response*, $p(t)$, as shown.

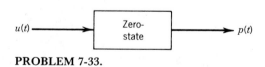

**PROBLEM 7-33.**

# 8

# CONVOLUTION
# AND SUPERPOSITION

In Chapter 1, we considered briefly the concept of superposition and linearity. In Chapter 7, the zero-state response[†] of a linear, constant network was developed via the concept of the network function $H(s)$:

$$R(s) = H(s)E(s) \tag{8-1}$$

In this chapter, we shall discuss some of these concepts in the time domain. Specifically, we shall consider the relations among the zero-state response $r(t)$, an arbitrary input $e(t)$, and the impulse response $h(t)$.

## 8-1 SUPERPOSITION

Let us forsake, for the time being, Laplace transform methods such as Eq. (8-1), and ask ourselves the following question: Since the network is linear, can we find a way to apply superposition in order to find the zero-state response to an arbitrary input? Continuing along this line of thought, consider the possibility of decomposing a given arbitrary input $e(t)$ into a sum of "simple" functions. If we know the zero-state response of the network to each one of those "simple" functions, then, by superposition, the response $r(t)$ will be the sum of the responses to each of the "simple" functions. Immediately, the impulse function suggests itself as a possible "simple" function, since we know the impulse response of the network: It is $h(t) = \mathscr{L}^{-1}H(s)$. The idea is illustrated in Fig. 8-1. The next problem is that of decomposing $e(t)$, an arbitrary input, into a sum of impulse functions. Let us illustrate the method by a preliminary example.

---

[†] In that chapter, a subscript 0 was used for the zero-state response to distinguish it from the zero-input response. In the present chapter, this subscript is omitted, because we deal here only with the zero-state response.

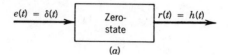

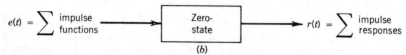

**FIGURE 8-1.** The use of superposition: (*a*) The unit impulse response; (*b*) Superposition of impulses.

## Example 1

Consider an input $e(t)$ shown in Fig. 8-2*a*. It has been approximated by a "staircase" of step functions, and these are shown separately in Fig. 7-2*b*. We can write

$$e(t) \approx u(t-1) - u(t-2) + 3[u(t-2) - u(t-3)] + 2.5[u(t-3) - u(t-4)]$$
$$+ 2[u(t-4) - u(t-5)] + u(t-5) - u(t-6)$$

where $\approx$ means "equals approximately." The separate step functions are shown in Fig. 8-2*b*. This is not a very satisfactory approximation. After all, the given $e(t)$ is continuous ("smooth"), while its approximation contains several discontinuities. However, we must remember that we took finite intervals of time, 1 sec. apart, for the

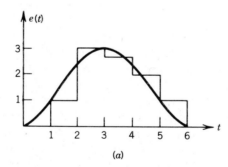

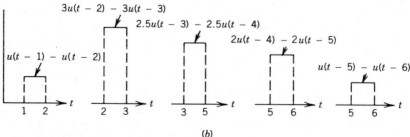

**FIGURE 8-2.** Example 1.

step functions in this example. In the general derivation which follows, we shall take infinitesimally small time intervals and, in the limit, such an approximation will be exact.  □

Consider, then, an input $e(x)$ as shown in Fig. 8-3. Here, we use a dummy variable $x$ to indicate the abscissa, and $t$ will be a particular point along this axis. We are interested in the input between 0 and $t$. As in the previous example, approximate $e(x)$ by a "staircase" of step functions, separated by intervals of $\Delta x$. A typical building block of this approximation, between the points $k\Delta x$ and $(k + 1)\,\Delta x$, is crosshatched in Fig. 8-3, and its expression is therefore $e(k\Delta x)\{u(x - k\Delta x) - u[x - (k + 1)\,\Delta x]\}$. Then,

$$e(t) \approx \sum_{k=0}^{k=n} e(k\Delta x)\{u(t - k\Delta x) - u[t - (k + 1)\,\Delta x]\} \tag{8-2}$$

is the approximation for the input between $x = 0$ and $x = t$. We have taken $n$, the upper limit of the summation, to be the total number of intervals between 0 and $t$. We can divide and multiply the summation by $\Delta x$

$$e(t) \approx \sum_{k=0}^{k=n} e(k\Delta x)\,\frac{\{u(t - k\Delta x) - u[t - (k + 1)\,\Delta x]\}}{\Delta x}\,\Delta x \tag{8-3}$$

and now let $\Delta x \to 0$. The points along the abscissa were previously *discrete*, i.e., 0, $\Delta x, 2\Delta x, \ldots, k\Delta x, (k + 1)\,\Delta x, \ldots, n\Delta x$. As $\Delta x \to 0$, these points get closer together, and $k\Delta x$ becomes the *continuous* variable $x$ along the abscissa. Also, by the definition of a derivative,

$$\lim_{\Delta x \to 0} \frac{u(t - k\Delta x) - u[t - (k + 1)\,\Delta x]}{\Delta x} = \delta(t - x) \tag{8-4}$$

while the summation becomes an integral. Thus,

$$e(t) = \int_{0_-}^{t} e(x)\delta(t - x)\,dx \tag{8-5}$$

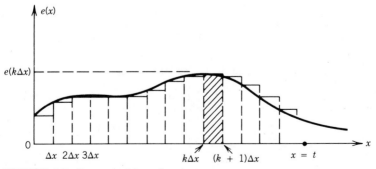

**FIGURE 8-3.** Decomposition of an input.

is the exact expression for the input, given as a sum (integral) of impulse functions. The result of Eq. (8-5) is also consistent with the definition of the impulse function and its sampling property. Each impulse function in Eq. (8-5) is multiplied by $e(x)$, the value of the input at that point. This multiplier is sometimes called the *strength* of the impulse.

The zero-state response of the linear, constant network to an input $\delta(t)$ is $h(t)$, the impulse response. Therefore, the zero-state response to $e(x)\delta(t - x)$ will be $e(x)h(t - x)$. By superposition, then, the zero-state response to $e(t)$, Eq. (8-5), is the sum of all the individual responses:

$$r(t) = \int_{0_-}^{t} e(x)h(t - x) \, dx \qquad (8\text{-}6)$$

The result in Eq. (8-6) is the *convolution integral* and, in words, we say: *The zero-state response of a linear network is obtained by convolving the input with the impulse response.* Such a convolution consists of three steps:

1. Substitute a dummy variable $x$ for $t$ in $e(t)$, obtaining $e(x)$.
2. Substitute $(t - x)$ for $t$ in the impulse response, obtaining $h(t - x)$.
3. Integrate the product $e(x)h(t - x) \, dx$ between $x = 0_-$ and $x = t$. The result of this integration will be $r(t)$. Note that since this is a definite integral with respect to $x$ between the indicated limits, the result will not depend on $x$.

---

### Example 2

Consider the $RC$ network shown in Fig. 8-4. The pertinent impulse response is easily calculated as

$$\mathscr{L}^{-1} \frac{1}{s + 1} = h(t) = e^{-t}u(t)$$

Let us compute, entirely in the time-domain, the zero-state response to an input $v_1(t) = 10e^{-3t}u(t)$. The convolution of $h(t)$ with $v_1(t)$ yields

$$r(t) = v_2(t) = \int_{0_-}^{t} 10e^{-3x}u(x)e^{-(t-x)}u(t - x) \, dx$$

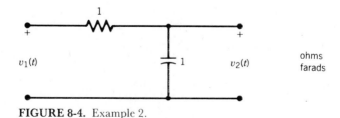

**FIGURE 8-4.** Example 2.

Since the integration is with respect to $x$, $u(x) = 1$ and $u(t - x) = 1$ between $x = 0_-$ and $x = t$. Also, terms that do not contain $x$ may be removed in front of the integral. Therefore,

$$v_2(t) = 10e^{-t} \int_{0_-}^{t} e^{-3x}e^x \, dx = 10e^{-t} \int_{0_-}^{t} e^{-2x} \, dx$$

$$= (5e^{-t} - 5e^{-3t})u(t)$$

As a quick check, the student should calculate

$$V_2(s) = H(s)V_1(s) = \frac{1}{s+1}\frac{10}{s+3} \qquad \square$$

---

The importance of the impulse response is again apparent: Once we know it, we can find the zero-state response to *any* input by convolving these two functions. This convolution, in the time domain, is a direct consequence of the linearity of the system. A basic physical significance is thus associated with the impulse response.

The convolution integral, Eq. (8-6), is often written as follows:

$$r(t) = e(t) * h(t) \equiv \int_{0_-}^{t} e(x)h(t - x) \, dx \tag{8-7}$$

where the symbol $*$ indicates convolution.[†] Let us prove that convolution is commutative; that is,

$$e(t) * h(t) = h(t) * e(t) = \int_{0_-}^{t} e(t - x)h(x) \, dx \tag{8-8}$$

which means, in essence, that in setting up the convolution integral, it does not matter whether the integrand is $e(x)h(t - x)$ or $e(t - x)h(x)$. If one function is in terms of $x$, the other will be in terms of $(t - x)$. The proof is quite simple. In Eq. (8-7), substitute a new variable

$$x = t - \eta \tag{8-9}$$

Then $dx = -d\eta$, the lower limit on the integral, $x = 0$, becomes $\eta = t$, and the upper limit $x = t$ becomes $\eta = 0$. Therefore,

$$\int_{0_-}^{t} e(x)h(t - x) \, dx = -\int_{t}^{0_-} e(t - \eta)h(\eta) \, d\eta = \int_{0_-}^{t} e(t - \eta)h(\eta) \, d\eta \tag{8-10}$$

and since $\eta$ is a dummy variable like $x$, the last result proves the commutative nature of convolution:

$$e(t) * h(t) = \int_{0_-}^{t} e(t - x)h(x) \, dx = \int_{0_-}^{t} e(x)h(t - x) \, dx = h(t) * e(t) \tag{8-11}$$

---

[†] It has nothing to do with the multiplication sign used in some computer languages!

## Example 3

In the previous example, we can write

$$r(t) = v_2(t) = \int_{0_-}^{t} 10e^{-3(t-x)}u(t-x)e^{-x}u(x)\,dx$$

$$= 10e^{-3t}\int_{0_-}^{t} e^{3x}e^{-x}\,dx = 5e^{-t} - 5e^{-3t}$$

as before.

☐ *Probs. 8-1*

*8-2*

*8-11*

## Example 4

Consider the $RC$ network shown previously in Fig. 8-4. This time, however, the resistor and capacitor are interchanged as shown in Fig. 8-5. The impulse response here is

$$h(t) = \mathcal{L}^{-1}\frac{s}{s+1} = \delta(t) - e^{-t}u(t),$$

and it contains an impulse, as can be readily seen from a physical point of view. It is required to find the zero-state response to an input $v_1(t) = 10e^{-3t}$, $t \geq 0$. The convolution integral reads:

$$r(t) = v_2(t) = \int_{0_-}^{t} [\delta(t-x) - e^{-(t-x)}u(t-x)]10e^{-3x}\,dx$$

$$= 10\int_{0_-}^{t} e^{-3x}\delta(t-x)\,dx - 10e^{-t}\int_{0_-}^{t} e^{-2x}\,dx$$

The first integral can be evaluated by using the sampling property of the impulse function, namely,

$$\int_{a}^{c} \delta(t-b)g(t)\,dt = g(b) \qquad a < b < c$$

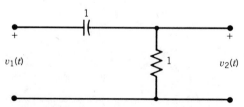

**FIGURE 8-5.** Example 4.

Therefore,

$$\int_{0_-}^{t} e^{-3x} \delta(t - x) \, dx = e^{-3t}$$

and the final answer is

$$v_2(t) = 15e^{-3t} - 5e^{-t} \qquad t \ge 0 \qquad \square \quad \textbf{\textit{Probs. 8-3}}$$
$$\textbf{\textit{8-4}}$$
$$\textbf{\textit{8-5}}$$

## 8-2   GRAPHICAL CONVOLUTION

It is instructive as well as of practical value to interpret graphically the convolution integral, Eq. (8-7) or (8-8). To be specific, let us consider Eq. (8-8), repeated here:

$$r(t) = e(t) * h(t) = \int_{0_-}^{t} e(t - x)h(x) \, dx \qquad (8\text{-}8)$$

If we interpret this integral graphically, $r(t_1)$ is the area (the integral between $x = 0_-$ and $x = t_1$) under the curve $e(t_1 - x)h(x)$. Similarly, $r(t_2)$ is the area under the curve $e(t_2 - x)h(x)$ between $x = 0_-$ and $x = t_2$, etc. In this way, $r(t)$ can be obtained for several points $t_1, t_2, \ldots$; in other words, it is a function of $t$. Let us illustrate with an example.

### Example 5

Let $h(t)$ and $e(t)$ be given as shown in Figs. 8-6a and 8-6b. The plot of $h(x)$ vs. $x$ will be identical to that of $h(t)$ vs. $t$, and is shown in Fig. 8-6c. The plot of $e(t - x)$, for a particular value of $t$, $t_1$, is done in two steps: (a) first, plot $e(-x)$ vs. $x$. This amounts to reflecting (or folding) the plot of $e(x)$ vs. $x$ about the vertical axis, as shown in Fig. 8-6d. For any positive $t_1$, the plot of $e(t_1 - x) = e(-x + t_1)$ is the plot shown in Fig. 8-6d *shifted to the right* by $t_1$ units along the abscissa. For example, $e(2 - x)$ is shown in Fig. 8-6e. To find $r(2)$, multiply the curve in part (c) by the curve in part (e). The product is shown in Fig. 8-6f. The area under the curve in part (f) is (numerically) the value of $r(t_1)$ for $t_1 = 2$. In Fig. 8-6g we show $h(x)$, $e(-x)$ and the

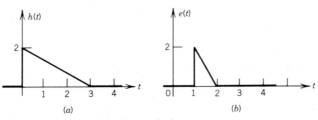

**FIGURE 8-6.** Graphical convolution.

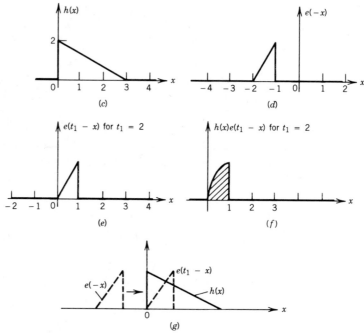

**FIGURE 8-6.** (*continued*)

shifted $e(t_1 - x)$. The entire process is then repeated for $t = t_2$, etc. The same results will be obtained with the roles of $h(t)$ and $e(t)$ exchanged, since convolution is commutative. $\square$

---

In summary, to evaluate a graphical convolution of $e(t)$ and $h(t)$, fold (reflect) one function about the vertical axis, shift it a distance $t_1$, and multiply by the other function. The area under the product curve is $e(t_1) * h(t_1)$. Repeat, for other values of $t$.

This graphical procedure of the convolution is suitable when, for example, $h(t)$ is given graphically (an oscilloscope display). Also, such numerical computations involving multiplication and integration can be programmed for a computer.

*Probs. 8-6*
*through*
*8-12*

### 8-3   NUMERICAL CONVOLUTION

Numerical convolution is convenient when either $h(t)$ and/or $e(t)$ are not given in analytical form. It is useful as an accurate method to obtain the zero-state output when $h(t)$ and $e(t)$ are given as sequences of numbers (in digitized form).

Let us develop this method with a specific example

---

## *Example 6*

Let $h(t)$ and $e(t)$ be given as shown in Fig. 8-7a,b as two sequences

$$h(t) = \{h(0), h(1), h(2), h(3), \ldots\}$$

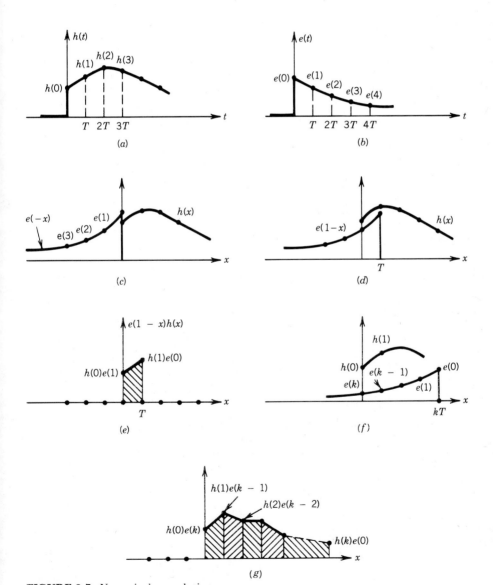

**FIGURE 8-7.** Numerical convolution.

and

$$e(t) = \{e(0), e(1), e(2), e(3), \ldots\}$$

where we use the simplified notation $h(k)$ instead of $h(kT)$. These sampled points are $T$ sec apart, and accuracy is improved as $T$ is made smaller. For $t = 0$, we fold (reflect) $e(x)$ about the vertical axis, while $h(x)$ remains unchanged, as shown in Fig. 8-7c. The area under the product curve is

$$r(t)]_{t=0} = \int_{0-}^{0} h(x)e(-x)\, dx = 0$$

provided $h$ or $e$ do not contain impulses at the origin. Next, shift $e(-x)$ to the right by one $T$, as in Fig. 8-7d. The product of $e(1 - x)h(x)$ is shown in Fig. 8-7e. In particular, at $x = 0$ the ordinate (height) of the product curve is $h(0)e(1)$, and at $x = T$ it is $h(1)e(0)$. All other points are of zero height. The area under this product curve is

$$r(t)]_{t=T} = \int_{0-}^{T} h(x)e(1 - x)\, dx = \frac{T}{2}[h(0)e(1) + h(1)e(0)]$$

where we have used the rule for the area of a trapezoid, that is, the height $(T)$ times one half the sum of the parallel sides.

At a general time $t = kT$, the curve $e(-x)$ has been shifted to the right by $k$ units, as shown in Fig. 8-7f. The product curve $h(x)e(k - x)$ is shown in Fig. 8-7g, with the ordinates $h(0)e(k)$, $h(1)e(k - 1)$, $h(2)e(k - 2), \ldots,$ $h(k)e(0)$. The area under this curve is

$$r(t)]_{t=kT} = \int_{0-}^{kT} h(x)e(k - x)\, dx$$

and is found as the sum of the areas of the trapezoids, that is,

$$r(t)]_{t=kT} = \frac{T}{2}[h(0)e(k) + 2h(1)e(k - 1) + 2h(2)e(k - 2)$$

$$+ \cdots + 2h(k - 1)e(1) + h(k)e(0)] \tag{8-12}$$

Notice that in the brackets we have the first parallel side $h(0)e(k)$, the last side $h(k)e(0)$, plus *twice* the internal sides, since each of these belongs to two adjacent trapezoids.

Equation (8-12) provides the numerical calculation for $r(kT)$, and the results, for various $k$, will yield the output sequence $r(t) = \{r(0), r(1), r(2), \ldots\}$[†]

☐   ***Probs. 8-13***

***8-14***

***8-15***

---

[†] See also Appendix E.

## 8-4 LAPLACE TRANSFORM OF THE CONVOLUTION

Let us return now to a consideration of the Laplace transform and make a comparison. On one hand, the zero-state response (in the time domain) is given by the convolution,

$$r(t) = h(t) * e(t) \tag{8-13}$$

On the other hand, $R(s)$, the Laplace transform of this zero-state response, is given by

$$R(s) = H(s)E(s) \tag{8-14}$$

Comparison of Eqs. (8-13) and (8-14) leads immediately to the conclusion that the latter is the Laplace transform of the former. That is,

$$\mathcal{L}[h(t) * e(t)] = H(s)E(s) \tag{8-15}$$

These results are shown in Fig. 8-8. In words: convolution in the time domain transforms into multiplication in the $s$ domain.

Although this result has been established through our study so far, let us prove Eq. (8-15) directly by the definitions of the convolution integral and the Laplace transform. In other words, we wish to establish that

$$\mathcal{L}\left[\int_{0-}^{t} e(t - x)h(x)\, dx\right] = E(s)H(s) \tag{8-16}$$

For this purpose, write out fully the left-hand side of Eq. (8-16)

$$\mathcal{L}\int_{0-}^{t} e(t - x)h(x)\, dx = \int_{0-}^{\infty}\left[\int_{0-}^{t} e(t - x)h(x)\, dx\right]e^{-st}\, dt$$

$$= \int_{0-}^{\infty}\int_{0-}^{t} e(t - x)h(x)e^{-st}\, dx\, dt \tag{8-17}$$

where the last double integral is with respect to two independent variables $x$ and $t$. We can change the upper limit on the first integral from $t$ to $\infty$ provided: 1) the integrand remains unchanged between 0 and $t$—the existing limits, and 2) the integrand is identically zero from $t$ to $\infty$. This can be achieved by multiplying the integrand by $u(t - x)$; as plotted vs. $x$, this unit step function is equal to unity between $0_-$ and $t$, and is zero from $t$ to $\infty$ (see Fig. 8-9). Therefore, Eq. (8-17) can be rewritten as follows:

$$\int_{0-}^{\infty}\int_{0-}^{t} e(t - x)h(x)e^{-st}\, dx\, dt = \int_{0-}^{\infty}\int_{0-}^{\infty} e(t - x)h(x)u(t - x)e^{-st}\, dx\, dt \tag{8-18}$$

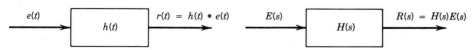

FIGURE 8-8. Input–output relations.

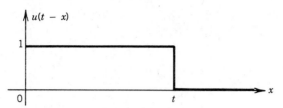

**FIGURE 8-9.** Plot of $u(t - x)$ versus $x$.

and we proceed to evaluate the double integral in Eq. (8-18)

$$\int_{0-}^{\infty} \int_{0-}^{\infty} e(t - x)h(x)u(t - x)e^{-st} \, dx \, dt$$

$$= \int_{0-}^{\infty} h(x) \left\{ \int_{0-}^{\infty} e(t - x)u(t - x)e^{-st} \, dt \right\} dx \quad (8\text{-}19)$$

where, again, we recall that $x$ and $t$ are totally independent variables, allowing us to perform the integration in this manner. A close examination of the bracketed integral in Eq. (8-19) tells us that this is precisely the Laplace transform of the shifted function $e(t - x)u(t - x)$. See Figs. 8-10 and 8-11. This Laplace transform is well-known (see Chap. 2)

$$\mathscr{L}e(t - x)u(t - x) = E(s)e^{-sx}$$

where $E(s) = \mathscr{L}e(t)$ is the transform of $e(t)$. Therefore, Eq. (8-19) reads

$$\int_{0-}^{\infty} h(x) \left\{ \int_{0-}^{\infty} e(t - x)u(t - x)e^{-st} \, dt \right\} dx = \int_{0-}^{\infty} h(x)E(s)e^{-sx} \, dx \quad (8\text{-}20)$$

Finally, since $s$ and $x$ are independent, $E(s)$ is a constant in the last integral and can be

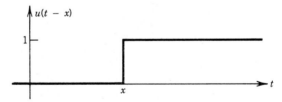

**FIGURE 8-10.** Plot of $u(t - x)$ versus $t$.

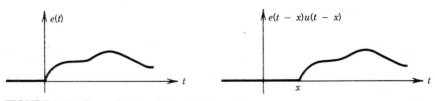

**FIGURE 8-11.** Plots of $e(t)$ and $e(t - x)u(t - x)$.

brought in front

$$\int_0^\infty h(x)E(s)e^{-sx}\,dx = E(s)\left\{\int_{0-}^\infty h(x)e^{-sx}\,dx\right\} = E(s)H(s) \qquad (8\text{-}21)$$

and the proof is completed.

In reviewing this proof, two important points should be mentioned: 1) the upper limit $t$ is changed to $\infty$ by a suitable introduction of a step function in Eq. (8-18), and 2) the same unit step function is re-interpreted (vs. $x$ formerly, vs. $t$ later) in Eq. (8-20). Indeed, the seemingly "innocent" step function is quite powerful!

The commutative nature of the convolution integral may be established easily from the previous proof. We have

$$\mathscr{L}[e(t) * h(t)] = E(s)H(s) \qquad (8\text{-}22)$$

but algebraic multiplication is certainly commutative. That is,

$$E(s)H(s) = H(s)E(s) \qquad (8\text{-}23)$$

Inverting Eq. (8-23), we obtain the desired result:

$$e(t) * h(t) = \mathscr{L}^{-1}E(s)H(s) = \mathscr{L}^{-1}H(s)E(s) = h(t) * e(t) \qquad (8\text{-}24)$$

The transform of the convolution can serve as a powerful tool to find the inverse Laplace of many functions. If we are given $F(s)$ which is a product of two simpler functions,

$$F(s) = F_1(s)F_2(s) \qquad (8\text{-}25)$$

and, furthermore, if we know

$$f_1(t) = \mathscr{L}^{-1}F_1(s) \qquad (8\text{-}26)$$

and

$$f_2(t) = \mathscr{L}^{-1}F_2(s) \qquad (8\text{-}27)$$

then we can find $\mathscr{L}^{-1}F(s)$ by convolution, that is,

$$\mathscr{L}^{-1}F(s) = f_1(t) * f_2(t) \qquad (8\text{-}28)$$

---

## Example 7

By convolution, find $\mathscr{L}^{-1}1/(s+1)(s+2)$. We write:

$$F_1(s) = \frac{1}{s+1} \qquad \therefore f_1(t) = e^{-t}u(t)$$

$$F_2(s) = \frac{1}{s+2} \qquad \therefore f_2(t) = e^{-2t}u(t)$$

And therefore:

$$\mathcal{L}^{-1} \frac{1}{(s+1)(s+2)} = e^{-t}u(t) * e^{-2t}u(t)$$

$$= \int_{0_-}^{t} e^{-x}u(x)e^{-2(t-x)}u(t-x)\,dx$$

$$= e^{-2t} \int_{0_-}^{t} e^{x}\,dx = e^{-t} - e^{-2t}$$

This result may be verified by a partial fraction expansion of $F(s)$.  ☐

---

The results of Eqs. (8-25) and (8-28) may be extended to several products. Suppose that we can write

$$F(s) = F_1(s)F_2(s)F_3(s) \tag{8-29}$$

then, treating $F_2(s)F_3(s)$ as a single function, say,

$$F_0(s) = F_2(s)F_3(s) \tag{8-30}$$

we have

$$\mathcal{L}^{-1}F(s) = \mathcal{L}^{-1}F_1(s)F_0(s) = f_1(t) * f_0(t) \tag{8-31}$$

But

$$f_0(t) = \mathcal{L}^{-1}F_2(s)F_3(s) = f_2(t) * f_3(t) \tag{8-32}$$

and therefore,

$$\mathcal{L}^{-1}F(s) = f_1(t) * f_2(t) * f_3(t) \tag{8-33}$$

In general, then,

$$\mathcal{L}^{-1}F_1(s)F_2(s)F_3(s) \cdots = f_1(t) * f_2(t) * f_3(t) * \cdots \tag{8-34}$$

*Probs. 8-16*
*8-17*
*8-18*
*8-19*

## PROBLEMS

8-1 Prove that convolution, as defined by Eq. (8-7) or (8-8), is a linear operation. *Hint*: Apply Eq. (1-4b).

8-2 The impulse response of a linear, time invariant network is the $10e^{-2t}u(t)$. By convolution, find the zero-state response of this network to an excitation $10e^{-2t}u(t)$. Sketch fairly accurately both responses.

8-3 The zero-state response of a network to a unit step input $e(t) = u(t)$ is given by

$3e^{-2t}u(t)$. What is the impulse response? Convolve the impulse response with an input $10e^{-3t}u(t)$ to obtain the zero-state response to this input. (See also Prob. 7-33).

8-4 A voltage source, given by

$$v(t) = \begin{cases} \dfrac{1}{\varepsilon}t & 0 < t < \varepsilon \\ 1 & t \geq \varepsilon \end{cases}$$

is applied to a series $RL$ circuit $(R = 1\,\Omega, L = 1\,H)$. The initial current in $L$ is zero.

a) Compute the impulse response of this network.

b) By convolution, find the zero-state current.

c) Sketch the results of (a) and (b).

d) Compare (b) and (c) with the results corresponding to $v(t) = u(t)$, the unit step.

*Note*: In part (b), set up carefully the limits on the integral, since $v(t)$ has two different expressions.

8-5 Repeat problem 8-4(a), (b), and (c) if the input is as shown below. *Hint*: use superposition.

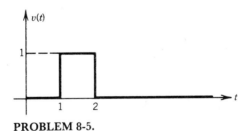

**PROBLEM 8-5.**

8-6 In the process of graphical convolution, sketch a typical $e(x)$ and show on it "the present value" $(x = 0)$ and "future values" $(x > 0)$. As you fold and shift $e(t_1 - x)$ to the right, indicate carefully which values ("past" or "future") get multiplied by the stationary function $h(x)$, to produce the response $r(t_1)$. Does this agree with the concept of a causal system, as discussed in Chapter 1?

8-7 Complete the graphical convolution of example 5 in section 8-2, by taking time increments of 0.5 sec. from 0 through 4.

8-8 Repeat problem 8-7 with $e(t)$ and $h(t)$ interchanged, that is, perform

$$\int_{0-}^{t} h(t-x)e(x)\,dx$$

graphically with the same increments from 0 through 4.

8-9 Solve problem 8-2 graphically, with time increments of 1 sec., from 0 through 5.

8-10 Solve problem 8-5 graphically, with time increments of 0.5 sec., from 0 through 3.

8-11 If $h(t) = \delta(t)$, find the zero-state response to the input $e(t)$.

8-12 Find graphically the zero-state response of a network for which $h(t)$ and $e(t)$ are given below.

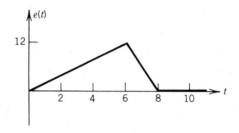

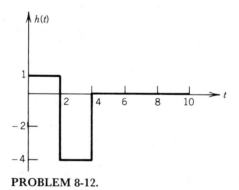

**PROBLEM 8-12.**

8-13 Different numerical approximations for $e(t)$ and $h(t)$ will yield different, but equally correct, results for the numerical convolution. Let $e(t)$ and $h(t)$ be

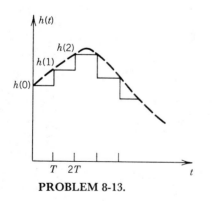

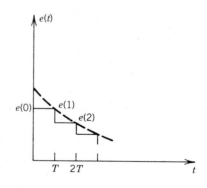

**PROBLEM 8-13.**

approximated by "staircase" shapes, as in Fig. 8-3 in the text, and as shown here. Prove that the numerical convolution here becomes

$$r(t)]_{t=kT} = (e * h)_{t=kT} = T \sum_{p=1}^{k} h(p)e(k - p) = T \sum_{p=1}^{k} h(k - p)e(p)$$

This form is called *the discrete convolution*.

8-14 Write a short program to implement the numerical convolution, either as in Eq. (8-12), or as in problem 8-13. Implement it on Example 2 using $T = 0.01$ sec and $0 \le t \le 2$ sec. Compare, with plots, with the analytical convolution.

8-15 Use numerical convolution to solve problem 8-12, with $T = 0.25$ sec.

8-16 By convolution, find

a) $\mathcal{L}^{-1} \dfrac{1}{s^2}$

b) $\mathcal{L}^{-1} \dfrac{1}{(s + a)^2}$

c) $\mathcal{L}^{-1} \dfrac{1}{s(s + a)}$

d) $\mathcal{L}^{-1} \dfrac{1}{(s^2 + 1)^2}$

e) $\mathcal{L}^{-1} \dfrac{s + 1}{(s + 2)(s + 3)}$

f) $\mathcal{L}^{-1} \dfrac{1}{(s + 1)(s + 2)(s + 3)}$

8-17 Prove that convolution, in addition to being linear (problem 8-1) and commutative, Eq. (8-8), is also distributive, i.e.,

$$f_1(t) * [f_2(t) + f_3(t)] = f_1(t) * f_2(t) + f_1(t) * f_3(t)$$

8-18 Use convolution to prove the well-known result

$$\mathcal{L} \int_{0_-}^{t} f(x)\, dx = \frac{1}{s} F(s)$$

*Hint*: Convolve $f(t)$ with $u(t)$.

8-19 The following equation (named after Volterra) occurs frequently in linear systems:

$$y(t) = g(t) + \int_{0_-}^{t} k(t - \xi)y(\xi)\, d\xi$$

This is an integral equation for the unknown $y(t)$. The functions $g(t)$ and $k(t)$ are given. Solve for $y(t)$, using Laplace transforms.

# 9

# FREQUENCY RESPONSE AND GRAPHICAL METHODS

In a previous chapter (Chapter 7, section 4) we discussed briefly some of the analytic properties of the network function for $s = j\omega$. In Chapter 8, we discussed the zero-state response in general. Here, we shall study some graphical methods associated with the evaluation of the *zero-state sinusoidal response* in the frequency domain. This response is commonly referred to as the *frequency response*. This response is

$$R(j\omega) = H(j\omega)E(j\omega) \tag{9-1}$$

where $E(j\omega)$ is the given transform excitation and $H(j\omega)$ is the appropriate network function, evaluated for $s = j\omega$. This equation implies that the magnitude of the response is

$$|R(j\omega)| = |H(j\omega)||E(j\omega)| \tag{9-2}$$

and that the phase of the response is

$$\theta_R(j\omega) = \theta_H(j\omega) + \theta_E(j\omega) \tag{9-3}$$

where the subscripts $R$, $E$ and $H$ identify the response, excitation and network function, respectively.

Since $E(j\omega)$ is given, we know $|E(j\omega)|$ and $\theta_E(j\omega)$. In order to find $|R(j\omega)|$ and $\theta_R(j\omega)$, we must investigate the behavior of $|H(j\omega)|$ and $\theta_H(j\omega)$. In the subsequent discussion, we shall omit the subscript $H$ on $\theta_H$, as this will be the only phase function under discussion. Our main problem will be, then: Given a network function $H(s)$, investigate various methods of plotting $H(j\omega)$ given by

$$H(j\omega) = |H(j\omega)|e^{j\theta(\omega)}. \tag{9-4}$$

In this process, we shall learn the important role played by poles and zeros of $H(s)$.

## Example 1

In this preliminary example, we review the phasor transform method which implements Eq. (9-1) at a *single* (fixed) $\omega$. For the circuit shown in Fig. 9-1a, we draw its phasor transform diagram as in Fig. 9-1b, where $\omega = 4$ rad/sec is fixed. From this circuit we obtain

$$\mathbf{V}_2 = \frac{-j3}{2 - j3}\,\mathbf{V}_1$$

knowing $\mathbf{V}_1 = 10\underline{/45°}$, we can calculate $|\mathbf{V}_2|$ and $\theta_{\mathbf{V}_2}$. Here, for $s = j\omega = j4$, we have

$$H(j4) = \frac{-j3}{2 - j3} = 0.832\,\underline{/-33.69°}$$

Therefore   $|\mathbf{V}_2| = (0.832)(10) = 8.32$   and   $\theta_{\mathbf{V}_2} = -33.69° + 45° = 11.31°$,
Finally, then

$$v_2(t) = 8.32\cos(4t + 11.31°)$$

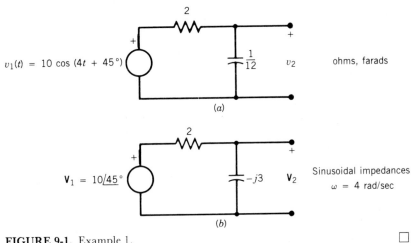

$v_1(t) = 10\cos(4t + 45°)$     $v_2$     ohms, farads

(a)

$\mathbf{V}_1 = 10\underline{/45°}$     $\mathbf{V}_2$     Sinusoidal impedances
$\omega = 4$ rad/sec

(b)

**FIGURE 9-1.** Example 1.   □

In the following discussion, we develop methods for a *generalized phasor* approach, good for *all* $\omega$.

## 9-1   MAGNITUDE AND PHASE: RECTANGULAR AND POLAR PLOTS

Given $H(s)$ as follows:

$$H(s) = K\frac{s^m + a_1 s^{m-1} + \cdots + a_{m-1}s + a_m}{s^n + b_1 s^{n-1} + \cdots + b_{n-1}s + b_n} = K\frac{P(s)}{Q(s)} \qquad (9\text{-}5)$$

we obtain directly

$$H(j\omega) = K \frac{(j\omega)^m + a_1(j\omega)^{m-1} + \cdots + a_{m-1}(j\omega) + a_m}{(j\omega)^n + b_1(j\omega)^{n-1} + \cdots + b_{n-1}(j\omega) + b_n} \qquad (9\text{-}6)$$

and we can write the numerator and the denominator in terms of their real and imaginary parts, say

$$H(j\omega) = \frac{A(\omega) + jB(\omega)}{C(\omega) + jD(\omega)} \qquad (9\text{-}7)$$

The magnitude function $|H(j\omega)|$ is then

$$|H(j\omega)| = \left| \frac{A(\omega) + jB(\omega)}{C(\omega) + jD(\omega)} \right| = \frac{|A(\omega) + jB(\omega)|}{|C(\omega) + jD(\omega)|}$$

$$= \sqrt{\frac{[A(\omega)]^2 + [B(\omega)]^2}{[C(\omega)]^2 + [D(\omega)]^2}} = \sqrt{H(j\omega)H(-j\omega)} \qquad (9\text{-}8)$$

and the phase function is

$$\theta(\omega) = \tan^{-1} \frac{B(\omega)}{A(\omega)} - \tan^{-1} \frac{D(\omega)}{C(\omega)} \qquad (9\text{-}9)$$

These are normally plotted for $0 \leq \omega \leq \infty$.

---

### Example 2

In Example 1 we have the following voltage transfer function

$$H(s) = \frac{12/s}{2 + (12/s)} = \frac{6}{s + 6}$$

Consequently,

$$H(j\omega) = \frac{6}{6 + j\omega} = \frac{6}{\sqrt{36 + \omega^2}} \underline{/-\tan^{-1}(\omega/6)}$$

for *all* $\omega$. In particular, for $\omega = 4$, we have

$$H(j4) = 0.832 \underline{/-33.69°}$$

as before. ☐

---

### Example 3

Plot $H(j\omega)$ and $\theta(\omega)$ for $H(s) = 2[(s-1)/(s+2)]$. Here $H(j\omega) =$

$[(2j\omega - 2)/(j\omega + 2)]$ with $A(\omega) = -2$, $B(\omega) = 2\omega$, $C(\omega) = 2$, and $D(\omega) = \omega$. Therefore,

$$|H(j\omega)| = \sqrt{\frac{4\omega^2 + 4}{\omega^2 + 4}}$$

and

$$\theta(\omega) = \tan^{-1}(-\omega) - \tan^{-1}\frac{\omega}{2}$$

Their *rectangular plots* are shown in Fig. 9-2.

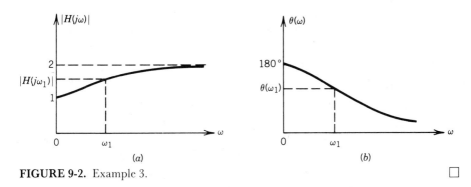

**FIGURE 9-2.** Example 3.

The two separate rectangular plots can be alternately shown as one *polar plot*: For any value of $\omega$, say $\omega_1$, we draw a directed line whose length is $|H(j\omega_1)|$ and whose angle is $\theta(\omega_1)$. One such directed line is shown in Fig. 9-3a for the previous example. We draw these lines for various values of $\omega$, $\omega = 0, \ldots, \infty$. The locus of the tips of these lines is the polar plot, also known as the *locus plot*. The locus plot shown in Fig. 9-3b is for $H(j\omega)$ of the previous example. It is customary to omit on the polar plot the directed lines themselves and show only the locus of their tips; also, the points

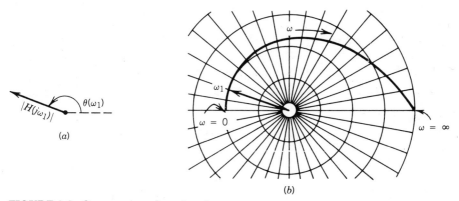

**FIGURE 9-3.** Construction of a polar plot.

corresponding to $\omega = 0$, $\omega = \infty$ are shown, together with the direction of increasing $\omega$.

As a useful aid for these plots (rectangular or polar), it is often convenient to consider the behavior of $H(j\omega)$ at very large frequencies ($\omega \to \infty$) and at very small frequencies ($\omega \to 0$). Using Eq. (9-6), we obtain

$$\lim_{\omega \to \infty} H(j\omega) = \lim_{\omega \to \infty} K(j\omega)^{m-n} \tag{9-10}$$

i.e., for large $\omega$, the highest powers of $\omega$ in the numerator and the denominator prevail. As a result, three cases occur:

a) $m > n$. The magnitude $H(j\omega)$ approaches infinity

$$\lim_{\omega \to \infty} |K(j\omega)^{m-n}| = \lim_{\omega \to \infty} K\omega^{m-n} \to \infty \tag{9-11}$$

which is to be expected because of the pole at infinity; the phase of $K(j\omega)^{m-n}$ is[†]

$$\lim_{\omega \to \infty} \theta(\omega) = (m - n)\frac{\pi}{2} \tag{9-12}$$

b) $m = n$. Here we obtain for the magnitude function at large frequencies

$$\lim_{\omega \to \infty} |H(j\omega)| = K \tag{9-13}$$

and its phase is[†]

$$\lim_{\omega \to \infty} \theta(\omega) = 0 \tag{9-14}$$

c) $m < n$. In this case

$$\lim_{\omega \to \infty} |H(j\omega)| = 0 \tag{9-15}$$

consistent with the zero at infinity, and [†]

$$\lim_{\omega \to \infty} \theta(\omega) = (m - n)\frac{\pi}{2} \tag{9-16}$$

These *high frequency asymptotes* are shown for the rectangular plots in Fig. 9-4a.

In a similar way, the low *frequency asymptotes* are determined by the lowest powers of $\omega$. Here, again, there are three possibilities, shown in Fig. 9-4b.

In summary, then, the high-frequency behavior is dictated by the highest powers in $P(s)$ and $Q(s)$; the low-frequency behavior is dictated by the lowest powers. No memorization of Eqs. (9-10) through (9-16) is ever needed!

---

### Example 4

A transistor amplifier has a voltage transfer function

$$G(s) = -2\,\frac{s - 10}{s^2 + 2s + 10}$$

[†] Assuming $K > 0$.

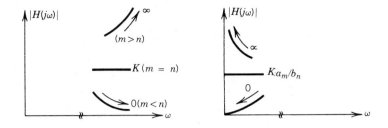

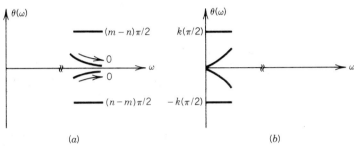

**FIGURE 9-4.** High-frequency and low-frequency asymptotes for $H(j\omega)$.

For $s = j\omega$ we get

$$G(j\omega) = -2 \frac{j\omega - 10}{-\omega^2 + j2\omega + 10}$$

At low frequencies

$$G(j\omega)]_{\omega \to 0} = (-2)\left(\frac{-10}{10}\right) = 2 \,\underline{/360°}$$

where each minus sign contributes $180°$. At high frequencies

$$G(j\omega)]_{\omega \to \infty} = -2 \frac{j\omega}{-\omega^2}\Bigg]_{\omega \to \infty} = 0 \,\underline{/90°}$$

At midpoints along $\omega$, we calculate magnitude and phase using Eqs. (9-8) and (9-9):

$$|G(j\omega)| = 2\sqrt{\frac{\omega^2 + 100}{(10 - \omega^2)^2 + 4\omega^2}}$$

$$\theta_G = 180° + \tan^{-1}\frac{\omega}{-10} - \tan^{-1}\frac{2\omega}{10 - \omega^2}$$

Notice that $180°$ is contributed by the negative $K(=-2)$. Also, the angle of the numerator, $\tan^{-1}(\omega/-10)$, is in the second quadrant, going from $180°$ to $90°$ as $\omega$ increases; the angle of the denominator is in the first quadrant $(0° \to 90°)$ for $0 \le \omega \le \sqrt{10}$, then in the second quadrant $(90° \to 180°)$ for $\omega > \sqrt{10}$.

The complete plots are shown in Fig. 9-5.

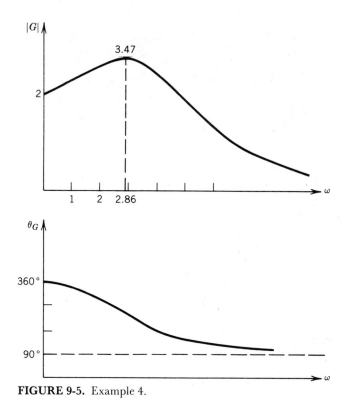

**FIGURE 9-5.**  Example 4.

□   **Prob. 9-1**
**9-2**
**9-6**

## 9-2   A GRAPHICAL METHOD: ZERO LINES AND POLE LINES

Let us now discuss a graphical method for evaluating $H(j\omega)$ at various frequencies $\omega = \omega_1, \omega_2, \ldots$ In the previous discussion, Eqs. (9-7) through (9-9), such an evaluation was done analytically. The graphical evaluation is based on the pole-zero configuration of $H(s)$. Write $H(s)$ in its factored form

$$H(s) = K \frac{(s - z_1)(s - z_2) \cdots (s - z_m)}{(s - p_1)(s - p_2) \cdots (s - p_n)} \qquad (9\text{-}17)$$

exhibiting the finite zeros $z_1, z_2, \ldots, z_m$ and the finite poles $p_1, p_2, \ldots, p_n$. These can be simple or multiple. A typical pole-zero configuration is shown in Fig. 9-6. To

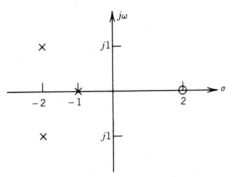

**FIGURE 9-6.** A typical pole-zero configuration of $H(s)$.

evaluate $H(s)$ for a specific $s = s_0$, write Eq. (9-17) as

$$H(s_0) = K \frac{(s_0 - z_1)(s_0 - z_2) \cdots (s_0 - z_m)}{(s_0 - p_1)(s_0 - p_2) \cdots (s_0 - p_n)} \qquad (9\text{-}18)$$

and consider the geometry of the term $(s_0 - z_1)$. Since $s_0$ is in general complex, as is $z_1$, we can represent them as directed lines in the $s$-plane. See Fig. 9-7. As shown, $s_0$ is a directed line from the origin to the point $s_0$, and $z_1$ is a directed line from the origin to the points $z_1$. Hence the complex number $(s_0 - z_1)$ is the directed line *from $z_1$ to $s_0$*. Similarly, $(s_0 - z_2)$ will be the directed line *from $z_2$ to $s_0$*, etc. Each will be called a *zero line* (from each zero to $s_0$). Similarly, the complex number $(s_0 - p_1)$ will be a directed line *from $p_1$ to $s_0$*, etc. Each of these will be called a *pole line* (from each pole to $s_0$). Therefore we can write Eq. (9-18) as

$$H(s_0) = K \frac{\text{product of all zero lines to } s_0}{\text{product of all pole lines to } s_0} \qquad (9\text{-}19)$$

Each zero line and pole line will be expressed in its polar form in magnitude and

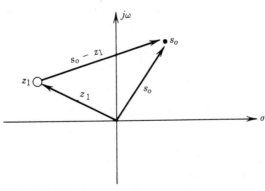

**FIGURE 9-7.** Geometry of $s_0 - z_1$.

angle. For example,

$$s_0 - z_1 = M_{z1}\underline{|\theta_{z1}}$$

$$\vdots \qquad \vdots$$

$$s_0 - z_m = M_{zm}\underline{|\theta_{zm}}$$

$$s_0 - p_1 = M_{p1}\underline{|\theta_{p1}} \tag{9-20}$$

$$\vdots \qquad \vdots$$

$$s_0 - p_n = M_{pn}\underline{|\theta_{pn}}$$

Where $M$ is the magnitude (length) of the line, $\theta$ is the angle (measured positive in the usual sense, that is, counterclockwise from the horizontal axis). The subscripts on $M$ and $\theta$ identify the respective zero- and pole lines.

---

## Example 5

For the network function

$$H(s) = 3\,\frac{s - 2}{s(s^2 + 4s + 5)}$$

find $H(j1)$. The pole-zero configuration for $H(s)$ is shown in Fig. 9-8. Here $s_0 = j1$, and therefore we draw the zero lines (only one in this example) from $z_1 = 2$ to $s_0 = j1$ and the pole lines from $p_1 = 0$ to $s_0 = j1$, from $p_2 = -2 + j1$ to $s_0 = j1$ $= -2 - j1$ to $s_0 = j1$ (see Fig. 9-8). We have here

$$s_0 - z_1 = \sqrt{5}\underline{|153.5°}$$

$$s_0 - p_1 = 1\underline{|90°}$$

$$s_0 - p_2 = 2\underline{|0°}$$

$$s_0 - p_3 = 2\sqrt{2}\underline{|45°}$$

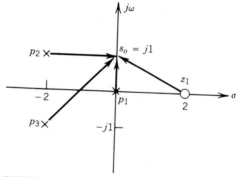

**FIGURE 9-8.** Example 5.

These lengths and angles may be measured directly (with a ruler and a protractor) if the plot is accurate, or else may be easily computed from the geometry of the plot. Hence

$$H(j1) = 3 \frac{\sqrt{5} \lfloor 153.5°}{1 \lfloor 90° \ 2 \lfloor 0° \ 2\sqrt{2} \lfloor 45°} = 1.19 \lfloor 18.5°$$

Such calculations are done repeatedly for successive values of $s_0$ along the $j\omega$ axis, and we obtain $H(j\omega)$ for these various values of $\omega$.   □

---

Finally, it should be noted from Eqs. (9-19) and (9-20), as well as from the previous example, that the magnitude function is given by

$$|H(j\omega_0)| = \frac{M_{z1} M_{z2} \cdots M_{zm}}{M_{p1} M_{p2} \cdots M_{pn}} \tag{9-21}$$

while the phase function is

$$\theta(\omega_0) = (\theta_{z_1} + \theta_{z_2} + \cdots + \theta_{zm}) - (\theta_{p_1} + \theta_{p_2} + \cdots + \theta_{pn}) \tag{9-22}$$

Also, recall that a pole or a zero at $s = \infty$ does not appear explicitly as a factor in Eq. (9-17). Therefore, only *finite* zeros and poles and plotted in this graphical procedure.

---

## Example 6

Use zero lines and pole lines to obtain the rectangular plots for

$$H(s) = \frac{s^2 + 1}{(s + 1)^2(s + 2)}$$

for $0 \le \omega \le \infty$. From these, sketch the polar plot. The pole-zero configuration is shown in Fig. 9-9a, together with zero-lines and pole-lines for evaluating $H(j0)$. For clarity, they are shown separated. Notice that there are two pole lines from the pole at $s = -1$ since this pole is of order 2. Thus we have

$$H(j0) = \frac{(1 \lfloor 90°)(1 \lfloor -90°)}{(1 \lfloor 0°)(1 \lfloor 0°)(2 \lfloor 0°)} = \tfrac{1}{2} \lfloor 0°$$

There is no difficulty in repeating this calculation until we reach $s = j1$. Due to the zero at $s = j1$, we know that the magnitude of $H(j1)$ will be zero. However, the angle variation is not too obvious. Therefore, let us evaluate graphically $H[j(1 - \varepsilon)]$, $\varepsilon > 0$, as shown in Fig. 9-9b. We get

$$H[j(1 - \varepsilon)] = \frac{(\varepsilon \lfloor -90°)[(2 - \varepsilon) \lfloor 90°]}{(\sqrt{2} \lfloor 45°)(\sqrt{2} \lfloor 45°(\sqrt{5} \lfloor 26.5°)}$$

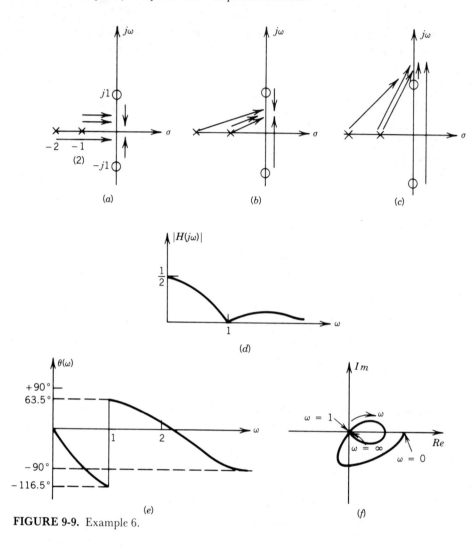

FIGURE 9-9. Example 6.

As $\varepsilon \to 0$, the magnitude approaches zero, as expected. The angle is $\theta[j(1 - \varepsilon)] = (-90° + 90°) - (45° + 45° + 26.5°) = -116.5°$. Now evaluate graphically $H[j(1 + \varepsilon)]$, as shown in Fig. 9-9c. Here we have

$$H[j(1 + \varepsilon)] = \frac{(\varepsilon \lfloor 90°)[(2 + \varepsilon) \lfloor 90°]}{(\sqrt{2}\lfloor 45°)(\sqrt{2}\lfloor 45°)(\sqrt{5}\lfloor 26.5°)}$$

and again the magnitude approaches zero. The angle, however is $\theta[j(1 + \varepsilon)] = (90° + 90°) - (45° + 45° + 26.5°) = 63.5°$. We note in this example that, in the range $0 \leq \omega < 1$, the two zero-lines contribute a net zero to the total angle (one zero-line contributes $90°$, the other $-90°$). In the range $1 < \omega \leq \infty$, however, these two zero-lines contribute $180°$ (each contributes $90°$). See also problems 9-3 and

9-4. Finally, rectangular plots are shown in Fig. 9-9d and e. From these, the polar plot in Fig. 9-9f is sketched.                                                             ☐ ***Prob. 9-3***
                                                                                                                        ***9-4***
                                                                                                                        ***9-5***
                                                                                                                        ***9-8***

## 9-3   BODE PLOTS

In many practical situations, the wide range of frequencies makes it necessary to use a logarithmic, rather than a linear, scale for $\omega$. These semi-logarithmic plots are called *Bode plots.*[†]

First, consider the logarithmic scale for $\omega$. In Fig. 9-10a, a linear scale is shown, labelled arbitrarily as $x$. Below it, and with corresponding intervals, is a logarithmic scale for $\omega$. Thus

$$x = \log \omega \tag{9-23}$$

and

$$\omega = 10^x \tag{9-24}$$

(all logarithms here are to the base of 10). We say that two frequencies $\omega_1$ and $\omega_2$ are *one decade* apart if

$$\omega_2 = 10\omega_1 \tag{9-25}$$

as shown in Fig. 9-10b. Another unit commonly used here is the *octave*. Two frequencies $\omega_1$ and $\omega_2$ are one ocatve apart if

$$\omega_2 = 2\omega_1 \tag{9-26}$$

A typical octave is shown in Fig. 9-10b. Note that the increment on the linear scale of $x$ for a decade is

$$x_2 = \log \omega_2 = \log(10\omega_1) = 1 + \log \omega_1 = 1 + x_1 \tag{9-27}$$

i.e., one unit; for an octave the increment is

$$x_2 = \log \omega_2 = \log(2\omega_1) = 0.3010 + \log \omega_1 \approx 0.3 + x_1 \tag{9-28}$$

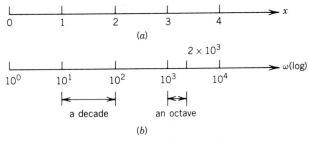

**FIGURE 9-10.** Linear and logarithmic scales.

[†] Other common names are *corner plots* and *break-frequency* plots.

i.e., approximately 0.3 units.

In plotting the magnitude, we shall not use $|H(j\omega)|$, but rather a related practical unit, the *gain*, $\alpha$, measured in *decibels* (dB) and defined by[†]

$$\alpha(\omega) = 20 \log|H(j\omega)| \qquad \text{dB} \tag{9-29}$$

Therefore, our Bode plots will be semilogarithmic: one of the gain (or magnitude in dB) versus a logarithmic $\omega$, the other of the phase (in degrees) versus the logarithmic scale of $\omega$.

In preparation for the Bode plots, consider again the most general form of $H(s)$, Eq. (9-5). In factored form, its numerator is a product of four types of factors:

1. $K$, a constant,

2. $s$, a zero at the origin,

3. $s + a$, a real zero, $\tag{9-30}$

4. $(s + b + jc)(s + b - jc) = s^2 + 2bs + b^2 + c^2$, a pair of complex conjugate zeros.

Any of these may be simple or multiple. A similar factorization holds, in principle, for the denominator. We can write therefore $H(s) = KP(s)/Q(s)$ where $P(s) = s \prod (s + a) \prod (s^2 + 2bs + b^2 + c^2)$ and $Q(s)$ is of the same general form. Here the symbol $\prod$ indicates "the product of." For $s = j\omega$, we write

$$H(j\omega) = |H(j\omega)|e^{j\theta(\omega)} \tag{9-31}$$

as before. We are interested in the gain, and so:

$$\alpha(\omega) = 20 \log|H(j\omega)| = 20\{\log|K| \pm \sum \log|j\omega| \pm \sum \log|j\omega + a|$$
$$\pm \sum \log|-\omega^2 + j2b\omega + b^2 + c^2|\}. \tag{9-32}$$

One advantage of this calculation becomes obvious now: By taking the logarithm, we obtain sums and differences of terms (instead of products). The plus sign ($+$) for the sums in Eq. (9-32) pertains to factors in the numerator of $H$, i.e., to zeros, and the minus sign ($-$) pertains to factors in the denominator, i.e., to poles.

---

[†] A larger unit, the *bel*, was originally introduced to measure the ratio of two powers $P_1$ and $P_2$ in a network, as follows:

$$\text{power gain} = \log \frac{P_2}{P_1} \qquad \text{bels}$$

where, for example, $P_1$ is the input power, and $P_2$ is the output power. In decibels, we have

$$\text{power gain} = 10 \log \frac{P_2}{P_1} \qquad \text{dB}$$

Since the ratio $P_2/P_1$, in the sinusoidal steady-state, is proportional to the *square* of the voltage ratio $V_2/V_1$, this concept leads to the voltage gain given by

$$\text{voltage gain} = 20 \log \frac{V_2}{V_1} \qquad \text{dB}$$

and, by extension, to all network functions (and not only $V_2/V_1$), as given in Eq. (9-29).

## Example 7

Let

$$H(s) = 3 \frac{(s + 2)}{s(s^2 + 4s + 100)}$$

$$H(j\omega) = 3 \frac{(j\omega + 2)}{j\omega(-\omega^2 + 4j\omega + 100)}$$

$$\therefore 20 \log|H(j\omega)| = 20\{\log 3 + \log|j\omega + 2| - \log|j\omega|$$

$$- \log|-\omega^2 + 4j\omega + 100|\} \qquad \square$$

It is now a relatively easy task to study the individual gain contributions and phase contribution of each of these four factors. We shall do so, and catalog them for general use.

1. *K, a constant.* The gain is

$$\alpha(\omega) = 20 \log|K| \qquad \text{dB} \qquad (9\text{-}33)$$

which is a constant (positive if $|K| > 1$, negative if $|K| < 1$). The phase contribution of $K$ is

$$\theta(\omega) = \begin{cases} 0° & K > 0 \\ 180° & K < 0 \end{cases} \qquad (9\text{-}34)$$

(see Fig. 9-11).

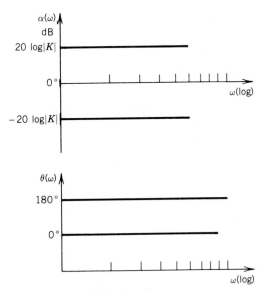

FIGURE 9-11. Bode plots for a constant.

2. $s^p$, *a zero or a pole at the origin (of order p)*. The gain is

$$\alpha(\omega) = \pm 20 \log|(j\omega)|^p = \pm 20p \log \omega \quad \text{dB} \quad (9\text{-}35)$$

In the notation of Eq. (9-23), with $\log \omega = x$, we recognize Eq. (9-35) to be a straight line with a slope of $\pm 20p$ dB/decade or $\pm 20p(0.3010) \approx \pm 6p$ dB/octave.

The phase of $(j\omega)^p$ is

$$\theta(\omega) = p\,90° \quad (9\text{-}36)$$

See Fig. 9-12.

3. $(s + a)^k$, *a real zero or pole of order k*. The gain is

$$\alpha(\omega) = \pm 20 \log|(j\omega + a)^k| = \pm 20k \log|(j\omega + a)| \quad (9\text{-}37)$$

Let us write the factor $(j\omega + a)$ in its standard form as follows:

$$(j\omega + a) = \frac{1}{\tau_a}(1 + j\omega\tau_a) \quad (9\text{-}38)$$

where $a = 1/\tau_a$. This merely serves to scale the frequency axis. Furthermore, $1/\tau_a$ in Eq. (9-38) can be absorbed in the constant multiplier $K$ earlier. We are essentially, then, interested in the gain of $(1 + j\omega\tau_a)$. This is given by

$$\alpha(\omega) = \pm 20 \log|1 + j\omega\tau_a| = \pm 20 \log\sqrt{1 + (\omega\tau_a)^2} \quad (9\text{-}39)$$

It is worthwhile to consider the asymptotic behavior of this gain first: When $\omega\tau_a \ll 1$, that is when $\omega \ll a$, Eq. (9-39) yields

$$\alpha(\omega) \approx 20 \log\sqrt{1} = 0 \quad \text{dB} \quad (9\text{-}40)$$

while for $\omega\tau_a \gg 1$, that is when $\omega \gg a$, we obtain

$$\alpha(\omega) \approx \pm 20 \log\sqrt{(\omega\tau_a)^2} = \pm 20 \log(\omega\tau_a) \quad \text{dB} \quad (9\text{-}41)$$

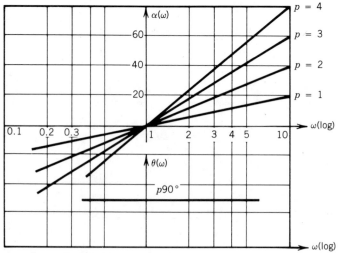

**FIGURE 9-12.** Bode plots for $(j\omega)^p$.

which is a straight line whose slope is $\pm 20$ dB/decade (Verify it!).

In summary, the low frequency ($\omega \ll a$) asymptote is a flat line of 0 dB; the high frequency asymptote ($\omega \gg a$) is a straight line whose slope is $\pm 20$ dB/decade. These two asymptotes meet when

$$0 = \pm 20 \log(\omega \tau_a) \tag{9-42}$$

that is, when

$$\omega \tau_a = 1 \quad \text{or} \quad \omega = a \tag{9-43}$$

This frequency is called the *break frequency* (or *cutoff* or *corner frequency*). To plot the exact $\alpha(\omega)$, all we need now are a few points in addition to these asymptotes. Convenient points are listed in Table 9-1 below.

**TABLE 9.1 Gain for $(1 + j\omega\tau_a)$**

| $\omega$ | Exact gain | Asymptote |
|---|---|---|
| $1/\tau_a$ | $\pm 20 \log\sqrt{2} \approx \pm 3$ dB | 0 dB |
| $1/2\tau_a$ | $\pm 20 \log\sqrt{5/4} \approx \pm 1$ dB | 0 dB |
| $2/\tau_a$ | $\pm 20 \log\sqrt{5} \approx \pm 7$ dB | $\pm 6$ dB |

Thus, at the break frequency the exact $\alpha(\omega)$ curve passes at 3 dB away from the asymptote, while at an octave above and below the break frequency it passes at 1 dB away from the asymptote. These are easily remembered, and the plot of $\alpha(\omega)$ can be sketched. See Fig. 9-13, where the asymptotes are shown in dashed lines and the actual curves in solid lines. The phase of $(1 + j\omega\tau_a)$ is

$$\theta(\omega) = \tan^{-1} \omega\tau_a \tag{9-44}$$

having $0°$ for the low frequency asymptote, going through $45°$ at the break frequency ($\tan^{-1} 1 = 45°$) and approaching $90°$ for $\omega \to \infty$.

4. $s^2 + 2bs + b^2 + c^2$, *a pair of complex conjugate zeros or poles.* Here, too, it will be useful to reduce this factor to its standard form. Define first

$$b^2 + c^2 = \omega_0^2 = \frac{1}{\tau_0^2} \tag{9-45}$$

where $\omega_0$ is called the *undamped frequency.* Also let

$$\frac{b}{\omega_0} = \frac{b}{\sqrt{b^2 + c^2}} = \zeta \tag{9-46}$$

be defined as the *damping factor.* In terms of the geometry of the two complex conjugate zeros (see Fig. 9-14), we see that $\omega_0$ is the distance from the origin to that zero and

$$\zeta = \cos\phi \tag{9-47}$$

and is therefore restricted to $0 \le \zeta \le 1$. With Eqs. (9-45) and (9-46), we obtain

$$s^2 + 2bs + b^2 + c^2 = \frac{1}{\tau_0^2}\left[\left(\frac{s}{\omega_0}\right)^2 + 2\zeta\frac{s}{\omega_0} + 1\right] \tag{9-48}$$

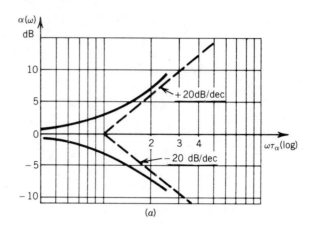

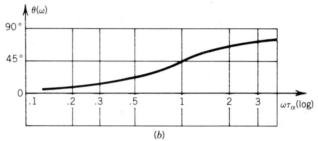

FIGURE 9-13. Bode plots for $(s + a)$: $(a)$ Gain; $(b)$ Phase.

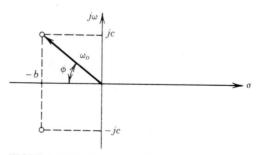

FIGURE 9-14. Geometry of complex conjugate zeros.

which, for $s = j\omega$, becomes

$$s^2 + 2bs + b^2 + c^2\Big]_{s=j\omega} = \frac{1}{\tau_0^2}\left[\frac{-\omega^2}{\omega_0^2} + j2\zeta\frac{\omega}{\omega_0} + 1\right]\tag{9-49}$$

Again, the factor $1/\tau_0^2$ can be absorbed in the constant multiplier $K$. Also, consider a new scaled frequency variable

$$\Omega = \frac{\omega}{\omega_0}\tag{9-50}$$

Then, the standard form of the factor under consideration is $(1 - \Omega^2 + j2\zeta\Omega)$. The gain is

$$\alpha(\omega) = \pm 20 \log|1 - \Omega^2 + j2\zeta\Omega| = \pm 20 \log\sqrt{(1 - \Omega^2)^2 + (2\zeta\Omega)^2} \quad (9\text{-}51)$$

Again, consider the asymptotic behavior first: For $\Omega \ll 1$, i.e., for $\omega \ll \omega_0$, we can neglect $\Omega^2$, $\Omega^4$, and $\zeta^2\Omega^2$ in Eq. (9-51). Therefore

$$\alpha(\omega) \approx \pm 20 \log 1 = 0 \qquad \text{dB} \quad (9\text{-}52)$$

while for $\Omega \gg 1$, i.e., for $\omega \gg \omega_0$, we can neglect 1, $\Omega^2$ and $\zeta^2\Omega^2$ in Eq. (9-51). The result is

$$\alpha(\omega) \approx \pm 20 \log\sqrt{(-\Omega^2)^2} = \pm 40 \log \Omega \qquad \text{dB} \quad (9\text{-}53)$$

which is a straight line whose slope is $\pm 40$ dB/decade. Briefly then, the low frequency asymptote ($\omega \ll \omega_0$) is a flat line of 0 dB; the high frequency asymptote ($\omega \gg \omega_0$) is a straight line with a slope of $\pm 40$ dB/decade. The two asymptotes meet when

$$0 = \pm 40 \log \Omega \quad (9\text{-}54)$$

that is, at

$$\Omega = 1 \quad \text{or} \quad \omega = \omega_0 \quad (9\text{-}55)$$

and so the undamped frequency $\omega_0 = \sqrt{b^2 + c^2}$ is the break frequency here (see Fig. 9-15).

When we come to the exact plot of $\alpha(\omega)$, Eq. (9-51), we see that it depends on the parameter $\zeta$. For each case of this damping factor ($0 \leq \zeta \leq 1$) a different

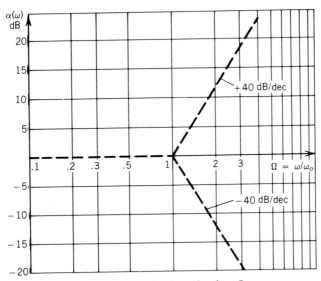

**FIGURE 9-15.** Asymptotes for the gain plot of $(s^2 + 2bs + b^2 + c^2)$.

curve exists. We can simply use Eq. (9-51), with the particular value of $\zeta$ at hand, to evaluate $\alpha(\omega)$ at several points. Or, alternately, we can use standard curves that are available, as shown in Fig. 9-16$a$. Note that these curves are for $1/[(s/\omega_0)^2 + 2\zeta(s/\omega_0) + 1]$, that is, a pair of complex conjugate poles. For a pair of complex conjugate *zeros*, the values of $\alpha$ are the negative of those shown.

The phase of $(1 - \Omega^2 + j2\zeta\Omega)$ is

$$\theta = \tan^{-1} \frac{2\zeta\Omega}{1 - \Omega^2} \tag{9-56}$$

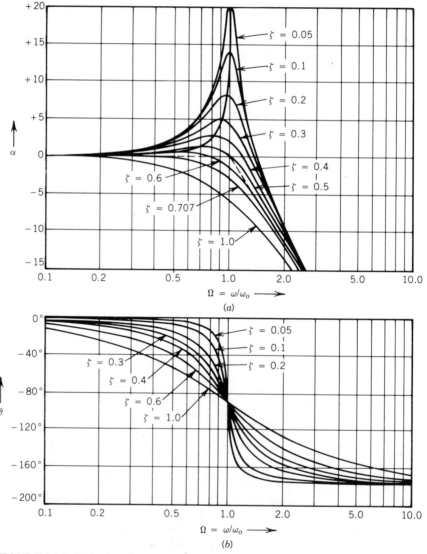

**FIGURE 9-16.** Bode plots for $1/[(s/\omega_o)^2 + 2\xi(s/\omega_o) + 1]$; (*a*) Gain; (*b*) Phase.

and again, depends on the parameter $\zeta$. See Fig. 9-16$b$. Note, however, that for all $\zeta$ the phase is $\pm 90°$ at the break frequency: $-90°$ for poles, $+90°$ for zeros. The low frequency asymptote $(\omega \to 0)$ is $0°$, while for high frequencies $\theta \to \pm 180°$.

In conclusion, let us outline the steps to be taken in preparing the Bode plots.

(a) Write all the first-order factors of $H$ in their standard form, Eq. (9-38). Identify the break frequencies.

(b) Write all the second-order factors in their standard form, Eq. (9-48). Identify the break frequencies and the various $\zeta$.

(c) For the gain plot, sketch the low- and high-frequency asymptotes of each first-order factor and each second-order factor. *Add* these asymptotes (see Eq. (9-32)) to obtain the *total* asymptote of the plot of $\alpha(\omega)$. In many cases, this asymptote will be sufficient. If not, add a few points, with the help of Table 9.1 (first-order factors) and with the help of Fig. 9-16 (second order factors).

(d) Repeat step (c) for the plot of $\theta(\omega)$, with the help of Fig. 9-13$b$ and Fig. 9-16$b$.

## Example 8

Let us obtain the Bode plots for

$$H(s) = 3 \frac{s + 2}{s(s^2 + 4s + 100)}$$

We rewrite $H(s)$, with all factors in standard form

$$H(s) = 3 \frac{2\left(\dfrac{s}{2} + 1\right)}{s \cdot 100\left[\left(\dfrac{s}{10}\right)^2 + \dfrac{4}{10} \cdot \dfrac{s}{10} + 1\right]}$$

$$= 0.06 \frac{\dfrac{s}{2} + 1}{s\left[\left(\dfrac{s}{10}\right)^2 + 2(0.2)\dfrac{s}{10} + 1\right]}$$

Therefore we have

a) a constant multiplier $K = 0.06$.

b) a first-order pole factor, $s$.

c) a first-order zero factor $(s/2) + 1$.

d) a second-order pole factor $(s/10)^2 + 2(0.2)(s/10) + 1$.

1. The gain of $K = 0.06$ is $20 \log 0.06 = -24.4$ dB and will be added to the total plot of $\alpha(\omega)$ at the end.

2. The break frequency of the pole factor $s$ is at $\omega = 1$. The high frequency asymptote has a slope of $-20$ dB/decade (or $-6$ dB/octave).

3. The break frequency of the zero factor $(s + 2)$ is at $\omega = 2$. The high frequency asymptote has a slope of $+20$ dB/decade ($+6$ dB/octave).

4. The break frequency of the pole factor $s^2 + 4s + 100$ is at $\omega = 10$. The value of $\zeta$ is $\zeta = 0.2$ and the high frequency asymptote has a slope of $-40$ dB/decade ($-12$ dB/octave).

These individual asymptotes are shown in Fig. 9-17a. The total asymptote is obtained as the sum of the three individual ones and is shown in dotted line in Fig. 9-17b. Up to $\omega = 2$, this sum is $-20 + 0 + 0 = -20$ dB/decade; for $2 < \omega < 10$, this sum is $-20 + 20 + 0 = 0$ dB/decade; for $\omega > 10$, this sum is $-20 + 20 - 40 = -40$ dB/decade. One exact point is shown at $\omega = 2$, where a correction of $+3$ dB is due to $(s + 2)$, while the correction due to $(s^2 + 4s + 10)$ here is negligible according to Fig. 9-16a. Another exact point is located at $\omega = 10$, where a correction of $+8$ dB is due to $(s^2 + 4s + 10)$ with $\zeta = 0.2$ from Fig. 9-16a. The correction due

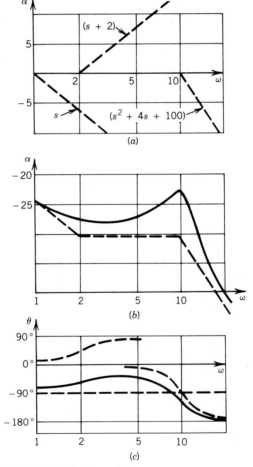

**FIGURE 9-17.** Example 7.

to $(s + 2)$ at $\omega = 10$ is also negligible. The total curve of $\alpha$ is shown in solid line in Fig. 9-17$b$, after we have added $-24.4$ dB to it, due to 20 log $K$.

The phase plot follows along similar lines.

a) The positive constant multiplier contributes nothing.

b) The first-order pole factor, $s$, contributes $-90°$.

c) The first-order zero factor $(s/2) + 1$ contributes an angle according to Fig. 9-13$b$.

d) The second-order pole factor contributes an angle according to Fig. 9-16$b$.

The individual plots (in dotted lines) and their sum, the total angle (in solid line), are shown in Fig. 9-17$c$. □ **Probs. 9-9 through 9-15**

---

In summary, we have learned several methods of computing and plotting frequency response curves. Many of the numerical calculations are adaptable to computer programming.

## PROBLEMS

9-1 For each of the following network functions, sketch the polar plot (see problem 7-20).

a) $\dfrac{s - 1}{s + 1}$

b) $\dfrac{1}{s^2 + s + 1}$

c) $\dfrac{s}{s + 1}$

d) $\dfrac{s}{s^2 + s + 1}$

9-2 Sketch the rectangular plots for $H(s) = \dfrac{s + 1}{s^2 + 1}$, and, from these, sketch the polar plot.

9-3 In the graphical method using zero lines and pole lines investigate the contribution of

a) a zero on the $+\sigma$ axis,

b) a zero on the $-\sigma$ axis,

c) a pair of imaginary (conjugate) zeros on the $j\omega$ axis,

d) a pair of complex conjugate zeros,

to the total magnitude $|H(j\omega)|$ and total angle $\theta(\omega)$, for $\omega \to 0$. Repeat, for a pole at these locations. Your results should agree, of course, with the analytical low frequency asymptotes shown in Fig. 9-4 in the text.

9-4 Repeat problem 9-3, for $\omega \to \infty$ and compare with the analytical high-frequency asymptotes in Fig. 9-4 of the text.

9-5 Sketch the rectangular and polar plots of $H(j\omega)$ for

$$H(s) = \frac{s(s^2 + 4)}{(s^2 + 1)(s^2 + 9)}$$

using zero lines and pole lines.

9-6 Sketch the polar plot of $Z_{dp}(j\omega)$ for a series $RL$ circuit. Next, sketch it for $Y_{dp}(j\omega)$ of the same circuit.

9-7 Use the idea of zero lines and pole lines to find graphically the residues in the partial fraction expansion of

$$F(s) = K \frac{(s - z_1)(s - z_2)}{(s - p_1)(s - p_2)(s - p_3)}$$

where $p_1 \neq p_2 \neq p_3$. *Hint*: Use Eq. (3-24) and interpret each term in it as zero lines or pole lines. Illustrate this procedure with

$$F(s) = \frac{s + 1}{(s^2 + 1)(s^2 + 2s + 2)}$$

9-8 Sketch $|H(j\omega)|$ for each one of the following network functions; accompany each sketch by the pole-zero configuration of $H(s)$.

a) $H(s) = \dfrac{1}{s + 1}$

b) $H(s) = \dfrac{1}{(s + 1)^2}$

c) $H(s) = \dfrac{1}{s^2 + 6s + 25}$

d) $H(s) = \dfrac{1}{s^2 + 2s + 17}$

e) $H(s) = \dfrac{1}{(s^2 + 2s + 17)^2}$

f) $H(s) = \dfrac{1}{(s^2 + s + 1)(s + 1)}$

g) $H(s) = \dfrac{1}{(s^2 + 0.765s + 1)(s^2 + 1.848s + 1)}$

From your results, note the effect of a multiple pole on the negative real axis as compared to a simple pole, cases (a) and (b), the effect of proximity to the $j\omega$ axis, cases (c) and (d), and the effect of equally spaced, constant locus poles, cases (f), (g) — the so-called Butterworth functions.

9-9 Verify the following procedure for obtaining graphically the break frequencies from the pole-zero diagram of $H(s)$: use a compass centered at the origin and draw an arc of a circle from each pole or zero until such arc intersects the $j\omega$ axis. Read the value of $\omega$ at this intersection as the break frequency corresponding to this pole or zero; these break frequencies are lined up, in increasing value, along the $j\omega$ axis.

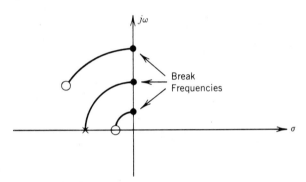

**PROBLEM 9-9.**

9-10 In the network shown, $R < 2\sqrt{L/C}$.

1. Find its oc voltage transfer function $G(s) = V_{out}(s)/V_{in}(s)$, and sketch its pole-zero configuration.

2. Let its two poles be at $s = -b \pm jc$. Find the expression for the two pole lines to any $\omega$ in terms of $b$, $c$, and $\omega$. Hence, write the expression for $|G(j\omega)|$.

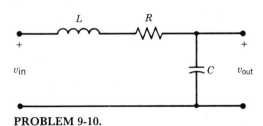

**PROBLEM 9-10.**

3. Using (2), show that $|G(j\omega)|$ attains its maximum at

$$\omega_{\text{max}} = \sqrt{c^2 - b^2}$$

and show this relation on the pole-zero configuration.

4. Also, since $\omega_{\text{max}}$ must be real, show that an equivalent condition to (3) is

$$\zeta \leq \tfrac{1}{2}\sqrt{2}$$

where $\zeta$ is the damping factor, Eq. (9-46).

5. Define *the Q factor* of the circuit as

$$Q = \frac{1}{2\zeta}$$

and sketch $|(Gj\omega)|$ for $b_1 = 1$, $c_1 = 4$, then for $b_2 = 0.5$, $c_2 = c_1$ and their respective $Q$ factors. Compare these magnitude plots with problem 9-8(c) and (d).

9-11 Draw the Bode plots (gain and phase) of:

a) $H(s) = 1.4 \dfrac{s^2(s + 2)}{(s^2 + 3s + 16)}$

b) $H(s) = 0.7 \dfrac{(s + 1)^2}{s(s^2 + 7s + 25)}$

c) $H(s) = \dfrac{0.04}{s^2 + s + 10}$

9-12 Investigate in detail the Bode plot (gain and phase) of a first order zero factor $(s - a)$ where $a > 0$. Follow the outline for $(s + a)$ in the text, Eq. (9-37) and following equations—to (9-44), and compare your results.

9-13 Use the results of problem 9-11(a) to obtain with minimum effort the Bode plots for

$$H(s) = 1.4 \frac{s^2(s - 2)}{s^2 + 3s + 16}$$

9-14 Draw the separate gain asymptotes, then the total asymptote for

$$H(s) = 2.7 \frac{(s - 1)(s + 2)^3}{(s^2 + 3s + 16)^2}$$

9-15 The Bode gain plot of a certain $H(s)$ is shown in the figure below. It is required that all the poles and zeros of $H(s)$ be simple. Find $H(s)$ within an arbitrary constant multiplier $K$. This type of problem is basic in network design; the procedure that you should apply is "in reverse" to that for plotting the gain $\alpha(\omega)$: Fit a total asymptote, recalling that the only slopes allowed are 0, $\pm 20$ and $\pm 40$ dB/decade. The "peak" in the plot should be accounted for by a second-order factor with a proper $\zeta$.

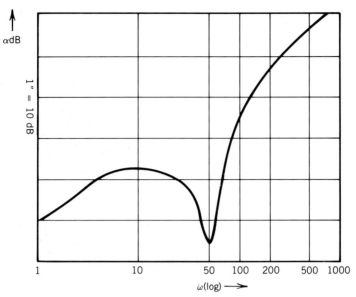

**PROBLEM 9-15.**

# 10

---

# STATE VARIABLES I: FORMULATION

Our previous studies of network characterization included loop analysis, node analysis (including the indefinite admittance matrix), and hybrid (mixed) variables. In this chapter, we shall study another method of analysis, the state variable approach. If it were only for a better understanding of network behavior, we should welcome yet another method of network analysis. However, the state variable analysis offers several additional important advantages:

1. It provides a better insight into the physical aspects of the problem.
2. It leads to a set of *first order* differential equations; the theory, properties, and solutions of such equations have been thoroughly studied.
3. These equations are suitable for solution by analog or digital computers.
4. The state variable approach is readily applicable to time-varying and nonlinear networks.

## 10-1 INTRODUCTORY EXAMPLE

Consider the network shown in Fig. 10-1. We could write two node equations, or one node equation and one loop equation—as studied earlier. In either case, we would

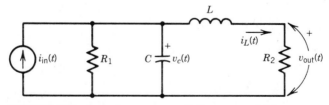

**FIGURE 10-1.** Introductory example for state variables.

have two simultaneous integro-differential equations which, after some manipulation, can be reduced to a second-order differential equation.

Instead, choose the voltage across the capacitance $v_c(t)$ and the current through the inductance $i_L(t)$, as the *state variables*. This choice seems quite reasonable: The initial conditions that are normally given are $v_c(0_-)$ and $i_L(0_-)$, and they constitute the *initial state* of the network. The current source $i_{in}(t)$ is the given input and, together with the initial state, it determines the solution of the network.

Let us write the network equations using $v_c(t)$ and $i_L(t)$ as variables. The current equation reads:[†]

$$C\frac{dv_c}{dt} = -\frac{1}{R_1}v_c + i_{in} - i_L \tag{10-1}$$

and the voltage equation is

$$L\frac{di_L}{dt} = v_c - R_2 i_L \tag{10-2}$$

Note, in passing, that by choosing $v_c(t)$ and $i_L(t)$ as variables, we have avoided an expression like $(1/L)\int v\,dt$ for the current through the inductance, and $(1/C)\int i\,dt$ for the voltage across the capacitance; and the two differential equations are of the first order.

For the desired output, we have the expression

$$v_{out} = R_2 i_L \tag{10-3}$$

After a slight rearrangement, Eqs. (10-1), (10-2), and (10-3) can be written in matrix form as:

$$\begin{bmatrix} \dfrac{dv_c}{dt} \\ \dfrac{di_L}{dt} \end{bmatrix} = \begin{bmatrix} -\dfrac{1}{R_1 C} & -\dfrac{1}{C} \\ \dfrac{1}{L} & -\dfrac{R_2}{L} \end{bmatrix} \begin{bmatrix} v_c \\ i_L \end{bmatrix} + \begin{bmatrix} \dfrac{1}{C} \\ 0 \end{bmatrix} i_{in} \tag{10-4}$$

and

$$v_{out} = \begin{bmatrix} 0 & R_2 \end{bmatrix} \begin{bmatrix} v_c \\ i_L \end{bmatrix} + [0]i_{in} \tag{10-5}$$

These are in the standard, or normal, form of the state equations:

$$\mathbf{\dot{x}} = \mathbf{Ax} + \mathbf{Be} \tag{10-6}$$

$$\mathbf{r} = \mathbf{Cx} + \mathbf{De} \tag{10-7}$$

where $\mathbf{x}(t)$ represents the state vector (i.e., the column matrix of the state variables), $\mathbf{e}(t)$ the input vector (here a single scalar $i_{in}$), and $\mathbf{r}(t)$ the output vector (here $v_{out}$).

$$\mathbf{x}(t) = \begin{bmatrix} v_c \\ i_L \end{bmatrix} \qquad \mathbf{e}(t) = i_{in} \qquad \mathbf{r}(t) = v_{out} \tag{10-8}$$

---

[†] Recall our convention: lowercase letters are functions of time $v_c = v_c(t)$, $i_L = i_L(t)$, etc.

The dot over **x** indicates differentiation with respect to time,

$$\dot{\mathbf{x}} = \frac{d}{dt}\mathbf{x}(t) \tag{10-9}$$

Of the two equations (10-6) and (10-7), the first one is usually called *the state equation* and the second one *the output equation*. Together, they are sometimes referred to as *the state equations*.[†] The solution of the state equations will yield, among other information, the zero-input response ($\mathbf{e}(t) = \mathbf{0}$), the natural frequencies of the network, the zero-state response ($\mathbf{x}(0_-) = \mathbf{0}$), convolution, and the network function **H**. In addition, several important properties and concepts related to these solutions will be brought into focus.

Conceptually, we may represent and think of the state equations as shown in Fig. 10-2 (see also Fig. 1-1 in Chapter 1). There are $k$ inputs $e_1, e_2, \ldots, e_k$ constituting the input vector **e**, the state vector **x** has $n$ components, and the output vector **r** has $p$ components. The state **x** is affected by the input **e** and the initial state through the *differential* state equation; the output equation relates *algebraically* the output to the input and state.

## 10-2 INDEPENDENT INITIAL CONDITIONS AND THE ORDER OF COMPLEXITY OF THE NETWORK

We recall (see Chapter 1) that a purely resistive network—consisting of resistors, sources (dependent and independent), ideal transformers and gyrators—is described by purely algebraic equations with no derivatives or integrals. No initial state is associated with such a network since it is *instantaneous*. A *dynamic* network, on the other hand, has one or more energy storage elements (inductors and capacitors), and the resulting differential equations will require for their solution some initial conditions—the initial state of the network. Are the state variables adequate for describing the other variables in a network? How many *independent* initial conditions are there for a given network? How do we recognize independent state variables? These important questions must be carefully answered before we embark on writing and solving state equations.

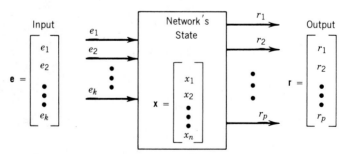

**FIGURE 10-2.** Representation of state variables.

---

[†] The matrices **A**, **B**, **C**, and **D** should not be confused with the chain parameters (scalars), A, B, C and D of a 2-P network!

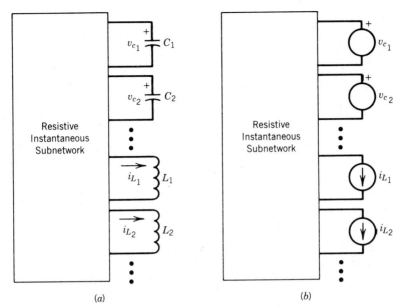

**FIGURE 10-3.** State variables solve the entire network.

Consider the first question. For convenience, let us draw a given dynamic network as shown in Fig. 10-3a, separating the energy storing elements ($L$'s and $C$'s) from the remaining resistive subnetwork ($R$'s, sources, etc.). The choice of state variables will be $i_{L_1}, i_{L_2}, \ldots, v_{c_1}, v_{c_2} \ldots$ . Suppose now we have found the solution for $\mathbf{x}(t)$, i.e., we have found $i_{L_1}(t), i_{L_2}(t), \ldots, v_{c_1}(t), v_{c_2}(t) \ldots$ . A little thought will verify that *as far as the resistive subnetwork is concerned, each capacitor can be replaced by a voltage source equal to the voltage $v_c(t)$ across it, and every inductor—by a current source equal to the current $i_L(t)$ through it.* See Fig. 10-3b. But now we have a resistive subnetwork with known sources, and therefore all the voltages and currents in this subnetwork can be found by pure algebraic calculations!

---

### Example 1

Consider the network discussed in the previous section. Having found $v_c(t)$ and $i_L(t)$, we can redraw the network as shown in Fig. 10-4. All the currents and voltages in

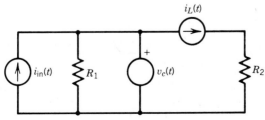

**FIGURE 10-4.** Example 1.

this resistive (instantaneous) network can be expressed in algebraic form (without differentiation or integration), in terms of $v_c(t)$, $i_L(t)$ and the input $i_{in}(t)$.    ☐

In summary, the state variables determine completely all the other variables in the network. Therefore, we define the *state of a network* (or, generally, of a system) as *the minimal amount of information necessary at any time to characterize completely the possible future behavior of the network*. At any time, $t = t_0$, the state $\mathbf{x}(t_0)$ and the input $\mathbf{e}(t)$, given for $t > t_0$, will determine the future values of the state for any $t > t_0$. In addition, the state and the input together determine any network variable.

Next, we turn our attention to the number of independent initial conditions. The natural initial conditions are initial capacitor voltages (or charges) and initial inductor currents (or fluxes). It would seem, therefore, that the number of independent initial conditions equals the number of energy-storing elements ($L$'s and $C$'s). This is true *provided* there are no constraints imposed by the topology of the network. For example, consider a network with a loop containing only capacitors and possibly voltage sources. Such a loop will be called an "all-capacitive" loop (see Fig. 10-5). Around this loop Kirchhoff's voltage law reads[†]

$$v_{c_1} + v_{c_2} + v_{c_3} + v_{c_4} = v_{in} \qquad t \geq 0 \tag{10-10}$$

(with proper reference polarities on these voltages). Therefore, the values at $t = 0_+$ of $v_{c_1}$, $v_{c_2}$, $v_{c_3}$ and $v_{c_4}$ *cannot* be specified independently: they are constrained by Eq. (10-10). Similarly, if there is a cut set consisting of only inductors and possibly current sources, Kirchhoff's current law for such an "all-inductive" cut set will place a constraint on the values of the inductor currents. In Fig. 10-5 we have[†]

$$i_{L_1} + i_{L_2} + i_{L_3} = 0 \qquad t \geq 0 \tag{10-11}$$

and therefore, the values of all three inductor currents at $t = 0_+$ *cannot* be specified independently.

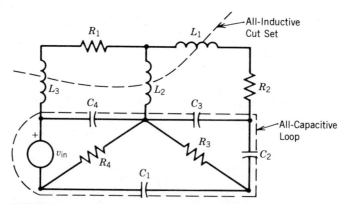

**FIGURE 10-5.** Constraints among state variables.

[†] At $t = 0_-$, initial conditions need not obey such restrictions. When $v_c(0_-) = v_c(0_+)$, or $i_L(0_-) = i_L(0_+)$, our Laplace transform takes care automatically with impulse functions. See Chapter 2, Example 9.

As a result, we define *the order of complexity of the network* as *the number of independent initial conditions that can be specified*. It is given by:

$$n = e_L + e_c - n_c - n_L \tag{10-12}$$

where

$e_L$ = number of inductors,
$e_c$ = number of capacitors,
$n_c$ = number of independent all-capacitive loops,
$n_L$ = number of independent all-inductive cut sets.

In counting $n_c$, the all-capacitive loops must be independent of each other; that is, a constraint like Eq. (10-10) must not be a linear combination of other such constraints. For this, it is sufficient that each all-capacitive loop contain at least one new capacitor. In the same way, the all-inductive cut set constraints must be independent, and each cut set should contain at least one new inductor.

---

### *Example 2*

Find the order of complexity of the network shown in Fig. 10-6. We count $e_L + e_c$ = 7. There are three all-capacitive loops ($v_{in} \to C_1 \to C_2 \to C_4$, $v_{in} \to C_3 \to C_4$, $C_1 \to C_2 \to C_3$) but only two are independent. Hence $n_c = 2$. Also, there are three all-inductive cut sets ($L_1, L_3$), ($L_1, L_2, i_{in}$), ($L_3, L_2, i_{in}$) but only two of them are independent, $n_L = 2$. Therefore,

$$n = 7 - 2 - 2 = 3.$$

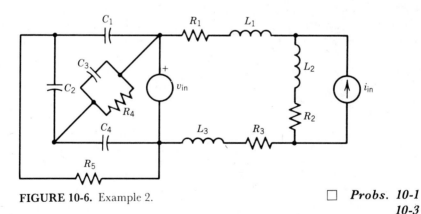

**FIGURE 10-6.** Example 2.                    ☐ *Probs.  10-1*
                                                  *10-3*

---

To obtain $n_c$ by inspection, open circuit all the resistors, inductors, and current sources. The remaining network will show clearly the all-capacitive loops. In a dual manner, short circuit all the resistors, capacitors, and voltage sources, to obtain by inspection the all-inductive cut sets.

Initial conditions serve to evaluate the arbitrary constants in the solution of a set of differential equations. That is, $n$ is also the number of these constants, and therefore also the degree of the characteristic equation of the network. But the degree of the characteristic equation is equal also to the number of natural frequencies (modes) of the network. In summary, the order of complexity of a network is also the number of independent initial conditions and is also the number of its natural frequencies.

___

### Example 3

Let

$$s^4 + 3s^3 + 5s^2 + 5s + 2 = (s + 1)^2(s^2 + s + 2) = 0$$

be the characteristic equation. The network has four natural frequencies, $s_{1,2} = -1$, $s_{3,4} = -\dfrac{1}{2} \pm j\dfrac{\sqrt{7}}{2}$; and its order of complexity is 4. A typical expression for its zero-input response will be of the form $K_1 e^{-t} + K_2 t e^{-t} + e^{-t/2}\left(K_3 \cos\dfrac{\sqrt{7}}{2}t + K_4 \sin\dfrac{\sqrt{7}}{2}t\right)$, with four arbitrary constants requiring four independent initial conditions.   □ **Prob. 10-2**

___

Finally, let us consider the natural frequency $s = 0$. As we recall, a constant current in an inductor has a natural frequency $s = 0$.[†] Such a constant current can flow in an "all-inductive" loop as shown in Fig. 10-7. In a similar fashion, a constant voltage across a capacitor, with a natural frequency $s = 0$, can exist for an "all-capacitive" cut set. Therefore, the number of *nonzero* natural frequencies is given by:

$$n' = n - m_L - m_c \tag{10-13}$$

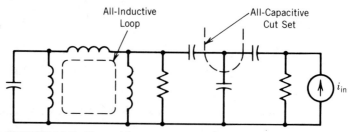

All-Inductive Loop

All-Capacitive Cut Set

**FIGURE 10-7.** Natural frequency at $s = 0$.

[†] $i_L = I_0 = I_0 e^{0t}$, or $I_L(s) = \dfrac{I_0}{s}$.

where

$n$ = total number of natural frequencies, Eq. (10-12),

$m_L$ = number of independent all-inductive loops,

$m_c$ = number of independent all-capacitive cut sets.

## 10-3 FORMULATION OF THE STATE EQUATIONS: LINEAR, CONSTANT, PASSIVE NETWORKS

We are now ready to discuss a systematic formulation of state equations. First, the networks under consideration will contain only linear, time invariant resistances, capacitances, inductances (self and mutual), and sources. It is clear that our independent state variables will be the independent capacitor voltages and inductor currents whose total number is $n$, the order of complexity of the network.

Consider first networks that do not contain any all-capacitive loops or all-inductive cut sets.[†] In this case the independent state variables will be *all* the capacitor voltages and *all* the inductor currents.

As with the formulation of loop and node equations, we make use of the topology of the network. Choose a tree containing all the capacitors and all the voltage sources, but no inductors or current sources. Such a choice is possible since, by definition, there cannot be a closed loop (all-capacitive) in a tree, nor can there be a cut set (all-inductive) in a co-tree. This tree is called a *proper tree*. Some resistors may be needed to complete this tree. Also, since identification of individual elements is important, a parallel or series combination must not be drawn as a single edge. Recall that branch voltages are independent (they determine the link voltages), and link currents are independent (they determine the branch currents). Next, we write current equations for fundamental cut sets and voltage equations for fundamental loops,[‡] retaining capacitor (branch) voltages and inductor (link) currents as variables.

---

### Example 4

Consider the network shown in Fig. 10-8a, where the edges of the graph are conveniently labelled with the numbers corresponding to the elements. The proper tree is shown in full lines, and its co-tree is in dashed lines in Fig. 10-8b. Our state variables are $v_3$, $v_5$ and $i_4$. The inputs are $v_1$ and $i_7$. That is,

$$\mathbf{x}(t) = \begin{bmatrix} v_3 \\ v_5 \\ i_4 \end{bmatrix} \qquad \mathbf{e}(t) = \begin{bmatrix} v_1 \\ i_7 \end{bmatrix} \qquad (10\text{-}14)$$

---

[†] Such an assumption is not unrealistic, since physical capacitors and inductors have inherent heat dissipation which can be represented by a series resistance (for a capacitor) and a shunt conductance (for an inductor)—thus eliminating all-capacitive loops and all-inductive cut sets.

[‡] For brevity, we refer to these as "cut set equations" and "loop equations" in the discussion that follows.

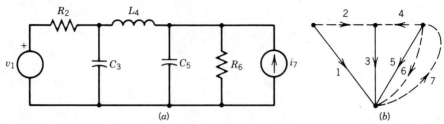

**FIGURE 10-8.** Example 4.

Two fundamental cut set equations are

$$i_3 = i_2 + i_4 \qquad \text{(for } C_3\text{)} \tag{10-15}$$

and

$$i_5 = -i_4 - i_6 + i_7 \qquad \text{(for } C_5\text{)} \tag{10-16}$$

A fundamental loop equation for $L_4$ is $v_4 = v_5 - v_3$. Next, we express $i_3$ in terms of $v_3$, and $i_2$ in terms of $v_1$ and $v_3$. Equation (10-15) becomes

$$C_3 \frac{dv_3}{dt} = \frac{1}{R_2}(v_1 - v_3) + i_4 \tag{10-17}$$

Similarly, $i_5$ is expressed in terms of $v_5$, and $i_6$ in terms of $v_5$; then Eq. (10-16) reads

$$C_5 \frac{dv_5}{dt} = -i_4 - \frac{1}{R_6}v_5 + i_7 \tag{10-18}$$

Finally, $v_4$ is expressed in terms of $i_4$, and then

$$L_4 \frac{di_4}{dt} = v_5 - v_3 \tag{10-19}$$

Equations (10-17), (10-18), and (10-19) are the desired equations. In matrix form they read:

$$
\begin{bmatrix} \dfrac{dv_3}{dt} \\[2mm] \dfrac{dv_5}{dt} \\[2mm] \dfrac{di_4}{dt} \end{bmatrix}
=
\begin{bmatrix} -\dfrac{1}{R_2 C_3} & 0 & \dfrac{1}{C_3} \\[2mm] 0 & -\dfrac{1}{R_6 C_5} & -\dfrac{1}{C_5} \\[2mm] -\dfrac{1}{L_4} & \dfrac{1}{L_4} & 0 \end{bmatrix}
\begin{bmatrix} v_3 \\[2mm] v_5 \\[2mm] i_4 \end{bmatrix}
+
\begin{bmatrix} \dfrac{1}{R_2 C_3} & 0 \\[2mm] 0 & \dfrac{1}{C_5} \\[2mm] 0 & 0 \end{bmatrix}
\begin{bmatrix} v_1 \\[2mm] i_7 \end{bmatrix} \tag{10-20}
$$

that is,

$$\dot{\mathbf{x}} = \mathbf{Ax} + \mathbf{Be}, \tag{10-21}$$

the state equation in normal form.

Using topological notation (see chapter 6), the above procedure may be summarized as follows: write the edge current matrix in a partitioned form as in

Eq. (6-9), namely,

$$i_e = \begin{bmatrix} i_{link} \\ \cdots\cdots \\ i_{branch} \end{bmatrix} \tag{10-22}$$

Similarly, write the edge voltage matrix as

$$v_e = \begin{bmatrix} v_{link} \\ \cdots\cdots \\ v_{branch} \end{bmatrix} \tag{10-23}$$

The fundamental cut set equations for the network are [see also Eq. (6-11)]

$$Q_f i_e = [Q_{f_{11}} \vdots U] \begin{bmatrix} i_{link} \\ \cdots\cdots \\ i_{branch} \end{bmatrix} = 0, \tag{10-24}$$

where $Q_f$ is the fundamental cut set matrix. Also the fundamental loop equations [see Eq. (6-21)] are

$$B_f v_e = [U \vdots B_{f_{12}}] \begin{bmatrix} v_{link} \\ v_{branch} \end{bmatrix} = 0 \tag{10-25}$$

Equations (10-24) and (10-25) are combined to yield

$$\begin{bmatrix} v_{link} \\ \cdots\cdots \\ i_{branch} \end{bmatrix} = \begin{bmatrix} 0 & \vdots & -B_{f_{12}} \\ \cdots\cdots & \cdots & \cdots\cdots \\ B_{f_{12}}^T & \vdots & 0 \end{bmatrix} \begin{bmatrix} i_{link} \\ \cdots\cdots \\ v_{branch} \end{bmatrix}, \tag{10-26}$$

where we have used Eq. (6-29), namely,

$$B_{f_{12}}^T = -Q_{f_{11}}. \tag{10-27}$$

Equation (10-26) expresses link voltages in terms of branch voltages that are our independent capacitor voltages and independent voltage sources; branch currents are expressed in terms of link currents that are our independent inductor currents and independent current sources.

Next, in the proper tree, the branch (capacitive) currents can be written as

$$i_c = \mathscr{C}\dot{v}_c, \tag{10-28}$$

where $\mathscr{C}$ is the capacitance matrix and $v_c$ is a submatrix of $v_{branch}$ containing the capacitor voltages. In a similar manner

$$v_L = \mathscr{L}\dot{i}_L, \tag{10-29}$$

where $\mathscr{L}$ is an inductance matrix and $i_L$ is a submatrix of $i_{link}$ containing the inductor currents.

The state equation (10-21) is obtained by eliminating all undesired voltages and currents (i.e., all except $v_c$, $i_L$ and the sources) between Eqs. (10-26), (10-28), and (10-29).   □

What if the network contains all-capacitive loops and all-inductive cut sets? Here the number of independent state variables still equals the order of complexity of the network. However, a proper tree cannot be drawn, because at least one capacitor must be excluded from the tree and included in the co-tree. Likewise, at least one inductor must be excluded from the co-tree and included in the tree. We build a modified proper tree (also called a *normal tree*) which contains all the voltage sources, the maximum number of capacitors, some resistors, and finally, the minimum number of inductors. All the current sources and a maximum number of inductors are assigned to the co-tree. The procedure is illustrated by an example.

---

### Example 5

Consider the network shown in Fig. 10-9(a). The order of complexity of this network is 2 and, hence, only two independent state variables are needed. The normal tree is shown in Fig. 10-9(b) in solid lines and the co-tree in dashed lines. The state vector and the input vector are

$$\mathbf{x} = \begin{bmatrix} v_7 \\ i_3 \end{bmatrix}, \qquad \mathbf{e} = \begin{bmatrix} v_1 \\ i_6 \end{bmatrix}.$$

We write the fundamental cut set equation for the branch $C_7$:[†]

$$C_7 \dot{v}_7 = i_6 - i_5 \tag{10-30}$$

Then write the fundamental loop equation for the link $L_3$:

$$L_3 \dot{i}_3 = -v_4 + v_1 + v_2 \tag{10-31}$$

The left-hand side of these two equations has the first derivative of $\mathbf{x}$, as required. All that remains to be done is to eliminate the undesired variables $i_5$, $v_4$ and $v_2$ from the right-hand side. To eliminate $i_5$, a link variable, write a fundamental loop

$$i_5 = C_5 \dot{v}_5 = C_5 (\dot{v}_7 - \dot{v}_1) \tag{10-32}$$

where $v_5$, a link voltage, is written in terms of the branch voltages in the all-capacitive

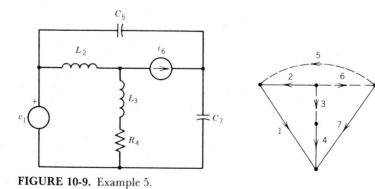

**FIGURE 10-9.** Example 5.

[†] We use the convenient notation $\dot{i}$ for $di/dt$, and $\dot{v}$ for $dv/dt$.

loop. To eliminate $v_4$, a branch variable, write a fundamental cut set

$$v_4 = R_4 i_4 = R_4 i_3 \tag{10-33}$$

where $i_4$, a branch current, is written in terms of link currents. To eliminate $v_2$ write, again, a fundamental cut set for this branch

$$v_2 = L_2 \dot{i}_2 = L_2(-\dot{i}_3 - \dot{i}_6) \tag{10-34}$$

when $i_2$, a branch current, is written in terms of link currents in the all-inductive cut set.

With these eliminations, Eqs. (10-30) and (10-31) become

$$C_7 \dot{v}_7 = i_6 - C_5(\dot{v}_7 - \dot{v}_1) \tag{10-35}$$

$$L_3 \dot{i}_3 = -R_4 i_3 + v_1 + L_2(-\dot{i}_3 - \dot{i}_6) \tag{10-36}$$

These are the state equations. After rearrangement, they read

$$
\begin{bmatrix} \dot{v}_7 \\ \dot{i}_3 \end{bmatrix} =
\begin{bmatrix} 0 & 0 \\ 0 & -\dfrac{R_4}{L_2 + L_3} \end{bmatrix}
\begin{bmatrix} v_7 \\ i_3 \end{bmatrix} +
\begin{bmatrix} 0 & \dfrac{1}{C_5 + C_7} \\ \dfrac{1}{L_2 + L_3} & 0 \end{bmatrix}
\begin{bmatrix} v_1 \\ i_6 \end{bmatrix}
$$

$$
+ \begin{bmatrix} \dfrac{C_5}{C_5 + C_7} & 0 \\ 0 & -\dfrac{L_2}{L_2 + L_3} \end{bmatrix}
\begin{bmatrix} \dot{v}_1 \\ \dot{i}_6 \end{bmatrix}. \tag{10-37}
$$

We observe that Eq. (10-37) is of the form

$$\dot{\mathbf{x}} = \mathbf{A}\mathbf{x} + \mathbf{B}\mathbf{e} + \mathbf{F}\dot{\mathbf{e}} \tag{10-38}$$

involving the additional matrix $\dot{\mathbf{e}}$, the derivative of the input. To reduce Eq. (10-37) to the standard form, substitute a new state vector $\boldsymbol{\xi}$ given by

$$\boldsymbol{\xi} = \mathbf{x} - \mathbf{F}\mathbf{e} \tag{10-39a}$$

where $\mathbf{F}$ is a constant matrix. Then

$$\dot{\boldsymbol{\xi}} = \dot{\mathbf{x}} - \mathbf{F}\dot{\mathbf{e}} \tag{10-39b}$$

Consequently Eq. (10-38) yields the normal form

$$\dot{\boldsymbol{\xi}} = \mathbf{A}\boldsymbol{\xi} + \mathbf{B}_1\mathbf{e} \tag{10-39c}$$

where

$$\mathbf{B}_1 = \mathbf{A}\mathbf{F} + \mathbf{B} \tag{10-39d}$$

☐ **Probs. 10-4**
**10-5**
**10-6**
**10-7**

## 10-4  NETWORKS CONTAINING TRANSFORMERS, GYRATORS AND CONTROLLED SOURCES

Rather than present general rules for writing the state equations when there is a coupling of currents or voltages in the network, let us outline the approach in each case by means of specific examples.

---

### Example 6

Consider the network shown in Fig. 10-10. The two coupled inductors represent a transformer (with resistances of the windings included in $R_1$ and $R_2$). The state variables are $i_{L_1}$ and $i_{L_2}$, and the two loop equations are written as usual:

$$R_1 i_{L_1} + L_1 \dot{i}_{L_1} + M \dot{i}_{L_2} = v_{\text{in}} \tag{10-40a}$$

$$L_2 \dot{i}_{L_2} + M \dot{i}_{L_1} + R_2 i_{L_2} = 0 \tag{10-40b}$$

($M$ can be positive or negative). These two can be solved *algebraically* for $\dot{i}_{L_1}$ and $\dot{i}_{L_2}$. The result, in matrix form, is indeed the normal state equation:

$$\begin{bmatrix} \dot{i}_{L_1} \\ \dot{i}_{L_2} \end{bmatrix} = \begin{bmatrix} \dfrac{-R_1 L_2}{\Delta} & \dfrac{R_2 M}{\Delta} \\ \dfrac{R_1 M}{\Delta} & \dfrac{-R_2 L_1}{\Delta} \end{bmatrix} \begin{bmatrix} i_{L_1} \\ i_{L_2} \end{bmatrix} + \begin{bmatrix} \dfrac{L_2}{\Delta} \\ \dfrac{-M}{\Delta} \end{bmatrix} v_{\text{in}} \tag{10-41}$$

where $\Delta = L_1 L_2 - M^2 \neq 0$. The output equation reads simply

$$v_{\text{out}} = -R_2 i_{L_2} \tag{10.42}$$

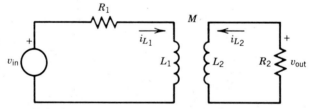

**FIGURE 10-10.**  State equations with a transformer.

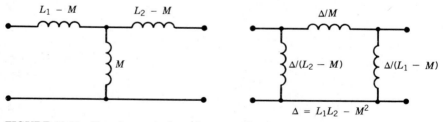

**FIGURE 10-11.**  *T*- and $\pi$-equivalents for a transformer.

The procedure, then, is essentially the loop analysis of mutually coupled inductors.

Alternately, we can replace the two mutually coupled inductors by their $T$- or $\pi$-equivalent (see Fig. 10-11) and proceed in the usual way outlined in the previous section. $\square$

---

### Example 7

An ideal transformer (see problem 4-14) is defined by the relations:

$$v_1 = \frac{1}{a}v_2, \qquad i_1 = -ai_2 \tag{10-43}$$

where $a$ is the *turns ratio*. See Fig. 10-12a. Consider the network shown in Fig. 10-12b and its graph in Fig. 10-12c. This graph has two separate parts because of the ideal transformer; it can be shown that one edge of this transformer, say edge 4, should be a branch while the other, say edge 5, should be a link. The state vector and the input are

$$\mathbf{x} = \begin{bmatrix} v_6 \\ i_2 \end{bmatrix}, \qquad \mathbf{e} = i_1 \tag{10-44}$$

The cut set equation for the capacitor is

$$C_6 \dot{v}_6 = -\frac{1}{R_7}v_6 - i_5 \tag{10-45}$$

and the loop equation for the inductor is

$$L_2 \dot{i}_2 = i_3 R_3 + v_4 \tag{10-46}$$

To eliminate $i_5$, $i_3$ and $v_4$, we write

$$i_5 = -\frac{1}{a}i_4 = -\frac{1}{a}i_3 \tag{10-47}$$

as in Eq. (10-43). Then we have

$$i_3 = i_1 - i_2 \tag{10-48}$$

Also

$$v_4 = \frac{1}{a}v_5 = \frac{1}{a}v_6 \tag{10-49}$$

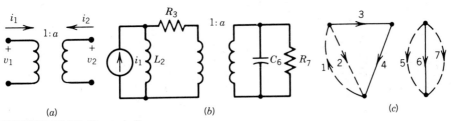

(a)  (b)  (c)

**FIGURE 10-12.** Example 7.

With these substitutions, Eqs. (10-45) and (10-46) become

$$
\begin{bmatrix} \dot{v}_6 \\ \dot{i}_2 \end{bmatrix} = \begin{bmatrix} -\dfrac{1}{R_7 C_6} & -\dfrac{1}{a} \\ \dfrac{1}{aL_2} & -\dfrac{R_3}{L_2} \end{bmatrix} \begin{bmatrix} v_6 \\ i_2 \end{bmatrix} + \begin{bmatrix} \dfrac{1}{a} \\ \dfrac{R_3}{} \end{bmatrix} i_1
\tag{10-50}
$$

the normal form of the state equation.

☐ **Probs. 10-8**
**10-9**
**10-10**

The development for a gyrator follows along similar lines, except that *both* its edges must be either branches or links. See problem 10-9.

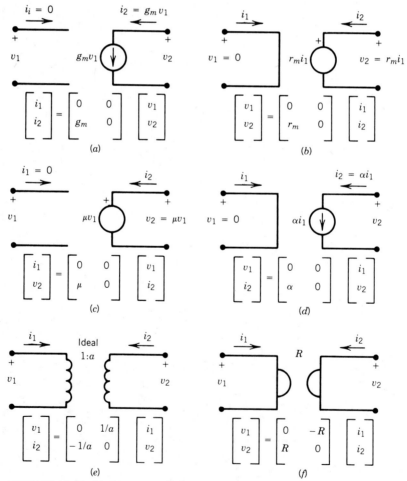

FIGURE 10-13. Various controlled sources.

Controlled (dependent) sources are either voltage controlled, as in Fig. 10-13a and c, or current controlled, as in Fig. 10-13b and d. Notice that all these common controlled sources may be considered as resistive coupling elements, because in all cases a controlled voltage or a current is proportional to a controlling voltage or current. In writing the state equations, these controlling voltages or currents may be the desired variables (elements of $\mathbf{x}$ and $\mathbf{e}$) or undesired variables—in which case an elimination is needed. The rule for eliminating undesired variables is similar to that of state variables: if the undesired variable is a tree branch, write a fundamental cut set for it; if it is a link, write a fundamental loop for it.

## Example 8

In the network shown in Fig. 10-14a, there are two controlled sources. A proper tree and its co-tree are shown, as usual, in Fig. 10-14b. The cut set equation for $C_3$ is

$$C_3 \dot{v}_3 = i_2 + i_6 \tag{10-51}$$

The loop equation for $L_6$ is

$$L_6 \dot{i}_6 = -v_3 + v_1 + v_4 + v_5 \tag{10-52}$$

Since

$$\mathbf{x} = \begin{bmatrix} v_3 \\ i_6 \end{bmatrix} \qquad \mathbf{e} = v_1 \tag{10-53}$$

we must eliminate $i_2$, $v_4$, and $v_5$ from Eqs. (10-51) and (10-52). For $i_2$, write a fundamental loop

$$R_2 i_2 + v_3 = v_1 \tag{10-54}$$

and therefore,

$$i_2 = \frac{1}{R_2} (v_1 - v_3) \tag{10-55}$$

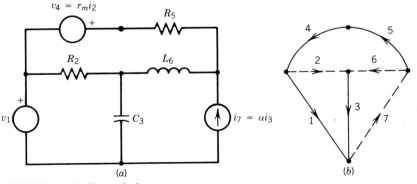

(a)                                        (b)

**FIGURE 10-14.** Example 8.

and

$$v_4 = r_m i_2 = \frac{r_m}{R_2} (v_1 - v_3) \tag{10-56}$$

Next we have a fundamental cut set for $i_5$

$$i_5 = (\alpha i_3 - i_6) = (\alpha C_3 \dot{v}_3 - i_6) \tag{10-57}$$

where we have substituted $i_3 = C_3 \dot{v}_3$. With these relationships, Eqs. (10-51) and (10-52) read

$$C_3 \dot{v}_3 = \frac{1}{R_2} (v_1 - v_3) + i_6 \tag{10-58}$$

and

$$L_6 \dot{i}_6 = -v_3 + v_1 + \frac{r_m}{R_2} (v_1 - v_3) + R_5 (\alpha C_3 \dot{v}_3 - i_6) \tag{10-59}$$

These last two equations must be considered as two simultaneous algebraic equations with the two unknowns $\dot{v}_3$ and $\dot{i}_6$. Solving for these unknowns will yield the desired state equations

$$\begin{bmatrix} \dot{v}_3 \\ \dot{i}_6 \end{bmatrix} = \begin{bmatrix} -\dfrac{1}{R_2 C_3} & \dfrac{1}{C_3} \\ -\dfrac{R_2 + \alpha R_5 + r_m}{R_2 L_6} & \dfrac{(\alpha - 1)R_5}{L_6} \end{bmatrix} \begin{bmatrix} v_3 \\ i_6 \end{bmatrix} + \begin{bmatrix} \dfrac{1}{R_2 C_3} \\ \dfrac{R_2 + \alpha R_5 + r_m}{R_2 L_6} \end{bmatrix} v_1 \tag{10-60}$$

The same results may be obtained easily by using equivalent sources together with superposition, as discussed in section 10-2 and in problem 10-7. ☐ **Probs. 10-11**
**10-12**
**10-13**

## 10-5 FORMULATION OF THE STATE EQUATIONS: TIME-VARYING AND NONLINEAR NETWORKS

As mentioned in the introduction, the state variable approach offers a great advantage in the analysis of linear time-varying networks. The topological relations, of course, remain the same regardless of the nature of the elements. The $v$–$i$ relation for a linear, time-varying resistor $R(t)$ is the same as for the constant resistor,

$$v_R(t) = R(t) i_R(t) \tag{10-61}$$

The only change occurs for the linear, time-varying inductors and capacitors:

$$v_L(t) = \frac{d\phi_L(t)}{dt} = \frac{d}{dt} \{L(t) i_L(t)\}$$

$$= \dot{L}(t) i_L(t) + L(t) \dot{i}_L(t) \tag{10-62}$$

and

$$i_c(t) = \frac{dq_c(t)}{dt} = \frac{d}{dt}\{C(t)v_c(t)\}$$

$$= \dot{C}(t)v_c(t) + C(t)\dot{v}_c(t) \qquad (10\text{-}63)$$

exhibiting the additional terms $\dot{L}(t)i_L(t)$ and $\dot{C}(t)v_c(t)$, respectively. These terms vanish in the constant, time-invariant, case.

## Example 9

Consider the introductory example of this chapter, Fig. 10-1, with all the elements linear and time varying: $R_1(t)$, $R_2(t)$, $L(t)$, $C(t)$. For convenience, we omit the parenthetical $t$ in the subsequent development.

The cut set equation for the capacitor is

$$\dot{C}v_c + C\dot{v}_c = -\frac{v_c}{R_1} - i_L + i_{\text{in}} \qquad (10\text{-}64)$$

and the loop equation for the inductor is

$$\dot{L}i_L + L\dot{i}_L = v_c - R_2i_L \qquad (10\text{-}65)$$

The output equation is

$$v_{\text{out}} = R_2i_L \qquad (10\text{-}66)$$

Compare these equations with Eqs. (10-1), (10-2), and (10-3).

Equations (10-64), (65), (67) after rearrangement, yield the equations

$$\dot{\mathbf{x}} = \begin{bmatrix} \dot{v}_c \\ \dot{i}_L \end{bmatrix} = \begin{bmatrix} -\dfrac{1 + R_1\dot{C}}{R_1C} & -\dfrac{1}{C} \\ \dfrac{1}{L} & -\dfrac{R_2 + \dot{L}}{L} \end{bmatrix} \begin{bmatrix} v_c \\ i_L \end{bmatrix} + \begin{bmatrix} \dfrac{1}{C} \\ 0 \end{bmatrix} i_{\text{in}} \qquad (10\text{-}67)$$

$$\mathbf{r} = v_{\text{out}} = \begin{bmatrix} 0 & R_2 \end{bmatrix} \begin{bmatrix} v_c \\ i_L \end{bmatrix} + [0]i_{\text{in}} \qquad (10\text{-}68)$$

That is,

$$\dot{\mathbf{x}}(t) = \mathbf{A}(t)\mathbf{x}(t) + \mathbf{B}(t)\mathbf{e}(t) \qquad (10\text{-}69)$$

$$\mathbf{r}(t) = \mathbf{C}(t)\mathbf{x}(t) + \mathbf{D}(t)\mathbf{e}(t) \qquad (10\text{-}70)$$

in the general case, with $\mathbf{A}$, $\mathbf{B}$, $\mathbf{C}$, and $\mathbf{D}$ being time-varying matrices. $\qquad \square$

In order to emphasize further the unified approach to constant and time-varying elements, we may choose as state variables *capacitor charges* and *inductor fluxes*:

$$q_c = Cv_c \qquad \phi_L = Li_L \qquad (10\text{-}71)$$

That is, the new state vector

$$\hat{\mathbf{x}} = \begin{bmatrix} q_{c_1} \\ q_{c_2} \\ \vdots \\ \phi_{L_1} \\ \phi_{L_2} \\ \vdots \end{bmatrix} \tag{10-72}$$

is equally suitable for constant *or* time-varying elements. In the case of constant elements, the difference between $\mathbf{x}$ (voltages and currents) and $\hat{\mathbf{x}}$ (charges and fluxes) is only a matter of constant multipliers. In the case of time-varying elements, the choice of $\hat{\mathbf{x}}$ avoids the derivatives of capacitances and inductances in Eqs. (10-62) and (10-63).

---

### Example 10

The previous example, with

$$\hat{\mathbf{x}} = \begin{bmatrix} q_c \\ \phi_L \end{bmatrix} \tag{10-73}$$

yields directly the following equations: the cut set equation,

$$\dot{q}_c = -\frac{q_c}{R_1 C} - \frac{\phi_L}{L} + i_{\text{in}} \tag{10-74}$$

the loop equation,

$$\dot{\phi}_L = \frac{q_c}{C} - R_2 \frac{\phi_L}{L} \tag{10-75}$$

and the output equation,

$$v_{\text{out}} = R_2 \frac{\phi_L}{L} \tag{10-76}$$

In matrix form, they read

$$\begin{bmatrix} \dot{q}_c \\ \dot{\phi}_L \end{bmatrix} = \begin{bmatrix} -\dfrac{1}{R_1 C} & -\dfrac{1}{L} \\ \dfrac{1}{C} & -\dfrac{R_2}{L} \end{bmatrix} \begin{bmatrix} q_c \\ \phi_L \end{bmatrix} + \begin{bmatrix} 1 \\ 0 \end{bmatrix} i_{\text{in}} \tag{10-77}$$

and

$$v_{\text{out}} = \begin{bmatrix} 0 & \dfrac{R_2}{L} \end{bmatrix} \begin{bmatrix} q_c \\ \phi_L \end{bmatrix} + [0] i_{\text{in}} \tag{10-78}$$

Thus, Eqs. (10-77) and (10-78) are valid for *either* the constant *or* time-varying network. Compare with Eqs. (10-4) and (10-5).  □ ***Probs. 10-14***
***10-15***

Finally, a brief comment on nonlinear elements. To be specific, consider the previous example, with all the elements nonlinear. That is, the defining relations of each element are given as

$$i_{R_1} = f_1(v_{R_1}, t)$$
$$i_{R_2} = f_2(v_{R_2}, t)$$
$$v_c = f_3(q_c, t)$$
$$i_L = f_4(\phi_L, t)$$

(10-79)

where the $f$'s designate nonlinear functions, given analytically or graphically. See Fig. 10-15. The cut set equation will be

$$\dot{q}_c = -i_{R_1} - i_L + i_{in} = -f_1(v_{R_1}, t) - f_4(\phi_L, t) + i_{in} \qquad (10\text{-}80)$$

and since $v_{R_1} = v_c = f_3(q_c, t)$ we rewrite Eq. (10-80),

$$\dot{q}_c = -f_1[f_3(q_c, t)] - f_4(\phi_L, t) + i_{in} \qquad (10\text{-}81)$$

The loop equation is

$$\dot{\phi}_L = v_c - v_{R_2} = f_3(q_c, t) - v_{R_2} \qquad (10\text{-}82)$$

In brief, Eqs. (10-81) and (10-82) are

$$\dot{\hat{x}} = F(\hat{x}, e(t), t) \qquad (10\text{-}83)$$

the normal form of the state equation, where $F$ designates a functional, nonlinear, relation.

Difficulties exist, however, with nonlinear networks. For example, since the second resistor is characterized by the relationship in Eq. (10-79) we have to solve for $v_{R_2}$ in terms of $i_{R_2}$ in order to use Eq. (10-82). In other words, we would need $f_2^{-1}$, the inverse of the nonlinear function $f_2$. This inverse may not exist, or may not be unique. If it exists, then we can express $v_{R_2}$ in terms of $\phi_L$.

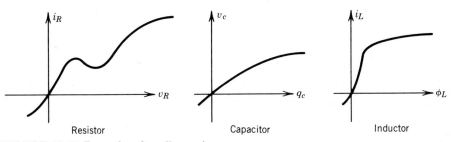

**FIGURE 10.15** Examples of nonlinear elements.

## PROBLEMS

10-1 Determine the order of complexity of each network shown.

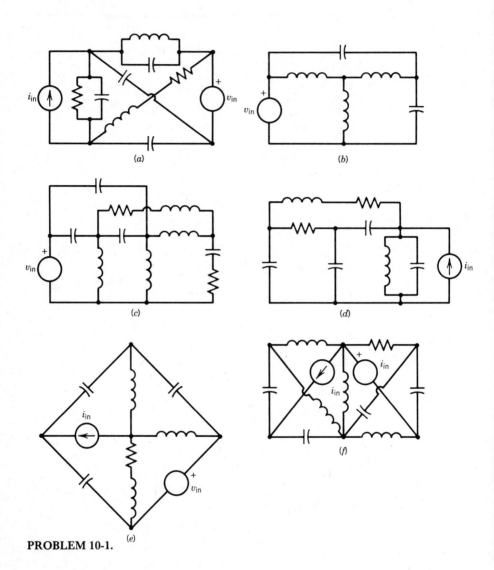

**PROBLEM 10-1.**

10-2 For each network in problem 10-1, write its characteristic equation as explicitly as possible.

10-3 In order to notice the effect of dependent sources on the order of complexity, consider the network shown. The current source $i_{in}(t)$ is independent, while $i_a$ is the controller for the dependent source. Write the state equation for the case $\alpha \neq 1$. What is the order of the network and what happens if $\alpha = 1$?

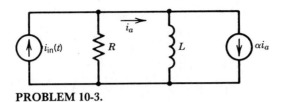

**PROBLEM 10-3.**

10-4 For each of the networks shown, obtain the state equation.

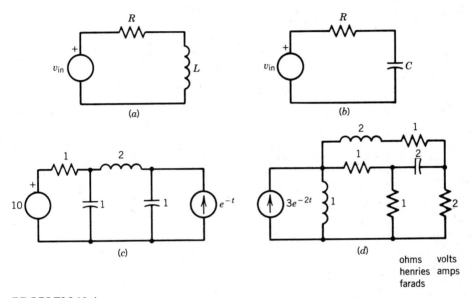

**PROBLEM 10-4.**

*10-5 In order to relate the derivation of state equations to topological considerations, we follow these steps:

a) Let the edge current matrix be partitioned into $i_{link}$ and $i_{branch}$, as in Eq. (6-9). Partition, in a similar way, the edge voltage matrix, Eq. (6-19).

b) Express $i_{branch}$ and $v_{link}$ in terms of the independent variables $v_{branch}$ and $i_{link}$, via the matrices $Q_f$ and $B_f$ and their appropriate submatrices.

c) The state variable matrix $x$ consists of $v_c$, a submatrix of $v_{branch}$, and of $i_L$, a submatrix of $i_{link}$.

d) In the proper tree, the branch currents are given by

$$i_c = \mathscr{C}\dot{v}_c$$

and the link voltages are given by

$$v_L = \mathscr{L}\dot{i}_L$$

Here, $\mathscr{C}$ is a capacitance matrix, and $\mathscr{L}$ is an inductance matrix. The state matrix equation is obtained by eliminating all undesired voltages and currents (i.e., all except $\mathbf{v}_c$, $\mathbf{i}_L$ and the sources) in steps b, c, and d. Use this method for the networks in problem 10-4.

10-6  Repeat problem 10-4 for the network shown.

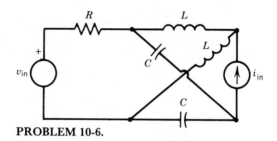

**PROBLEM 10-6.**

10-7  Use the method of equivalent sources as shown in Fig. 10-3 in the text to solve problem 10-4. *Hint*: After replacing the inductors and capacitors by their equivalent sources, use superposition (one source at a time) to solve the resistive network. Then find the expressions for capacitor currents and inductor voltages.

10-8  Obtain the normal state equation in matrix form for the network shown.

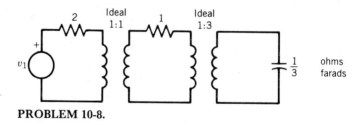

**PROBLEM 10-8.**

10-9  Replace the ideal transformer in Fig. 10-12b of the text by a gyrator (see Fig. 10-13f) then obtain the state equation in matrix form. In forming the tree, include *both* edges of the gyrator either as branches or links.

10-10  Obtain the normal state equation in matrix form for the network shown.

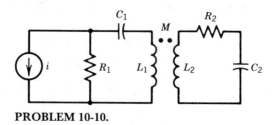

**PROBLEM 10-10.**

10-11  The network shown is an equivalent circuit for a transistor amplifier. Obtain its normal state equations, including the output equation. The parameters $h_{11}, h_{12}, h_{21}$, and $h_{22}$ are given hybrid parameters; the notations "$b$," "$c$," and "$e$" refer to the base, the collector, and the emitter terminals, respectively. Note the dependent sources.

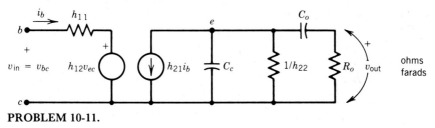

**PROBLEM 10-11.**

10-12  Repeat problem 10-10 for the network shown.

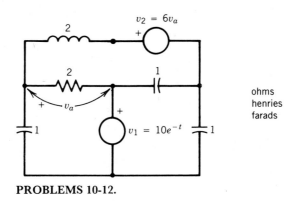

**PROBLEMS 10-12.**

10-13  Develop fully the idea illustrated in Fig. 10-3a for a network whose order of complexity is 2. Specifically, consider a purely resistive 2-P network, as shown, with an inductor $L$ and a capacitor $C$ at the ports. The 2-P itself is characterized by its open-circuit impedance parameters (here purely resistive). Derive the state equation in matrix form for this network.[1]

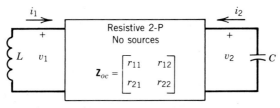

**PROBLEM 10-13.**

10-14 Obtain the state equations in matrix form for the network shown in Fig. 10-8 with all the linear elements time varying.

    a) Use the capacitor voltages and inductor currents as state variables.

    b) Use charges and fluxes as state variables.

10-15 Repeat problem 10-14 for the network in Fig. 10-9.

## REFERENCES

1. W. R. Lepage, "State variable approach to a second-order system," *IEEE Trans. on Education.* E-9, June 1966, pp. 80–90.

# 11

# STATE VARIABLES II: SOLUTION

We are now ready to study the solution of the state equations. We shall concentrate on the linear, constant case. The first section of this chapter considers time-domain solutions, and the second section—solutions by Laplace transforms. At the end of the chapter we discuss certain topics associated with signal flow graphs of state equations.

## 11-1  INTRODUCTORY EXAMPLE

In order to gain confidence and, at the same time, introduce some of the notation, let us solve completely the introductory example in Chapter 10, using the Laplace transform. Specifically, consider the network in Fig. 10-1, with $R_1 = 0.5$ ohms, $R_2 = 2.5$ ohms, $L = 0.5$ henry, and $C = 1$ farad. Also let the initial state be given by $v_c(0_-)$ and $i_L(0_-)$. The state equation, Eq. (10-4), becomes

$$\begin{bmatrix} \dfrac{dv_c}{dt} \\[2mm] \dfrac{di_L}{dt} \end{bmatrix} = \begin{bmatrix} -2 & -1 \\ 2 & -5 \end{bmatrix}\begin{bmatrix} v_c \\ i_L \end{bmatrix} + \begin{bmatrix} 1 \\ 0 \end{bmatrix} i_{\text{in}} \tag{11-1a}$$

that is,

$$\dot{\mathbf{x}}(t) = \mathbf{A}\mathbf{x}(t) + \mathbf{B}e(t) \tag{11-1b}$$

It is important to note that, using the Laplace transform, we have to take the transform of first-order differential equations only—a much easier and simpler task than with integro-differential loop or node equations. Taking the Laplace transform of each equation in Eq. (11-1a) we obtain

$$sV_c(s) - v_c(0_-) = -2V_c(s) - I_L(s) + I_{\text{in}}(s)$$

and

$$sI_L(s) - i_L(0_-) = 2V_c(s) - 5I_L(s) \tag{11-2a}$$

These are ready now for algebraic solution. Collecting terms and rearranging, we get

$$(s + 2)V_c(s) + I_L(s) = v_c(0_-) + I_{in}(s)$$

$$-2V_c(s) + (s + 5)I_L(s) = i_L(0_-)$$

(11-2b)

or, in matrix form

$$\begin{bmatrix} s + 2 & 1 \\ -2 & s + 5 \end{bmatrix} \begin{bmatrix} V_c(s) \\ I_L(s) \end{bmatrix} = \begin{bmatrix} v_c(0_-) \\ i_L(0_-) \end{bmatrix} + \begin{bmatrix} 1 \\ 0 \end{bmatrix} I_{in}(s)$$

(11-2c)

It is worthwhile to recognize this form as being

$$\left\{ \begin{bmatrix} s & 0 \\ 0 & s \end{bmatrix} - \begin{bmatrix} -2 & -1 \\ 2 & -5 \end{bmatrix} \right\} \begin{bmatrix} V_c(s) \\ I_L(s) \end{bmatrix} = \begin{bmatrix} v_c(0_-) \\ i_L(0_-) \end{bmatrix} + \begin{bmatrix} 1 \\ 0 \end{bmatrix} I_{in}(s)$$

(11-2d)

that is,

$$(s\mathbf{U} - \mathbf{A})\mathbf{X}(s) = \mathbf{x}(0_-) + \mathbf{B}\mathbf{E}(s)$$

(11-2e)

The algebraic solution for $V_c(s)$ and $I_L(s)$ proceeds from Eq. (11-2b) by determinants, or from Eq. (11-2d) by matrix inversion. We obtain

$$V_c(s) = \frac{(s + 5)v_c(0_-) - i_L(0_-)}{s^2 + 7s + 12} + \frac{s + 5}{s^2 + 7s + 12} I_{in}(s)$$

and

(11-3a)

$$I_L(s) = \frac{(s + 2)i_L(0_-) + 2v_c(0_-)}{s^2 + 7s + 12} + \frac{2}{s^2 + 7s + 12} I_{in}(s)$$

Each answer exhibits clearly the zero-input part and the zero-state part. With specific values for $v_c(0_-)$, $i_L(0_-)$ and $I_{in}(s)$, we can invert Eq. (11-3a) to obtain the time domain answers $v_c(t)$ and $i_L(t)$. Let us do it here for the zero-input answers only. We have then

$$V_c(s) = \frac{s + 5}{(s + 3)(s + 4)} v_c(0_-) - \frac{1}{(s + 3)(s + 4)} i_L(0_-)$$

$$I_L(s) = \frac{2}{(s + 3)(s + 4)} v_c(0_-) + \frac{s + 2}{(s + 3)(s + 4)} i_L(0_-)$$

(11-3b)

Expansion by partial fractions yields

$$V_c(s) = \left( \frac{2}{s + 3} + \frac{-1}{s + 4} \right) v_c(0_-) + \left( \frac{-1}{s + 3} + \frac{1}{s + 4} \right) i_L(0_-)$$

$$I_L(s) = \left( \frac{2}{s + 3} + \frac{-2}{s + 4} \right) v_c(0_-) + \left( \frac{-1}{s + 3} + \frac{2}{s + 4} \right) i_L(0_-)$$

(11-3c)

and finally, be inversion and in matrix form,

$$\begin{bmatrix} v_c(t) \\ i_L(t) \end{bmatrix} = \begin{bmatrix} 2e^{-3t} - e^{-4t} & -e^{-3t} + e^{-4t} \\ 2e^{-3t} - 2e^{-4t} & -e^{-3t} + 2e^{-4t} \end{bmatrix} \begin{bmatrix} v_c(0_-) \\ i_L(0_-) \end{bmatrix}$$

(11-3d)

## *11-2 TIME DOMAIN SOLUTIONS OF THE STATE EQUATIONS

As mentioned earlier, the output equation, Eq. (10-7), is an algebraic equation giving directly the output **r** in terms of the state **x** and the input **e**. We shall therefore concentrate only on the solution of the state equation, Eq. (10-6), repeated here for convenience

$$\dot{\mathbf{x}} = \mathbf{A}\mathbf{x} + \mathbf{B}\mathbf{e} \tag{11-4}$$

For the type of networks considered (linear, constant), matrices **A** and **B** are constant. The order of **A** is $(n \times n)$, where $n$ is the number of independent state variables, while the order of **B** is $(n \times k)$ with **e** being of order $(k \times 1)$ as in Fig. 10-2.

### The Scalar State Equation

Consider first the simplest case, that of a network whose order of complexity is $n = 1$. Here the state equation becomes a single *scalar* differential equation

$$\dot{x}(t) = ax(t) + be(t) \tag{11-5}$$

or

$$\dot{x}(t) - ax(t) = be(t) \tag{11-6}$$

with a given initial value $x(t_0)$.

The classical solution to this equation may be obtained with an integrating factor. The integrating factor for Eq. (11-6) is $e^{-at}$, and when we multiply that equation by it we get[†]

$$e^{-at}\dot{x} - ae^{-at}x = e^{-at}be(t) \tag{11-7}$$

that is,

$$\frac{d}{dt}(xe^{-at}) = e^{-at}be(t) \tag{11-8}$$

Integrating both sides from $t_0$, the initial time, to $t$ yields

$$\int_{t_0}^{t} \frac{d}{dt} x(\tau)e^{-a\tau} \, d\tau = \int_{t_0}^{t} e^{-a\tau}be(\tau) \, d\tau \tag{11-9}$$

where $\tau$ is a dummy variable of integration. Finally, the evaluation of the left-hand side of Eq. (11-9) yields the answer,

$$x(t) = e^{a(t-t_0)}x(t_0) + \int_{t_0}^{t} e^{a(t-\tau)}be(\tau) \, d\tau \tag{11-10}$$

This *complete state response* consists of two terms. First, *the zero-input state response*:

$$x_{zi}(t) = e^{a(t-t_0)}x(t_0) \tag{11-11}$$

---

[†] Do not confuse $e = 2.71828\ldots$, the base of natural logarithms, with $e(t)$, our symbol for input!

obtained, as its name implies, when $e(t) = 0$. This corresponds to the *homogeneous solution* of Eq. (11-6); to emphasize it, the subscript $zi$ is placed here. The second term is the *zero-state state response*,

$$x_{zs}(t) = \int_{t_0}^{t} e^{a(t-\tau)} be(\tau) \, d\tau \tag{11-12}$$

corresponding to a zero initial state $x(t_0) = 0$. Notice that if $t_0 = 0$ (as is usually the case), the zero-state solution is

$$x_{zs}(t) = \int_{0}^{t} e^{a(t-\tau)} be(\tau) \, d\tau \tag{11-13}$$

the usual convolution integral.

Other remarks pertaining to the dependence of the zero-input response on the initial state, the dependence of the zero-state response on the input, etc., were discussed in Chapter 7 (section 7-7) and the reader is urged to review them.

## The Homogeneous State Equation

Since there is no loss of generality in assuming $t_0 = 0$ for linear constant networks, we shall do so in discussing the solution of the matrix state equation. We begin with the homogeneous, zero-input case. The equation to solve is

$$\dot{\mathbf{x}} = \mathbf{A}\mathbf{x} \qquad \mathbf{x}(0_-) \text{ given} \tag{11-14}$$

By analogy to the scalar case, Eq. (11-11), we expect an exponential solution,

$$\mathbf{x}(t) = e^{\mathbf{A}t} \mathbf{x}(0_-) \tag{11-15}$$

where $e^{\mathbf{A}t}$ is an exponential matrix function defined as follows:

$$e^{\mathbf{A}t} = \mathbf{U} + \mathbf{A}t + \mathbf{A}^2 \frac{t^2}{2!} + \mathbf{A}^3 \frac{t^3}{3!} + \cdots = \sum_{k=0}^{\infty} \mathbf{A}^k \frac{t^k}{k!} \tag{11-16}$$

where $\mathbf{U}$ is the unit matrix. Notice the similarity to the scalar series for $e^{at}$

$$e^{at} = 1 + at + a^2 \frac{t^2}{2!} + a^3 \frac{t^3}{3!} + \cdots = \sum_{k=0}^{\infty} a^k \frac{t^k}{k!} \tag{11-17}$$

It can be shown that the series in Eq. (11-16) converges for all $\mathbf{A}$ and for all finite $t$. That Eq. (11-15) is indeed the solution of Eq. (11-14) can be verified by a direct substitution: first, for $t = 0_-$, Eq. (11-15) is satisfied since $\mathbf{x}(0_-) = \mathbf{U}\mathbf{x}(0_-)$. Next differentiate Eq. (11-16) to yield

$$\frac{d}{dt}(e^{\mathbf{A}t}) = \mathbf{A} + \mathbf{A}^2 t + \mathbf{A}^3 \frac{t^2}{2!} + \cdots = \mathbf{A}e^{\mathbf{A}t} \tag{11-18}$$

and therefore the derivative of $\mathbf{x}$ in Eq. (11-15) becomes

$$\dot{\mathbf{x}} = \mathbf{A}e^{\mathbf{A}t}\mathbf{x}(0_-) = \mathbf{A}\mathbf{x} \tag{11-19}$$

as required. The exponential matrix function $e^{At}$ is usually denoted by $\phi(t)$.

$$\phi(t) = e^{At} \tag{11-20}$$

and is called the *state transition matrix* (also *the fundamental matrix*). Thus, the zero-input state response is

$$x(t) = \phi(t)x(0_-) \tag{11-21}$$

which explains the term "state transition": The initial state $x(0_-)$ undergoes a change, a transition, to another state at time $t$, according to Eq. (11-21).

## Example 1

Consider the introductory example where the homogeneous state equation (see Eq. (11-1a)) becomes

$$\begin{bmatrix} \dot{v}_c \\ \dot{i}_L \end{bmatrix} = \begin{bmatrix} -2 & -1 \\ 2 & -5 \end{bmatrix} \begin{bmatrix} v_c \\ i_L \end{bmatrix} \tag{11-22}$$

that is, $\dot{x} = Ax$. The state transition matrix will be computed according to Eq. (11-16). First we have

$$A = \begin{bmatrix} -2 & -1 \\ 2 & -5 \end{bmatrix} \qquad A^2 = A \cdot A = \begin{bmatrix} 2 & 7 \\ -14 & 23 \end{bmatrix}$$

$$A^3 = A \cdot A^2 = \begin{bmatrix} 10 & -37 \\ 74 & -101 \end{bmatrix}, \dots \tag{11-23}$$

And therefore,

$$\phi(t) = \begin{bmatrix} 1 & 0 \\ 0 & 1 \end{bmatrix} + \begin{bmatrix} -2 & -1 \\ 2 & -5 \end{bmatrix} t + \begin{bmatrix} 2 & 7 \\ -14 & 23 \end{bmatrix} \frac{t^2}{2!} + \begin{bmatrix} 10 & -37 \\ 74 & -101 \end{bmatrix} \frac{t^3}{3!} + \cdots$$

$$= \begin{bmatrix} 1 - 2t + t^2 + \dfrac{5}{3}t^3 + \cdots & -t + \dfrac{7}{2}t^2 - \dfrac{37}{6}t^3 + \cdots \\ 2t - 7t^2 + \dfrac{37}{3}t^3 + \cdots & 1 - 5t + \dfrac{23}{2}t^2 - \dfrac{101}{6}t^3 + \cdots \end{bmatrix} \tag{11-24}$$

The infinite series in $\phi(t)$ can be shown to be

$$\phi(t) = \begin{bmatrix} 2e^{-3t} - e^{-4t} & -e^{-3t} + e^{-4t} \\ 2e^{-3t} - 2e^{-4t} & -e^{-3t} + 2e^{-4t} \end{bmatrix} \tag{11-25}$$

and therefore the solution of Eq. (11-22) is (compare Eq. (11-3d))

$$\begin{bmatrix} v_c(t) \\ i_L(t) \end{bmatrix} = \begin{bmatrix} 2e^{-3t} - e^{-4t} & -e^{-3t} + e^{-4t} \\ 2e^{-3t} - 2e^{-4t} & -e^{-3t} + 2e^{-4t} \end{bmatrix} \begin{bmatrix} v_c(0_-) \\ i_L(0_-) \end{bmatrix} \tag{11-26}$$

It appears that the transition matrix $\phi(t)$ is of great importance. However, its evaluation by Eq. (11-16), as illustrated in this example, is tedious. The repeated multiplications of $\mathbf{A}$, the infinite series representations, and the obvious question of how to recognize the closed form of the series are good reasons for seeking more efficient ways for evaluating $\boldsymbol{\phi}(t)$. $\square$

## Evaluation of $\boldsymbol{\phi}(t)$ from the Characteristic Values of $\mathbf{A}$

The classical approach to the solution of the scalar differential equation

$$\dot{x} = ax \tag{11-27}$$

is to assume a scalar exponential solution

$$x = Ke^{\lambda t} \tag{11-28}$$

and substitute it into Eq. (11-27). The result is

$$K\lambda e^{\lambda t} = aKe^{\lambda t} \tag{11-29a}$$

that is,

$$\lambda = a \tag{11-29b}$$

*the characteristic equation* giving the value of $\lambda$. If such a solution is assumed for each component $x_j$ of the state vector $\mathbf{x}$, then the matrix differential equation,

$$\dot{\mathbf{x}} = \mathbf{Ax} \tag{11-30}$$

will have the assumed solution

$$\mathbf{x} = \mathbf{K}e^{\lambda t} \tag{11-31}$$

where the constant matrix $\mathbf{K}$ replaces the scalar $K$ of Eq. (11-28). Substitute Eq. (11-31) into Eq. (11-30) to obtain

$$\lambda \mathbf{K}e^{\lambda t} = \mathbf{AK}e^{\lambda t} \tag{11-32}$$

that is,

$$\lambda \mathbf{x} = \mathbf{Ax} \tag{11-33a}$$

which is a set of linear algebraic equations. Written fully, they read:

$$
\begin{aligned}
a_{11}x_1 + a_{12}x_2 + \cdots + a_{1n}x_n &= \lambda x_1 \\
a_{21}x_1 + a_{22}x_2 + \cdots + a_{2n}x_n &= \lambda x_2 \\
\vdots \qquad\qquad \vdots \qquad \vdots \\
a_{n1}x_1 + a_{n2}x_2 + \cdots + a_{nn}x_n &= \lambda x_n
\end{aligned}
\tag{11-33b}
$$

or

$$
\begin{aligned}
(\lambda - a_{11})x_1 - a_{12}x_2 - \cdots - a_{1n}x_n &= 0 \\
-a_{21}x_1 + (\lambda - a_{22})x_2 - \cdots - a_{2n}x_n &= 0 \\
\vdots \qquad\qquad \vdots \\
-a_{n1}x_1 - a_{n2}x_2 - \cdots + (\lambda - a_{nn})x_n &= 0
\end{aligned}
\tag{11-34}
$$

In matrix form, these equations become

$$(\lambda \mathbf{U} - \mathbf{A})\mathbf{x} = \mathbf{0} \qquad (11\text{-}35)$$

where $\mathbf{U}$ is the unit matrix. Compare with Eq. (11-2e).

A nontrivial solution for $\mathbf{x}$ exists $(\mathbf{x} \neq \mathbf{0})$ provided the determinant of the coefficients is zero, that is,

$$\begin{vmatrix} (\lambda - a_{11}) & -a_{12} & \cdots & -a_{1n} \\ -a_{21} & (\lambda - a_{22}) & \cdots & -a_{2n} \\ & \vdots & & \\ -a_{n1} & -a_{n2} & \cdots & (\lambda - a_{nn}) \end{vmatrix} = \det(\lambda \mathbf{U} - \mathbf{A}) = 0 \qquad (11\text{-}36)$$

Equation (11-36) is *the characteristic equation* of the network.[†] When the determinant is expanded, we obtain

$$\lambda^n + \alpha_1 \lambda^{n-1} + \alpha_2 \lambda^{n-2} + \cdots + \alpha_n = 0 \qquad (11\text{-}37)$$

The values of $\lambda$ that satisfy this characteristic equation are the *characteristic values* or *eigenvalues* of $\mathbf{A}$, and they are the *natural frequencies* of the network. The same results were obtained in the introductory example, with $s$ replacing $\lambda$. The characteristic equation there is

$$\det(s\mathbf{U} - \mathbf{A}) = 0 \qquad (11\text{-}38)$$

This derivation emphasizes the important feature of the matrix $\mathbf{A}$: It serves to describe completely the natural (free) response of the network.

---

## Example 2

The eigenvalues of $\mathbf{A}$ in Eq. (11-22) are found from

$$\det(\lambda \mathbf{U} - \mathbf{A}) = \begin{vmatrix} \lambda + 2 & 1 \\ -2 & \lambda + 5 \end{vmatrix} = \lambda^2 + 7\lambda + 12 = 0 \qquad (11\text{-}39)$$

or $\lambda_1 = -3$, $\lambda_2 = -4$. These are the natural frequencies of the network in that example. ☐

---

## Example 3

Find the eigenvalues of

$$\mathbf{A} = \begin{bmatrix} 1 & 1 & 2 \\ 0 & 2 & 2 \\ -1 & 1 & 3 \end{bmatrix} \qquad (11\text{-}40)$$

---

[†] See also section 7-5 in Chapter 7.

The characteristic equation is

$$\begin{vmatrix} \lambda - 1 & -1 & -2 \\ 0 & \lambda - 2 & -2 \\ 1 & -1 & \lambda - 3 \end{vmatrix} = 0 \qquad (11\text{-}41)$$

or

$$\lambda^3 - 6\lambda^2 + 11\lambda - 6 = 0$$

giving

$$\lambda_1 = 1 \qquad \lambda_2 = 2 \qquad \lambda_3 = 3 \qquad \square$$

How will the eigenvalues of **A** help in finding the transition matrix $\boldsymbol{\phi}(t)$? Here we shall make use of the *Cayley–Hamilton theorem* which can be stated as follows: If the characteristic equation of any square matrix **A** is

$$\lambda^n + \alpha_1 \lambda^{n-1} + \alpha_2 \lambda^{n-2} + \cdots + \alpha_n = 0 \qquad (11\text{-}42)$$

then **A** satisfies the matrix equation,

$$\mathbf{A}^n + \alpha_1 \mathbf{A}^{n-1} + \alpha_2 \mathbf{A}^{n-2} + \cdots + \alpha_n \mathbf{U} = \mathbf{0} \qquad (11\text{-}43)$$

That is, every square matrix satisfies its own characteristic equation.

## Example 4

The characteristic equation of **A** in example 2 is $\lambda^2 + 7\lambda + 12 = 0$. Therefore, substituting **A** we get

$$\mathbf{A}^2 + 7\mathbf{A} + 12\mathbf{U} = \begin{bmatrix} 2 & 7 \\ -14 & 23 \end{bmatrix} + 7\begin{bmatrix} -2 & -1 \\ 2 & -5 \end{bmatrix} + 12\begin{bmatrix} 1 & 0 \\ 0 & 1 \end{bmatrix} = \begin{bmatrix} 0 & 0 \\ 0 & 0 \end{bmatrix}$$

as expected. $\square$

The usefulness of the Cayley–Hamilton theorem for our purpose lies in the fact that Eq. (11-43) can be written as

$$\mathbf{A}^n = -\alpha_1 \mathbf{A}^{n-1} - \alpha_2 \mathbf{A}^{n-2} - \cdots - \alpha_n \mathbf{U} \qquad (11\text{-}44)$$

that is, the $n$th power of **A** can be expressed in terms of its lower powers $\mathbf{A}^{n-1}$, $\mathbf{A}^{n-2}$, etc. Postmultiply now Eq. (11-44) by **A**:

$$\mathbf{A}^n \mathbf{A} = \mathbf{A}^{n+1} = -\alpha_1 \mathbf{A}^n - \alpha_2 \mathbf{A}^{n-1} - \cdots - \alpha_n \mathbf{A} \qquad (11\text{-}45)$$

and substitute for $\mathbf{A}^n$ from Eq. (11-44) to get

$$\mathbf{A}^{n+1} = -\alpha_1(-\alpha_1 \mathbf{A}^{n-1} - \alpha_2 \mathbf{A}^{n-2} - \cdots) - \alpha_2 \mathbf{A}^{n-1} \cdots \qquad (11\text{-}46)$$

This scheme may be repeated for $\mathbf{A}^{n+2}$, $\mathbf{A}^{n+3}$, etc. Briefly, then: *all* the powers of $\mathbf{A}$ from $\mathbf{A}^n$ and up can be expressed in terms of the powers of $\mathbf{A}^{n-1}$ and down.

---

## Example 5

Evaluate $\mathbf{A}^2$, $\mathbf{A}^3$, and $\mathbf{A}^4$ if

$$\mathbf{A} = \begin{bmatrix} -3 & 2 \\ 1 & -2 \end{bmatrix}$$

The characteristic equation of $\mathbf{A}$ is

$$\begin{vmatrix} \lambda + 3 & -2 \\ -1 & \lambda + 2 \end{vmatrix} = \lambda^2 + 5\lambda + 4 = 0 \tag{11-47}$$

Therefore,

$$\mathbf{A}^2 + 5\mathbf{A} + 4\mathbf{U} = 0 \tag{11-48}$$

that is,

$$\mathbf{A}^2 = -5\mathbf{A} - 4\mathbf{U} = -5\begin{bmatrix} -3 & 2 \\ 1 & -2 \end{bmatrix} - \begin{bmatrix} 4 & 0 \\ 0 & 4 \end{bmatrix} = \begin{bmatrix} 11 & -10 \\ -5 & 6 \end{bmatrix} \tag{11-49}$$

Next,

$$\mathbf{A}^3 = \mathbf{A}^2\mathbf{A} = (-5\mathbf{A} - 4\mathbf{U})\mathbf{A} = -5\mathbf{A}^2 - 4\mathbf{A} = -5(-5\mathbf{A} - 4\mathbf{U}) - 4\mathbf{A}$$

$$= 21\mathbf{A} + 20\mathbf{U} = \begin{bmatrix} -43 & 42 \\ 21 & -22 \end{bmatrix} \tag{11-50}$$

And

$$\mathbf{A}^4 = \mathbf{A}^3\mathbf{A} = (25\mathbf{A} + 20\mathbf{U})\mathbf{A} = 25\mathbf{A}^2 + 20\mathbf{A} = 25(-5\mathbf{A} - 4\mathbf{U}) + 20\mathbf{A}$$

$$= -105\mathbf{A} - 100\mathbf{U} = \begin{bmatrix} 215 & -210 \\ -105 & 110 \end{bmatrix} \qquad \square \tag{11-51}$$

---

In particular, this method can be used to evaluate the powers of $\mathbf{A}$ in $e^{\mathbf{A}t}$. To illustrate the evaluation of the transition matrix $\boldsymbol{\phi}(t)$, consider a specific example.

---

## Example 6

Let us use again $\mathbf{A}$ of example 1. Since $\mathbf{A}$ is of order two, its characteristic equation is of that order. In other words, $\mathbf{A}^2$, $\mathbf{A}^3$, $\mathbf{A}^4$, etc., can be expressed as linear combinations of $\mathbf{A}$ and $\mathbf{U}$ with constant coefficients. If, instead of $\mathbf{A}$, we consider $\mathbf{A}t$ ($t$ is a scalar), then $(\mathbf{A}t)^2$, $(\mathbf{A}t)^3$, etc., are linear combinations of $\mathbf{A}$ and $\mathbf{U}$—only here the coefficients of these combinations are *dependent on t*. That is,

$$e^{\mathbf{A}t} = k_0(t)\mathbf{U} + k_1(t)\mathbf{A} \tag{11-52}$$

It can be shown that a scalar equation, equivalent to Eq. (11-52), is satisfied when $\mathbf{A}$ is replaced by $\lambda$; that is,

$$e^{\lambda t} = k_0(t) + k_1(t)\lambda \tag{11-53}$$

where the $\lambda$'s are the eigenvalues of $\mathbf{A}$. For $\lambda = \lambda_1 = 3$, we have

$$e^{-3t} = k_0(t) - 3k_1(t) \tag{11-54}$$

and for $\lambda = \lambda_2 = -4$, we have

$$e^{-4t} = k_0(t) - 4k_1(t) \tag{11-55}$$

Solving Eqs. (11-54) and (11-55) simultaneously, we get

$$k_0(t) = 4e^{-3t} - 3e^{-4t} \qquad k_1(t) = e^{-3t} - e^{-4t} \tag{11-56}$$

and Eq. (11-52) yields the transition matrix,

$$\begin{bmatrix} 4e^{-3t} - 3e^{-4t} & 0 \\ 0 & 4e^{-3t} - 3e^{-4t} \end{bmatrix} + \begin{bmatrix} -2(e^{-3t} - e^{-4t}) & -(e^{-3t} - e^{-4t}) \\ 2(e^{-3t} - e^{-4t}) & -5(e^{-3t} - e^{-4t}) \end{bmatrix}$$

$$= \begin{bmatrix} 2e^{-3t} - e^{-4t} & -e^{-3t} + e^{-4t} \\ 2e^{-3t} - 2e^{-4t} & -e^{-3t} + 2e^{-4t} \end{bmatrix} \tag{11-57}$$

as before.   □

---

In case of a repeated eigenvalue, we differentiate Eq. (11-53) with respect to $\lambda$ to provide a second relationship between $k_0(t)$ and $k_1(t)$, as illustrated in example 7.

---

### Example 7

Consider

$$\mathbf{A} = \begin{bmatrix} -1 & -2 \\ 2 & -5 \end{bmatrix} \tag{11-58}$$

Here the characteristic equation is $\lambda^2 + 6\lambda + 9 = 0$ and yields a repeated eigenvalue $\lambda_1 = \lambda_2 = -3$. Then Eq. (11-53) provides one relation,

$$e^{-3t} = k_0(t) - 3k_1(t) \tag{11-59}$$

A second relation between $k_0(t)$ and $k_1(t)$ is obtained by differentiation of Eq. (11-53):

$$\frac{d}{d\lambda}(e^{\lambda t})\Big]_{\lambda = -3} = k_1(t) \tag{11-60}$$

that is,

$$te^{-3t} = k_1(t) \tag{11-61}$$

Finally, from Eq. (11-59) we get

$$k_0(t) = e^{-3t} + 3k_1(t) = e^{-3t} + 3te^{-3t} \tag{11-62}$$

With $k_0(t)$ and $k_1(t)$ known, use Eq. (11-52) to obtain the transition matrix $\phi(t)$.

□ **Prob. 11-1**

More than one differentiation with respect to $\lambda$ may be needed. Consider, for example, $\mathbf{A}$ of order three with a cubic characteristic equation. The corresponding Eqs. (11-52) and (11-53) become

$$e^{\mathbf{A}t} = k_0(t)\mathbf{U} + k_1(t)\mathbf{A} + k_2(t)\mathbf{A}^2 \tag{11-63}$$

and

$$e^{\lambda t} = k_0(t) + k_1(t)\lambda + k_2(t)\lambda^2 \tag{11-64}$$

If the three eigenvalues are distinct $\lambda_1 \neq \lambda_2 \neq \lambda_3$, Eq. (11-64) will provide the three simultaneous equations for $k_0(t)$, $k_1(t)$, and $k_2(t)$. If one eigenvalue is repeated, $\lambda_1 = \lambda_2 \neq \lambda_3$, then Eq. (11-64) will provide only two equations (one with $\lambda_1$, another with $\lambda_3$), and we have to differentiate it once before setting $\lambda = \lambda_2$, that is,

$$\frac{d}{d\lambda}e^{\lambda t}\bigg]_{\lambda = \lambda_2} = k_1(t) + 2k_2(t)\lambda_2 \tag{11-65}$$

If all three eigenvalues are identical $\lambda_1 = \lambda_2 = \lambda_3$, then Eq. (11-64) will have to be differentiated twice in order to provide the two additional equations for $k_0(t)$, $k_1(t)$ and $k_2(t)$. We have then

$$e^{\lambda_1 t} = k_0(t) + k_1(t)\lambda_1 + k_2(t)\lambda_1^2 \tag{11-66}$$

$$\frac{d}{d\lambda}e^{\lambda t}\bigg]_{\lambda = \lambda_1} = k_1(t) + 2k_2(t)\lambda_1 \tag{11-67}$$

$$\frac{d^2}{d\lambda^2}e^{\lambda t}\bigg]_{\lambda = \lambda_1} = 2k_2(t) \tag{11-68}$$

Finally, it should be pointed out that this differentiation with respect to $\lambda$ is familiar to us in a similar situation; namely, in the solution of a homogeneous differential equation when the natural frequencies are repeated. For example, the third-order differential equation,

$$\frac{d^3i}{dt^3} + 3\frac{d^2i}{dt^2} + 3\frac{di}{dt} + i = 0 \tag{11-69}$$

has the solution:

$$i(t) = K_1e^{-t} + K_2te^{-t} + K_3t^2e^{-t} \tag{11-70}$$

where the second and third terms, $te^{-t}$ and $t^2e^{-t}$, are obtained (within the arbitrary constant multipliers) by differentiating the first term, $e^{\lambda t}(\lambda = -1)$, with respect to $\lambda$ once and twice, respectively.

## The Zero-State Response

The zero-state response is obtained when the initial state is zero, $\mathbf{x}(0_-) = \mathbf{0}$. Generalizing the result obtained in Eq. (11-13), we write

$$
\begin{aligned}
\mathbf{x}_{zs}(t) &= \int_0^t e^{\mathbf{A}(t-\tau)}\mathbf{Be}(\tau)d\tau \\
&= \int_0^t \boldsymbol{\phi}(t-\tau)\mathbf{Be}(\tau)d\tau
\end{aligned}
\tag{11-71}
$$

as the zero-state state response. The zero-state output response is

$$
\mathbf{r}_{zs}(t) = \mathbf{Cx}_{zs}(t) + \mathbf{De}(t)
\tag{11-72}
$$

where $zs$ stands for "zero-state."

The importance of the state transition matrix is evident again, as it appears in Eqs. (11-71) and (11-72). The integration of a matrix, as in Eq. (11-71), implies integrating every element of that matrix.

---

### Example 8

Let us continue with example 1, using the network in Fig. 10-1. For convenience, we repeat Eqs. (10-4) and (10-5), the state and output equations for that network:

$$
\begin{bmatrix} \dot{v}_c \\ \dot{i}_L \end{bmatrix} = \begin{bmatrix} -\dfrac{1}{R_1 C} & -\dfrac{1}{C} \\ \dfrac{1}{L} & -\dfrac{R_2}{L} \end{bmatrix} \begin{bmatrix} v_c \\ i_L \end{bmatrix} + \begin{bmatrix} \dfrac{1}{C} \\ 0 \end{bmatrix} i_{in}
\tag{11-73}
$$

and

$$
[v_{out}] = [0 \quad R_2] \begin{bmatrix} v_c \\ i_L \end{bmatrix} + [0]i_{in}
\tag{11-74}
$$

With the element values $R_1 = 0.5$ ohm, $R_2 = 2.5$ ohm, $L = 0.5$ henry, and $C = 1$ farad, we have

$$
\mathbf{A} = \begin{bmatrix} -2 & -1 \\ 2 & -5 \end{bmatrix} \quad \mathbf{B} = \begin{bmatrix} 1 \\ 0 \end{bmatrix} \quad \mathbf{C} = [0 \quad 2.5] \quad \mathbf{D} = [0] \tag{11-75}
$$

The state transition matrix is [see Eq. (11-25)]

$$
\boldsymbol{\phi}(t) = e^{\mathbf{A}t} = \begin{bmatrix} 2e^{-3t} - e^{-4t} & -e^{-3t} + e^{-4t} \\ 2e^{-3t} - e^{-4t} & -e^{-3t} + 2e^{-4t} \end{bmatrix}
\tag{11-76}
$$

Therefore, the zero-state state response is

$$\mathbf{x}_{zs}(t) = \int_0^t \left\{ \begin{bmatrix} 2e^{-3(t-\tau)} - e^{-4(t-\tau)} \\ 2e^{-3(t-\tau)} - e^{-4(t-\tau)} \end{bmatrix} i_{in}(\tau) \right\} d\tau \qquad (11\text{-}77)$$

and the zero-state output is

$$\mathbf{r}_{zs}(t) = v_{out}(t) = 5 \int_0^t \{e^{-3(t-\tau)} - e^{-4(t-\tau)}\} i_{in}(\tau)\, d\tau \qquad (11\text{-}78)$$

With a unit step input $i_{in}(t) = u(t)$, Eq. (11-77) yields

$$\mathbf{x}_{zs}(t) = \int_0^t \left\{ \begin{matrix} 2e^{-3(t-\tau)}u(\tau) - e^{-4(t-\tau)}u(t) \\ 2e^{-3(t-\tau)}u(\tau) - e^{-4(t-\tau)}u(t) \end{matrix} \right\} d\tau$$

$$= \begin{bmatrix} \dfrac{2}{3}(1 - e^{-3t}) - \dfrac{1}{4}(1 - e^{-4t}) \\[2ex] \dfrac{2}{3}(1 - e^{-3t}) - \dfrac{1}{2}(1 - e^{-4t}) \end{bmatrix} u(t) \qquad (11\text{-}79)$$

and Eq. (11-78) becomes

$$\mathbf{r}_{zs}(t) = v_{out}(t) = \frac{5}{3}(1 - e^{-3t}) - \frac{5}{4}(1 - e^{-4t}) \qquad (11\text{-}80)$$

The present analysis, then, gives a complete picture of the network's behavior due to the external excitations only.     ☐   ***Probs. 11-2***
***through***
***11-8***

In summary, the *complete response* is the sum of the zero-input response and the zero-state response. The basic analysis involves the computation of the state transition matrix $e^{\mathbf{A}t}$; from this, any output can be obtained by simple operations. The outstanding feature of this method of analysis is that it gives complete information on the behavior of the network in terms of the energy storing elements, as well as the desired outputs.

## 11-3   FREQUENCY DOMAIN SOLUTION OF THE STATE EQUATIONS

So far we have discussed at some length the time domain solutions of the state equations. Throughout that discussion the important features, such as zero-input and zero-state responses, convolution (superposition), etc., were salient. We are now going to consider the Laplace transform approach. Although this approach may seem repetitive at times, it will be very instructive nonetheless: Many relations between time-domain and frequency-domain functions will become clear. Review here the introductory example, section 11-1.

## The Scalar State Equation

The Laplace transform of Eq. (11-6) yields

$$sX(s) - x(0_-) - aX(s) = bE(s) \tag{11-81}$$

Solving for $X(s)$ we obtain

$$X(s) = \frac{x(0_-)}{s-a} + \frac{bE(s)}{s-a} \tag{11-82}$$

The inversion of Eq. (11-82) is straightforward,

$$x(t) = e^{at}x(0_-) + e^{at} * be(t) \tag{11-83}$$

exactly as in Eq. (11-10) (with $t_0 = 0$). The first term on the right-hand side of Eq. (11-83) is the zero-input state response $x_{zi}(t)$, the homogeneous solution; the state transition matrix is here the scalar function $e^{at}$. The second term on the right-hand side is the zero-state state response, $x_{zs}(t)$, obtained by convolution.

## The Homogeneous State Equation: Evaluation of $\phi(t)$ by Laplace Transform

Taking the Laplace transform of

$$\dot{\mathbf{x}} = \mathbf{A}\mathbf{x} \tag{11-84}$$

we obtain

$$s\mathbf{X}(s) - \mathbf{x}(0_-) = \mathbf{A}\mathbf{X}(s) \tag{11-85}$$

This is a set of linear algebraic equations for $X_1(s)$, $X_2(s),\ldots$, the components of $\mathbf{X}(s)$. Equation (11-85) can be written as follows:

$$(s\mathbf{U} - \mathbf{A})\mathbf{X}(s) = \mathbf{x}(0_-) \tag{11-86}$$

Solving for $\mathbf{X}(s)$, we have

$$\mathbf{X}(s) = (s\mathbf{U} - \mathbf{A})^{-1}\mathbf{x}(0_-) \tag{11-87}$$

and by comparison with Eq. (11-15) or (11-21) we have the important result

$$\mathbf{\Phi}(s) = \mathscr{L}\phi(t) = (s\mathbf{U} - \mathbf{A})^{-1} \tag{11-88}$$

or,

$$\phi(t) = \mathscr{L}^{-1}(s\mathbf{U} - \mathbf{A})^{-1} \tag{11-89}$$

which provides a method for computing the state transition matrix via the Laplace transformation.

## Example 9

Consider again the introductory example and example 1 of the previous section. With the element values, matrix $\mathbf{A}$ is given as:

$$\mathbf{A} = \begin{bmatrix} -2 & -1 \\ 2 & -5 \end{bmatrix} \tag{11-90}$$

$$\therefore s\mathbf{U} - \mathbf{A} = \begin{bmatrix} s + 2 & 1 \\ -2 & s + 5 \end{bmatrix} \tag{11-91}$$

and[†]

$$(s\mathbf{U} - \mathbf{A})^{-1} = \frac{1}{\det (s\mathbf{U} - \mathbf{A})} \operatorname{Adj} (s\mathbf{U} - \mathbf{A}) \tag{11-92}$$

$$= \frac{1}{s^2 + 7s + 12} \begin{bmatrix} s + 5 & -1 \\ 2 & s + 2 \end{bmatrix}$$

$$= \begin{bmatrix} \dfrac{s + 5}{(s + 3)(s + 4)} & \dfrac{-1}{(s + 3)(s + 4)} \\ \dfrac{2}{(s + 3)(s + 4)} & \dfrac{s + 2}{(s + 3)(s + 4)} \end{bmatrix} \tag{11-93}$$

Finally

$$\boldsymbol{\phi}(t) = \mathcal{L}^{-1}(s\mathbf{U} - \mathbf{A})^{-1} = \mathcal{L}^{-1} \begin{bmatrix} \dfrac{s + 5}{(s + 3)(s + 4)} & \dfrac{-1}{(s + 3)(s + 4)} \\ \dfrac{2}{(s + 3)(s + 4)} & \dfrac{s + 2}{(s + 3)(s + 4)} \end{bmatrix}$$

$$= \begin{bmatrix} 2e^{-3t} - e^{-4t} & -e^{-3t} + e^{-4t} \\ 2e^{-3t} - 2e^{-4t} & -e^{-3t} + 2e^{-4t} \end{bmatrix} \tag{11-94}$$

which checks with Eq. (11-25). The characteristic polynomial, $\det (s\mathbf{U} - \mathbf{A})$, appears in the denominator of $\boldsymbol{\Phi}(s)$, and the natural frequencies are therefore the poles of $\boldsymbol{\Phi}(s)$.  ☐  **Prob. 11-9**

## Partial Fraction Expansion of $\boldsymbol{\Phi}(s)$

The previous example suggests that there may be an extension of the partial fraction expansion method, used for a scalar rational function $F(s)$, to the matrix function $\boldsymbol{\Phi}(s)$. Indeed, there is an analogous expansion, best illustrated by an example.

[†] See Appendix A.

## Example 10

Let **A** be given as

$$\mathbf{A} = \begin{bmatrix} 0 & 1 & 0 \\ 0 & 0 & -1 \\ -6 & -11 & -6 \end{bmatrix} \tag{11-95}$$

Then

$$\mathbf{\Phi}(s) = (s\mathbf{U} - \mathbf{A})^{-1} = \begin{bmatrix} s & -1 & 0 \\ 0 & s & 1 \\ 6 & 11 & s+6 \end{bmatrix}^{-1} \tag{11-96}$$

$$= \frac{\text{Adj}\,(s\mathbf{U} - \mathbf{A})}{(s+1)(s+2)(s+3)}$$

Now assume the expansion

$$\frac{\text{Adj}\,(s\mathbf{U} - \mathbf{A})}{(s+1)(s+2)(s+3)} = \frac{\mathbf{K}_1}{s+1} + \frac{\mathbf{K}_2}{s+2} + \frac{\mathbf{K}_3}{s+3} \tag{11-97}$$

where $\mathbf{K}_1$, $\mathbf{K}_2$ and $\mathbf{K}_3$ are constant, unknown $(3 \times 3)$ residue matrices. To find $\mathbf{K}_1$, multiply both sides of Eq. (11-97) by $(s+1)$, then set $s = -1$:

$$\mathbf{K}_1 = \frac{\text{Adj}\,(s\mathbf{U} - \mathbf{A})}{(s+2)(s+3)}\Bigg]_{s=-1} \tag{11-98}$$

Similarly,

$$\mathbf{K}_2 = \frac{\text{Adj}\,(s\mathbf{U} - \mathbf{A})}{(s+1)(s+3)}\Bigg]_{s=-2} \tag{11-99}$$

and

$$\mathbf{K}_3 = \frac{\text{Adj}\,(s\mathbf{U} - \mathbf{A})}{(s+1)(s+2)}\Bigg]_{s=-3} \tag{11-100}$$

in complete analogy to the expansion of a scalar rational function. The computations yield

$$\mathbf{K}_1 = \begin{bmatrix} 3 & 2.5 & 0.5 \\ -3 & -2.5 & -0.5 \\ 3 & 2.5 & 0.5 \end{bmatrix} \quad \mathbf{K}_2 = \begin{bmatrix} -3 & -4 & -1 \\ 6 & 8 & 2 \\ -12 & -16 & -4 \end{bmatrix}$$

$$\mathbf{K}_3 = \begin{bmatrix} 1 & 1.5 & 0.5 \\ -3 & -4.5 & -1.5 \\ 9 & 13.5 & 4.5 \end{bmatrix} \tag{11-101}$$

and so

$$\boldsymbol{\phi}(t) = \mathbf{K}_1 e^{-t} + \mathbf{K}_2 e^{-2t} + \mathbf{K}_3 e^{-3t} \tag{11-102}$$

It can be shown that[1]

$$\sum \mathbf{K}_k = \mathbf{U} \tag{11-103}$$

that is, the sum of all the residue matrices is the unit matrix. This important result can serve as a check on the correctness of the computations, or else may be used to evaluate one residue matrix when the others are already known.

□  **Prob. 11-10**

## Input–Output Relations: The Matrix Network Function

The Laplace transform of the state equations, Eqs. (10-6) and (10-7) yields

$$s\mathbf{X}(s) - \mathbf{x}(0_-) = \mathbf{A}\mathbf{X}(s) + \mathbf{B}\mathbf{E}(s) \tag{11-104}$$

$$\mathbf{R}(s) = \mathbf{C}\mathbf{X}(s) + \mathbf{D}\mathbf{E}(s) \tag{11-105}$$

where $\mathbf{x}(0_-)$ is the initial state vector, the given initial conditions at $t = 0_-$.

The elimination of $\mathbf{X}(s)$—a purely algebraic step—from these equations gives the input-output relation, i.e., the complete output response is

$$\mathbf{R}(s) = [\mathbf{C}(s\mathbf{U} - \mathbf{A})^{-1}\mathbf{B} + \mathbf{D}]\mathbf{E}(s) + \mathbf{C}(s\mathbf{U} - \mathbf{A})^{-1}\mathbf{x}(0_-) \tag{11-106}$$

The first term on the right-hand side of Eq. (11-106) is the (transform) zero-state output, while the second term is the zero-input output.

The classical definition of a matrix network function $\mathbf{H}(s)$ relates the zero-state response $\mathbf{R}_{zs}(s)$ to the excitation $\mathbf{E}(s)$; that is

$$\mathbf{R}_{zs}(s) = \mathbf{H}(s)\mathbf{E}(s) \tag{11-107}$$

Therefore, Eqs. (11-106) and (11-107) provide the matrix network function as

$$\mathbf{H}(s) = \mathbf{C}(s\mathbf{U} - \mathbf{A})^{-1}\mathbf{B} + \mathbf{D} = \mathbf{C}\boldsymbol{\Phi}(s)\mathbf{B} + \mathbf{D} \tag{11-108}$$

## *Example 11*

Consider again the example in Fig. 10-1. The matrices $\mathbf{A}$, $\mathbf{B}$, $\mathbf{C}$, and $\mathbf{D}$ are given in Eq. (11-75). Therefore, we have here

$$\mathbf{C}(s\mathbf{U} - \mathbf{A})^{-1}\mathbf{B} + \mathbf{D} = [0 \quad 2.5] \begin{bmatrix} s + 2 & 1 \\ -2 & s + 5 \end{bmatrix}^{-1} \begin{bmatrix} 1 \\ 0 \end{bmatrix} + [0]$$

$$= \frac{5}{s^2 + 7s + 12}$$

which is, in this example, the (scalar) transfer impedance $V_{out}(s)/I_{in}(s)$.

□  **Probs. 11-11**
**11-12**
**11-13**
**11-14**

As expected, the poles of $\mathbf{H}(s)$ are eigenvalues of $\mathbf{A}$. The converse, however, is not necessarily true: Some eigenvalues of $\mathbf{A}$ may not appear as poles of $\mathbf{H}(s)$. This is illustrated in the next example.

---

### Example 12

Consider the network shown in Fig. 11-1. The state equation is (verify!)

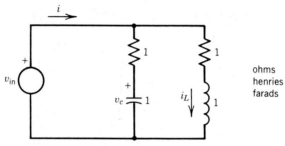

**FIGURE 11-1.** Example 12.

$$\begin{bmatrix} \dot{v}_c \\ \dot{i}_L \end{bmatrix} = \begin{bmatrix} -1 & 0 \\ 0 & -1 \end{bmatrix} \begin{bmatrix} v_c \\ i_L \end{bmatrix} + \begin{bmatrix} 1 \\ 1 \end{bmatrix} v_{in}$$

and if the output is $\mathbf{r} = i$ then we have

$$\mathbf{r} = [-1 \quad 1] \begin{bmatrix} v_c \\ i_L \end{bmatrix} + [1] v_{in}$$

with the matrices $\mathbf{A}, \mathbf{B}, \mathbf{C},$ and $\mathbf{D}$ clearly identified. The characteristic equation of $\mathbf{A}$ is

$$\begin{vmatrix} \lambda + 1 & 0 \\ 0 & \lambda + 1 \end{vmatrix} = \lambda^2 + 2\lambda + 1 = 0$$

giving a multiple eigenvalue $\lambda_1 = \lambda_2 = -1$. If we compute the network function according to Eq. (11-108), we get here $Y_{dp}(s)$, the driving-point admittance,

$$\mathbf{C}(s\mathbf{U} - \mathbf{A})^{-1}\mathbf{B} + \mathbf{D} = [-1 \quad 1] \begin{bmatrix} s + 1 & 0 \\ 0 & s + 1 \end{bmatrix}^{-1} \begin{bmatrix} 1 \\ 1 \end{bmatrix} + 1 = 1$$

Here, then, an eigenvalue of $\mathbf{A}$ does not appear as a pole of $\mathbf{H}(s)$.

☐ **Probs. 11-15**
**11-16**

---

## *11-4 SIMULATION DIAGRAMS

As mentioned in Chapter 10, state equations are very suitable for analog computer calculations. Let us discuss briefly several simulation diagrams associated with these equations.

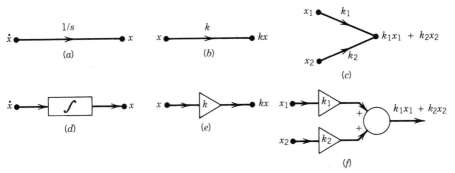

**FIGURE 11-2.** Basic operations of an analog computer.

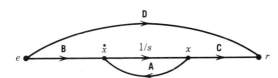

**FIGURE 11-3.** Simulation diagram for the state equations.

Three basic operations of an analog computer are: (a) integration, (b) amplification (multiplication by a constant), and (c) addition. These are shown symbolically in Fig. 11-2a, b, and c, respectively, using the notation of signal flow graphs.[†] Another notation, using "boxes," is shown in (d), (e), and (f). The actual implementation of an integrator, and amplifier, and an adder can be done with operational amplifier circuits, as we will see in Appendix B.

The simulation of the state equations, Eqs. (10-6) and (10-7), is shown in Fig. 11-3. Here the nodes must be considered as having several "components." For example, the node $e$ represents actually $k$ inputs, $e_1, e_2, \ldots, e_k$, and node $x$ represents $n$ states, $x_1, x_2, \ldots, x_n$. Similarly, a branch represents several component branches and may be thought of as a "cable" containing many individual "wires," each with its own transmittance. For a specific network, such "cables" can, of course, be shown in detail.

---

## Example 13

Obtain the state equations and draw the simulation diagram for a network characterized by the differential equation,

$$\frac{dr(t)}{dt} + br(t) = e(t) \qquad (11\text{-}109)$$

[†] See Appendix D.

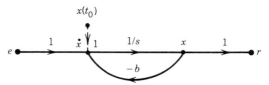

**FIGURE 11-4.** Example 13.

The order of complexity of the network is unity, hence the state vector has one element; let us choose it as

$$\mathbf{x} = x = r \tag{11-110}$$

Then from Eq. (11-109), we have

$$\dot{x} = \dot{r} = -br + e = -bx + e \tag{11-111}$$

which is the state equation. Here

$$\mathbf{A} = -a \qquad \mathbf{B} = 1 \tag{11-112}$$

The output equation is Eq. (11-110), yielding

$$\mathbf{C} = 1 \qquad \mathbf{D} = 0 \tag{11-113}$$

The simulation diagram is shown in Fig. 11-4. Notice that the state variable is the output of the integrator. Also, the initial state $x(t_0)$ must be given, and it is treated as a source shown in dotted lines.[†]    □

## Example 14

Draw the simulation diagram for example 1 in Fig. 10-1. Its matrices are given by Eq. (11-75), repeated here:

$$\dot{\mathbf{x}} = \begin{bmatrix} -2 & -1 \\ 2 & -5 \end{bmatrix} \mathbf{x} + \begin{bmatrix} 1 \\ 0 \end{bmatrix} e \tag{11-114}$$

$$\mathbf{r} = \begin{bmatrix} 0 & 2.5 \end{bmatrix} \mathbf{x} + \begin{bmatrix} 0 \end{bmatrix} e \tag{11-115}$$

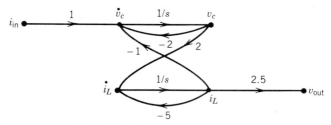

**FIGURE 11-5.** Example 14.

[†] Sources due to initial states are omitted for clarity in all subsequent diagrams.

with $\mathbf{e} = i_{in}$, $x_1 = v_c$, $x_2 = i_L$, $\mathbf{r} = v_{out}$. A simulation diagram is shown in Fig. 11-5. Using Mason's formula (see Appendix D), we can obtain the transfer function as follows:

$$G_1 = (1)\left(\frac{1}{s}\right)(2)\left(\frac{1}{s}\right)\left(2\frac{1}{2}\right) = \frac{5}{s^2} \qquad (11\text{-}116)$$

$$G_2 = G_3 = \cdots = 0, \qquad (11\text{-}117)$$

$$\sum P_{m1} = -\frac{2}{s} - \frac{5}{s} - \left(\frac{1}{s}\right)(2)\left(\frac{1}{s}\right) = -\frac{7}{s} - \frac{2}{s^2} \qquad (11\text{-}118)$$

$$\sum P_{m2} = \left(-\frac{2}{s}\right)\left(-\frac{5}{s}\right) = \frac{10}{s^2} \qquad (11\text{-}119)$$

$$\Delta_1 = \Delta_2 = 1 \qquad (11\text{-}120)$$

$$\therefore G = \frac{\dfrac{5}{s^2}}{1 + \dfrac{7}{s} + \dfrac{2}{s^2} + \dfrac{10}{s^2}} = \frac{5}{s^2 + 7s + 12} \qquad (11\text{-}121)$$

as before. □

The previous example suggests a method of using Mason's formula in order to construct a simulation diagram from a given transfer (gain) function. The following example illustrates this approach.

**Example 15**

Let

$$\frac{R(s)}{E(s)} = \frac{K}{s^3 + 2s^2 + 7s + 3} \qquad (11\text{-}122a)$$

Dividing by $s^3$, we obtain

$$\frac{R(s)}{E(s)} = \frac{\dfrac{K}{s^3}}{1 + \dfrac{2}{s} + \dfrac{7}{s} + \dfrac{3}{s^3}} = \frac{\dfrac{K}{s^3}}{1 - \left(-\dfrac{2}{s} - \dfrac{7}{s^2} - \dfrac{3}{s^3}\right)} \qquad (11\text{-}122b)$$

and by comparison with Mason's formula, we can consider the numerator $K/s^3$ as the gain of the forward path, and the denominator as one-minus-the-sum of the various loop gains. We have therefore three state variables (the order of complexity of the network is 3), shown in Fig. 11-6 as the outputs of three integrators; the three loops have the required gains $-2/s$, $-7/s^2$ and $-3/s^3$, respectively.

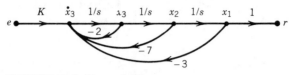

**FIGURE 11-6.** Example 15.                                                        ☐

## Simulation Diagrams for Higher-Order Networks, Scalar $H(s)$

Consider next a generalization of a second-order network. Let the scalar network function be

$$\frac{R(s)}{E(s)} = H(s) = \frac{K}{s^2 + b_1 s + b_2} \tag{11-123}$$

that is,

$$(s^2 + b_1 s + b_2)R(s) = KE(s) \tag{11-124}$$

Since $H(s)$ is defined for zero initial conditions, Eq. (11-124) is simply the Laplace transform of the differential equation

$$\frac{d^2 r(t)}{dt^2} + b_1 \frac{dr(t)}{dt} + b_2 r(t) = Ke(t) \tag{11-125}$$

To obtain the state equation corresponding to Eq. (11-125), let the state vector be

$$\mathbf{x} = \begin{bmatrix} x_1 \\ x_2 \end{bmatrix} = \begin{bmatrix} r \\ \dot{r} \end{bmatrix} \tag{11-126}$$

where the dot indicates differentiation with respect to time, $\dot{r} = dr/dt$. We have therefore from Eqs. (11-126) and (11-125)

$$\dot{x}_1 = x_2 \tag{11-127}$$

$$\dot{x} = \ddot{r} = -b_1\dot{r} - b_2 r + Ke = -b_1 x_2 - b_2 x_1 + Ke \tag{11-128}$$

These are the state equations

$$\begin{bmatrix} \dot{x}_1 \\ \dot{x}_2 \end{bmatrix} = \begin{bmatrix} 0 & 1 \\ -b_2 & -b_1 \end{bmatrix}\begin{bmatrix} x_1 \\ x_2 \end{bmatrix} + \begin{bmatrix} 0 \\ K \end{bmatrix} e \tag{11-129}$$

and

$$r = \begin{bmatrix} 1 & 0 \end{bmatrix}\begin{bmatrix} x_1 \\ x_2 \end{bmatrix} + \begin{bmatrix} 0 \end{bmatrix} e \tag{11-130}$$

with the matrices

$$\mathbf{A} = \begin{bmatrix} 0 & 1 \\ -b_2 & -b_1 \end{bmatrix} \quad \mathbf{B} = \begin{bmatrix} 0 \\ K \end{bmatrix} \quad \mathbf{C} = \begin{bmatrix} 1 & 0 \end{bmatrix} \quad \mathbf{D} = \begin{bmatrix} 0 \end{bmatrix} \tag{11-131}$$

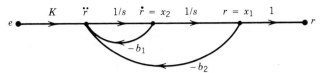

**FIGURE 11-7.** Simulation diagram for Eq. (11-123).

A simulation diagram is shown in Fig. 11-7, where the state variables are the outputs of the integrators.

Let us now consider an $n$th order network function with no finite zeros, i.e.,

$$\frac{R(s)}{E(s)} = H(s) = \frac{K}{s^n + b_1 s^{n-1} + b_2 s^{n-2} + \cdots + b_{n-1}s + b_n} \qquad (11\text{-}132)$$

Again, the corresponding differential equation (with zero initial conditions) is

$$\frac{d^n r}{dt^n} + b_1 \frac{d^{n-1}r}{dt^{n-1}} + b_2 \frac{d^{n-2}r}{dt^{n-2}} + \cdots + b_{n-1}\frac{dr}{dt} + b_n r = Ke \qquad (11\text{-}133)$$

An appropriate choice for state variables would be here

$$x_1 = r$$

$$x_2 = \frac{dr}{dt}$$

$$\vdots \qquad\qquad (11\text{-}134)$$

$$x_n = \frac{d^{n-1}r}{dt^{n-1}}$$

With this choice, we have

$$\dot{x}_1 = \dot{r} = x_2$$
$$\dot{x}_2 = \ddot{r} = x_3 \qquad\qquad (11\text{-}135)$$
$$\vdots$$

and Eq. (11-133) provides the relationship

$$\frac{d^n r}{dt^n} = \dot{x}_n = -b_n x_1 - b_{n-1}x_2 - \cdots - b_1 x_n + Ke \qquad (11\text{-}136)$$

Equations (11-135) and (11-136) are the state equation

$$\dot{\mathbf{x}} = \begin{bmatrix} 0 & 1 & 0 & \cdots & 0 \\ 0 & 0 & 1 & \cdots & 0 \\ & & \vdots & & \\ -b_n & b_{n-1} & -b_{n-2} & \cdots & -b_1 \end{bmatrix} \mathbf{x} + \begin{bmatrix} 0 \\ 0 \\ \vdots \\ K \end{bmatrix} e \qquad (11\text{-}137)$$

while the output equation is simply $r = x_1$, that is,

$$r = [1 \quad 0 \quad 0 \quad \cdots \quad 0]\mathbf{x} + [0]e \qquad (11\text{-}138)$$

The matrices $\mathbf{A}$, $\mathbf{B}$, $\mathbf{C}$, and $\mathbf{D}$ are clearly identified in Eqs. (11-137) and (11-138).

Consider next the general network function where the constant multiplier $K$ has been absorbed in the numerator,

$$H(s) = \frac{a_0 s^m + a_1 s^{m-1} + a_2 s^{m-2} + \cdots + a_{m-1} s + a_m}{s^n + b_1 s^{n-1} + b_2 s^{n-2} + \cdots + b_{n-1} s + b_n} \qquad (11\text{-}139)$$

With all initial conditions zero, this may be identified with the Laplace transform of the differential equation

$$\frac{d^n r}{dt^n} + b_1 \frac{d^{n-1} r}{dt^{n-1}} + \cdots + b_{n-1} \frac{dr}{dt} + b_n r$$

$$= a_0 \frac{d^m e}{dt^m} + a_1 \frac{d^{m-1} e}{dt^{m-1}} + \cdots + a_{m-1} \frac{de}{dt} + a_m e \qquad (11\text{-}140)$$

From our knowledge of physical, realizable networks, $m$ cannot exceed $n$ by more than 1. If $m = n$, the response contains an impulse. Let us consider the practical cases for which $m \le n$.

The difficulty that arises in trying to fit Eq. (11-139) into a diagram similar to Figs. 11-3, 11-5, or 11-6 is the presence of the *derivatives* of the input. Since differentiators are not allowed, we must modify our approach somewhat.

---

## *Example 16*

Consider

$$\frac{d^2 r}{dt^2} + b_1 \frac{dr}{dt} + b_2 r = a_0 \frac{d^2 e}{dt^2} + a_1 \frac{de}{dt} + a_2 e \qquad (11\text{-}141)$$

Let us try and modify Fig. 11-6 to accommodate the terms containing the first and second derivatives of $e$. This is done by *assuming* a simulation diagram (as shown in Fig. 11-8) which contains, in addition to the basic structure of Fig. 11-6, branches

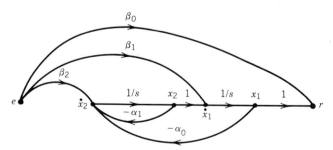

**FIGURE 11-8.** Simulation for Eq. (11-141).

from the input to three nodes in the diagram. The $\alpha$'s and $\beta$'s are to be determined to satisfy Eq. (11-141). From the diagram we write

$$r = x_1 + \beta_0 e \tag{11-142}$$

$$\dot{x}_1 = x_2 + \beta_1 e \tag{11-143}$$

$$\dot{x}_2 = -(\alpha_0 x_1 + \alpha_1 x_2) + \beta_2 e \tag{11-144}$$

Differentiation of $r$ once yields

$$\dot{r} = \dot{x}_1 + \beta_0 \dot{e} \tag{11-145}$$

And using Eq. (11-143), we get

$$\dot{r} = x_2 + \beta_1 e + \beta_0 \dot{e} \tag{11-146}$$

Similarly we can write

$$\ddot{r} = \dot{x}_2 + \beta_1 \dot{e} + \beta_0 \ddot{e} \tag{11-147}$$

and with Eq. (11-144) the result is

$$\ddot{r} = -(\alpha_0 x_1 + \alpha_1 x_2) + \beta_2 e + \beta_1 \dot{e} + \beta_0 \ddot{e} \tag{11-148}$$

Now substitute these expressions into Eq. (11-141) and compare terms on both sides. We obtain:

$$\begin{aligned}
\alpha_0 &= b_2 \\
\alpha_1 &= b_1 \\
\beta_0 &= a_0 \\
\beta_1 &= a_1 - b_1 \beta_0 \\
\beta_2 &= a_2 - b_1 \beta_1 - b_2 \beta_0
\end{aligned} \tag{11-149}$$

□ **Probs. 11-20**
**11-21**
**11-22**
**11-23**

---

This approach can be generalized now for Eq. (11-140), with $m = n$. The simulation diagram is shown in Fig. 11-9. From this diagram we write immediately

$$r = x_1 + \beta_0 e \tag{11-150}$$

$$\dot{x}_p = x_{p+1} + \beta_p e \qquad p = 1, 2, \ldots, m - 1 \tag{11-151}$$

$$\dot{x}_n = -(\alpha_0 x_1 + \alpha_1 x_2 + \cdots + \alpha_{n-1} x_n) + \beta_n e \tag{11-152}$$

As before, we differentiate Eq. (11-150) once to obtain $\dot{r}$, then substitute in it Eq. (11-151) for $\dot{x}_1$ to get

$$\dot{r} = x_2 + \beta_1 + \beta_0 \dot{e} \tag{11-153}$$

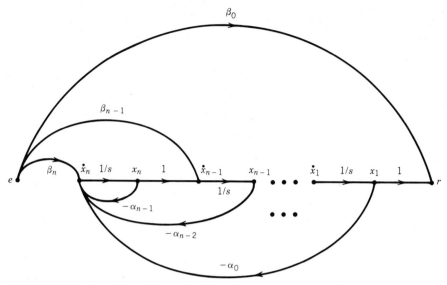

**FIGURE 11-9.** Simulation diagram for Eq. (11-140), with $m = n$.

The same method is followed to obtain the higher derivatives of $r$.

$$\ddot{r} = \dot{x}_2 + \beta_1 \dot{e} + \beta_0 \ddot{e} = x_3 + \beta_2 e + \beta_1 \dot{e} + \beta_0 \ddot{e} \qquad (11\text{-}154)$$

$$\vdots$$

$$\frac{d^{n-1} r}{dt^{n-1}} = x_n + \beta_{n-1} e + \beta_{n-2} \dot{e} + \cdots + \beta_0 \frac{d^{n-1} e}{dt^{n-1}}, \qquad (11\text{-}155)$$

$$\frac{d^n r}{dt^n} = -(\alpha_0 x_1 + \alpha_1 x_2 + \cdots + \alpha_{n-1} x_n) + \beta_n e + \beta_{n-1} \dot{e} + \cdots + \beta_0 \frac{d^n e}{dt^n} \qquad (11\text{-}156)$$

Substitution into Eq. (11-140) and comparison of terms yields

$$\alpha_k = b_{n-k} \qquad k = 0, 1, 2, \ldots, n - 1 \qquad (11\text{-}157)$$

$$\beta_0 = a_0$$

$$\beta_1 = a_1 - b_1 \beta_0$$

$$\beta_2 = a_2 - b_1 \beta_1 - b_2 \beta_0$$

$$\vdots$$

$$\beta_n = a_n - b_1 \beta_{n-1} - b_2 \beta_{n-2} - \cdots - b_n \beta_0 \qquad (11\text{-}158)$$

Equation (11-157) gives all the $\alpha$'s. The second one gives $\beta_0$. Each subsequent relation gives a new $\beta$ in terms of known $a$'s and $b$'s and in terms of previously calculated $\beta$'s. Alternately, Eq. (11-158) may be written in matrix form as

$$
\begin{bmatrix} a_0 \\ a_1 \\ a_2 \\ \vdots \\ a_n \end{bmatrix} = \begin{bmatrix} 1 & 0 & 0 & \cdots & 0 \\ b_1 & 1 & 0 & \cdots & 0 \\ b_2 & b_1 & 1 & \cdots & 0 \\ & \cdots & & & \cdots \\ b_n & b_{n-1} & \cdots & b_1 & 1 \end{bmatrix} \begin{bmatrix} \beta_0 \\ \beta_1 \\ \beta_2 \\ \vdots \\ \beta_n \end{bmatrix}
\tag{11-159}
$$

and then, by inversion, we can solve for the $\beta$'s,

$$
\begin{bmatrix} \beta_0 \\ \beta_1 \\ \beta_2 \\ \vdots \\ \beta_n \end{bmatrix} = \begin{bmatrix} 1 & 0 & 0 & \cdots & 0 \\ b_1 & 1 & 0 & \cdots & 0 \\ b_2 & b_1 & 1 & \cdots & 0 \\ & \cdots & & & \\ b_n & b_{n-1} & & b_1 & 1 \end{bmatrix}^{-1} \begin{bmatrix} a_0 \\ a_1 \\ a_2 \\ \vdots \\ a_n \end{bmatrix}
\tag{11-160}
$$

Finally, if we compare Fig. 11-9 and its equation, Eq. (11-145), with Fig. 11-3, we conclude that for this choice of state variables, the matrices **A**, **B**, **C**, and **D** are:

$$
\mathbf{A} = \begin{bmatrix} 0 & 1 & 0 & 0 & \cdots & 0 \\ 0 & 0 & 1 & 0 & \cdots & 0 \\ & \cdots & & & & \cdots \\ -b_n & -b_{n-1} & -b_{n-2} & \cdots & \cdots & -b_1 \end{bmatrix} \qquad \mathbf{B} = \begin{bmatrix} \beta_1 \\ \beta_2 \\ \vdots \\ \beta_n \end{bmatrix}
\tag{11-161}
$$

$$
\mathbf{C} = [1 \quad 0 \quad 0 \quad 0 \quad \cdots \quad 0], \qquad \mathbf{D} = \beta_0
$$

Other simulation diagrams, with different matrices, are also available.

---

## Example 17

Obtain a simulation diagram and the state equations for

$$
\frac{d^3 r}{dt^3} + 6\frac{d^2 r}{dt^2} + 11\frac{dr}{dt} + 6r = \frac{d^3 e}{dt^3} + 9\frac{d^2 e}{dt^2} + 2\frac{de}{dt} + 4e
\tag{11-162}
$$

Here we have $b_1 = 6$, $b_2 = 11$, $b_3 = 6$, and $a_0 = 1$, $a_1 = 9$, $a_2 = 2$, $a_3 = 4$. From Eq. (11-157) we obtain

$$
\alpha_0 = b_3 = 6, \; \alpha_1 = b_2 = 11, \; \alpha_2 = b_1 = 6
\tag{11-163}
$$

and Eq. (11-158) yields

$$
\begin{aligned}
\beta_0 &= a_0 = 1 \\
\beta_1 &= a_1 - b_1\beta_0 = 9 - 6 = 3 \\
\beta_2 &= a_2 - b_1\beta_1 - b_2\beta_0 = 2 - 18 - 11 = -27 \\
\beta_3 &= a_3 - b_1\beta_2 - b_2\beta_1 - b_3\beta_0 = 4 - (-162) - 33 - 6 = 127
\end{aligned}
\tag{11-164}
$$

The simulation diagram corresponding to these values can be drawn easily. The state equations are obtained with Eq. (11-161) as

$$\dot{\mathbf{x}} = \begin{bmatrix} 0 & 1 & 0 \\ 0 & 0 & 1 \\ -6 & -11 & -6 \end{bmatrix} \mathbf{x} + \begin{bmatrix} 3 \\ -27 \\ 127 \end{bmatrix} e \tag{11-165}$$

$$\mathbf{r} = [1 \quad 0 \quad 0]\mathbf{x} + [1]e \qquad \square \quad \textbf{\textit{Probs. 11-24}}$$
$$\textbf{\textit{11-25}}$$

## Parallel Decomposition by Partial Fractions

Earlier we discussed the derivation of a simulation diagram from the network function $H(s)$ by using Mason's formula. Another approach is provided by the partial fraction expansion of $H(s)$, Eq. (11-139). Again, assume that $m \le n$. If the poles of $H(s)$ are all simple, then the partial fraction expansion will be

$$H(s) = A_0 + \frac{A_1}{s - \lambda_1} + \frac{A_2}{s - \lambda_2} + \cdots + \frac{A_n}{s - \lambda_n} \tag{11-166}$$

Consider again Eq. (11-109) and Fig. 11-3: The network function corresponding to that case is

$$\frac{R(s)}{E(s)} = H(s) = \frac{1}{s + b} \tag{11-167}$$

which is essentially the same form as the term $A_k/(s - \lambda_k)$ in our partial fraction expansion. In other words, Fig. 11-3 is the basic building block for our purpose here.

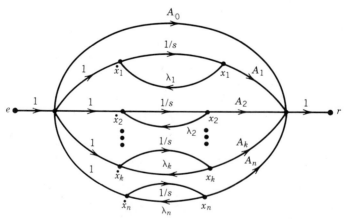

**FIGURE 11-10.** Parallel decomposition by partial fractions (simple eigenvalues).

The response $R(s) = H(s)E(s)$ corresponding to Eq. (11-166) is therefore

$$R(s) = A_0 E(s) + \frac{A_1}{s - \lambda_1} E(s) + \frac{A_2}{s - \lambda_2} E(s) + \cdots + \frac{A_n}{s - \lambda_n} E(s) \quad (11\text{-}168)$$

and a simulation diagram is shown in Fig. 11-10. The state variables are shown as the outputs of the integrators, and we write by inspection

$$\dot{x}_1 = \lambda_1 x_1 + e$$
$$\dot{x}_2 = \lambda_2 x_2 + e$$
$$\vdots \qquad \vdots \qquad\qquad (11\text{-}169)$$
$$\dot{x}_n = \lambda_n x_n + e$$

and

$$r = A_1 x_1 + A_2 x_2 + \cdots + A_n x_n + A_0 e \quad (11\text{-}170)$$

In matrix form these state equations read

$$\dot{\mathbf{x}} = \begin{bmatrix} \lambda_1 & 0 & & \cdots & 0 \\ 0 & \lambda_2 & 0 & \cdots & 0 \\ & \cdots & & \cdots & \\ 0 & & & \ddots & 0 \\ 0 & & & & \lambda_n \end{bmatrix} \mathbf{x} + \begin{bmatrix} 1 \\ 1 \\ \vdots \\ 1 \end{bmatrix} e \quad (11\text{-}171)$$

$$r = \begin{bmatrix} A_1 & A_2 & \cdots & A_n \end{bmatrix} \mathbf{x} + A_0 e \quad (11\text{-}172)$$

Among the advantages of this method, then, is the special form of the matrix **A**: It is diagonal, with the (simple) eigenvalues as its elements. In other words, each state variable is independent of the other state variables, since its equation is uncoupled from the others. Also note that **B** is a column unit matrix and **C** is a row residue matrix.

---

## Example 18

Obtain the state equations and the state transition matrix $\boldsymbol{\phi}(t)$ for

$$H(s) = \frac{2s}{(s + 1)(s + 2)} \quad (11\text{-}173)$$

The partial fraction expansion yields

$$H(s) = \frac{-2}{s + 1} + \frac{4}{s + 2} \quad (11\text{-}174)$$

Therefore Eq. (11-171) reads

$$\dot{\mathbf{x}} = \begin{bmatrix} -1 & 0 \\ 0 & -2 \end{bmatrix} \mathbf{x} + \begin{bmatrix} 1 \\ 1 \end{bmatrix} e \quad (11\text{-}175)$$

$$r = \begin{bmatrix} -2 & 4 \end{bmatrix} \mathbf{x} \quad (11\text{-}176)$$

To find $\boldsymbol{\phi}(t)$ we write first

$$\boldsymbol{\Phi}(s) = (s\mathbf{U} - \mathbf{A})^{-1} = \begin{bmatrix} s+1 & 0 \\ 0 & s+2 \end{bmatrix}^{-1} = \begin{bmatrix} \dfrac{1}{s+1} & 0 \\ 0 & \dfrac{1}{s+2} \end{bmatrix} \qquad (11\text{-}177)$$

recalling that the inverse of a diagonal matrix is obtained by simply inverting every element. Finally

$$\boldsymbol{\phi}(t) = \begin{bmatrix} e^{-t} & 0 \\ 0 & e^{-2t} \end{bmatrix} \qquad (11\text{-}178) \qquad \square \qquad \textbf{\textit{Prob. 11-26}} \atop \textbf{\textit{11-27}}$$

---

With a proper modification, this method is applicable to multiple eigenvalues. Let $\lambda_1$ be an eigenvalue of multiplicity $r$. Then,

$$H(s) = A_0 + \frac{A_{1,r}}{(s-\lambda_1)^r} + \frac{A_{1,r-1}}{(s-\lambda_1)^{r-1}} + \cdots + \frac{A_{1,1}}{(s-\lambda_1)} + \sum \frac{A_k}{s-\lambda_k} \qquad (11\text{-}179)$$

where $\lambda_k$ $(k = 2, 3, \ldots, n)$ are the simple eigenvalues. Since we know how to treat $A_0$ and simple eigenvalues (Fig. 11-10), let us consider only those terms due to $\lambda_1$. That is, let

$$H_1(s) = \frac{A_{1,r}}{(s-\lambda_1)^r} + \frac{A_{1,r-1}}{(s-\lambda_1)^{r-1}} + \cdots + \frac{A_{1,1}}{(s-\lambda_1)} \qquad (11\text{-}180)$$

The response corresponding to $H_1(s)$ is

$$R_1(s) = H_1(s) E(s) = \frac{A_{1,r}}{(s-\lambda_1)^r} E(s) + \frac{A_{1,r-1}}{(s-\lambda_1)^{r-1}} E(s) + \cdots + \frac{A_{1,1}}{(s-\lambda_1)} E(s),$$
$$(11\text{-}181)$$

which can be written as

$$R_1(s) = A_{1,r} X_1(s) + A_{1,r-1} X_2(s) + \cdots + A_{1,1} X_r(s) \qquad (11\text{-}182)$$

where

$$X_1(s) = \frac{1}{(s-\lambda_1)^r} E(s) = \frac{1}{s-\lambda_1} X_2(s)$$

$$X_2(s) = \frac{1}{(s-\lambda_1)^{r-1}} E(s) = \frac{1}{s-\lambda_1} X_2(s)$$

$$\vdots \qquad \vdots \qquad\qquad\qquad\qquad (11\text{-}183)$$

$$X_{r-1}(s) = \frac{1}{(s-\lambda_1)^2} E(s) = \frac{1}{s-\lambda_1} X_r(s)$$

$$X_r(s) = \frac{1}{s-\lambda_1} E(s)$$

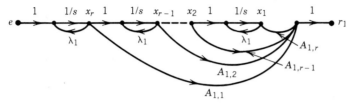

**FIGURE 11-11.** Parallel decomposition for a multiple eigenvalue.

A simulation diagram corresponding to Eqs. (11-180) through (11-183) is shown in Fig. 11-11. The additional branches, needed for $A_0$ and the simple eigenvalues in Eq. (11-179) are not shown. From this diagram, together with Eq. (11-179), we can write for the multiple eigenvalue $\lambda_1$:

$$\dot{x}_1 = \lambda_1 x_1 + x_2$$

$$\dot{x}_2 = \lambda_1 x_2 + x_3$$

$$\vdots \qquad \vdots \tag{11-184}$$

$$\dot{x}_{r-1} = \lambda_1 x_{r-1} + x_r$$

$$\dot{x}_r = \lambda_1 x_r + e$$

while for $\lambda_2, \lambda_3, \ldots, \lambda_n$ we have Eq. (11-169), as usual. Therefore the state equations here are

$$
\dot{\mathbf{x}} = 
\begin{bmatrix}
\dot{x}_1 \\
\dot{x}_2 \\
\vdots \\
\dot{x}_r \\
\vdots \\
\dot{x}_n
\end{bmatrix}
=
\left[
\begin{array}{cccc:ccc}
\lambda_1 & 1 & 0 & \cdots & & & \\
0 & \lambda_1 & 1 & \cdots & & \mathbf{0} & \\
0 & \cdots & \cdots & \lambda_1 & & & \\
\hdashline
& & & & \lambda_2 & & \mathbf{0} \\
& \mathbf{0} & & & & \ddots & \\
& & & & \mathbf{0} & & \lambda_n
\end{array}
\right]
\begin{bmatrix}
x_1 \\
x_2 \\
\vdots \\
x_r \\
\vdots \\
x_n
\end{bmatrix}
+
\begin{bmatrix}
0 \\
0 \\
\vdots \\
1 \\
1 \\
\vdots \\
1
\end{bmatrix}
e \tag{11-185}
$$

and

$$r = [A_{1,r} A_{1,r-1} \cdots A_{1,1} A_2 \cdots A_n]\mathbf{x} + A_0 e \tag{11-186}$$

The matrices **A**, **B**, **C**, and **D** are clearly exhibited in Eqs. (11-185), (11-186). Matrix **A** here has a special form, known as the *Jordan form*.

---

*Example 19*

Let

$$H(s) = \frac{s}{(s+1)(s+3)^2} \tag{11-187}$$

Its partial fraction expansion is

$$H(s) = \frac{3/2}{(s+3)^2} + \frac{1/4}{(s+3)} + \frac{-1/4}{s+1}. \tag{11-188}$$

Therefore,

$$\mathbf{A} = \begin{bmatrix} -3 & 1 & 0 \\ 0 & -3 & 0 \\ 0 & 0 & -1 \end{bmatrix} \quad \mathbf{B} = \begin{bmatrix} 0 \\ 1 \\ 1 \end{bmatrix} \quad \mathbf{C} = \begin{bmatrix} \frac{3}{2} & \frac{1}{4} & -\frac{1}{4} \end{bmatrix} \quad \mathbf{D} = 0$$

□  **Prob. 11-28**

## Chain (Cascade) Decomposition by Factorization

In addition to the decomposition by Mason's formula and by partial fractions, we can decompose a given $H(s)$ into a product of first-order factors, and use basic structures as Fig. 11-4. Let us illustrate by an example.

### Example 20

Consider $H(s)$ as in example 18 previously, and write it as follows:

$$H(s) = \frac{2s}{(s+1)(s+2)} = \left(\frac{s}{s+1}\right)\left(\frac{2}{s+2}\right) \tag{11-189}$$

A simulation diagram for each of the first-order factors is easily drawn. The chain connection of these is the required simulation diagram and is shown in Fig. 11-12. From this diagram we have the state equations,

$$\begin{bmatrix} \dot{x}_1 \\ \dot{x}_2 \end{bmatrix} = \begin{bmatrix} -2 & -1 \\ 0 & -1 \end{bmatrix} \begin{bmatrix} x_1 \\ x_2 \end{bmatrix} + \begin{bmatrix} 1 \\ 1 \end{bmatrix} e \tag{11-190}$$

and

$$r = \begin{bmatrix} 2 & 0 \end{bmatrix} \begin{bmatrix} x_1 \\ x_2 \end{bmatrix} \tag{11-191}$$

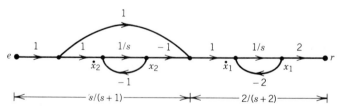

**FIGURE 11-12.** Example 20.

Different grouping of the first-order factors will lead to different simulation diagrams. The main advantage of this method is in the flexibility that it provides in the design of networks.    ☐    ***Prob. 11-29***

Several comments are appropriate at this point about simulation on analog computers. Their accuracy and the range of signals are rather limited (recall the small linear range of op amps!). A digital computer simulation affords a much better accuracy. On the other hand, a simulation on an analog computer is much more interactive: we can adjust continuously the values of the parameters and note the results instantly. A digital computer, by comparison, require new entries for such adjustments and a re-run of the program or a portion of it. However, very good interactive simulation can be done nowadays with improved graphical displays and computational speeds of digital computers.

## PROBLEMS

11-1  Obtain in closed form the transition matrix for

$$\mathbf{A} = \begin{bmatrix} -1 & -2 \\ 2 & -5 \end{bmatrix}$$

by using the Cayley–Hamilton theorem.

11-2  Solve in the time domain the state equation,

$$\begin{bmatrix} \dot{v}_c \\ \dot{i}_L \end{bmatrix} = \begin{bmatrix} -2 & 0 \\ -3 & -1 \end{bmatrix} \begin{bmatrix} v_c \\ i_L \end{bmatrix}$$

given $v_c(0_-)$, $i_L(0_-)$.

11-3  Solve, with zero input, the state equation of the network shown in Fig. 10-14 in the text. The given values are $R_2 = 0.5$ ohm, $C_3 = 1$ farad, $R_5 = 1$ ohm, $L_6 = 2$ henries, $\alpha = 1$, $r_m = 0.5$.

11-4  Obtain in closed form the state transition matrix for each network in problem 10-4.

11-5  Prove that:

a)  $e^{(\mathbf{A} + \mathbf{B})} = e^{\mathbf{A}} e^{\mathbf{B}}$ only if $\mathbf{AB} = \mathbf{BA}$.

b)  $[e^{\mathbf{A}}]^{-1} = e^{-\mathbf{A}}$. *Hint*: Set $\mathbf{A} = -\mathbf{B}$ in (a).

11-6  Prove the following properties of the state transition matrix

a)  $\boldsymbol{\phi}(0) = \mathbf{U}$

b)  $\boldsymbol{\phi}(t_2 - t_0) = \boldsymbol{\phi}(t_2 - t_1)\boldsymbol{\phi}(t_1 - t_0)$

c)  $\boldsymbol{\phi}(t_2 - t_1) = [\boldsymbol{\phi}(t_1 - t_2)]^{-1}$

Interpret each property in term of the transition of the network from one state to another.

11-7 Find the eigenvalues of **A** and then the transition matrix for

a) $\mathbf{A} = \begin{bmatrix} 0 & 2 & 0 \\ 0 & 0 & 2 \\ -12 & -22 & -12 \end{bmatrix}$

b) $\mathbf{A} = \begin{bmatrix} 6 & -2 & 2 \\ 4 & 0 & 2 \\ 2 & -2 & 4 \end{bmatrix}$

11-8 Use the results of problem 11-3 to find the complete state response for the given input $v_1(t) = 10e^{-t}$.

11-9 Use partial fractions to find $\boldsymbol{\phi}(t) = \mathcal{L}^{-1}(s\mathbf{U} - \mathbf{A})^{-1}$ for the matrix **A**:

a) in problem 10-4 (compare with problem 11-4)
b) in problem 10-8
c) in problem 11-1
d) in problem 11-2.

11-10 Prove Eq. (11-103) which states

$$\sum \mathbf{K}_k = \mathbf{U}$$

that is, the sum of the residue matrices in the partial fraction expansion of $\boldsymbol{\Phi}(s)$ equals the unit matrix. *Hint*: Write $\boldsymbol{\Phi}(s)$ in its assumed expansion (as for example in Eq. (11-97), then multiply both sides by the characteristic polynomial det $(s\mathbf{U} - \mathbf{A})$ and equate the coefficients of $s^{n-1}$ on both sides.

11-11 Using Eq. (11-108), find the matrix network function for problem 10-4(a), considering the voltage across the resistor and the current through the inductor as the outputs.

11-12 Repeat problem 11-11 for the network in problem 10-4(c). The output is $v_L$.

11-13 Repeat problem 11-11 for problem 10-12. The output is the current through the resistor. Remember: in defining a network function, only the initial state is set to zero. Dependent sources remain intact!

11-14 Consider the solution obtained in Eq. (11-26) of the text, and let $v_c(0_-) = 1$, $i_L(0_-) = -1$. Write fully the expressions for $v_c(t)$ and for $i_L(t)$, and plot each one versus $t$. Next, consider $(v_c(t), i_L(t))$ as the coordinates of the state vector,

$$\mathbf{x} = \begin{bmatrix} v_c \\ i_L \end{bmatrix}$$

for every $t \geq 0$. These coordinates establish a point on the $v_c$-$i_L$ plane for every $t \geq 0$. As $t$ changes, these points form a curve called the *trajectory* of the state.

The $v_c$-$i_L$ plane is called the *state space*. Compute and plot the trajectory of **x** using $t = 0, 0.1, 0.2, \ldots, 1$.

11-15  Compute and plot the trajectory of **x** for problem 11-14, with

$$\mathbf{x}(0_-) = \begin{bmatrix} 2 \\ 1 \end{bmatrix}$$

11-16  Compute and plot the trajectory of **x** for problem 11-14 with

$$\mathbf{x}(0_-) = \begin{bmatrix} -1 \\ -2 \end{bmatrix}$$

11-17  Draw the simulation diagram for problem 10-4(a). See also problem 11-11.

11-18  Repeat problem 11-17 for problem 10-4(c). See also problem 11-12.

11-19  For each of the simulation diagrams shown, write the state equations. Then, using Mason's formula, obtain the network function.

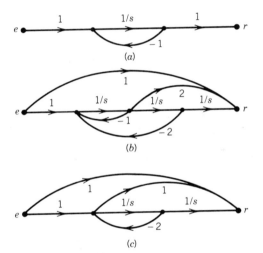

**PROBLEM 11-19.**

11-20  For each of the networks described by the given differential equation, draw a simulation diagram:

a) $\dfrac{d^3r}{dt^3} + 2\dfrac{d^2r}{dt^2} + \dfrac{dr}{dt} + r = 3e$

b) $\dfrac{d^2r}{dt^2} + 5\dfrac{dr}{dt} + 6r = 2e + \dfrac{de}{dt}$

11-21  Obtain a simulation diagram for

$$H(s) = \frac{R(s)}{E(s)} = \frac{s^3 + a_1 s^2 + a_2 s + a_3}{s^4 + b_1 s^3 + b_2 s^2 + b_3 s + b_4}$$

**11-22** Obtain a simulation diagram, the matrices **A**, **B**, **C**, and **D**, and $\Phi(s) = (s\mathbf{U} - \mathbf{A})^{-1}$ for

$$\frac{d^3r}{dt^3} + 3\frac{d^2r}{dt^2} + 2\frac{dr}{dt} + 2r = \frac{de}{dt} - 2e$$

**11-23** Repeat problem 11-22 for

$$2\frac{dr}{dt} + r = 3\frac{de}{dt} + e$$

**11-24** Another possible simulation diagram for Eq. (11-140) is shown here.[2] Follow a derivation based on this diagram and obtain the values of the $\alpha$'s and $\beta$'s in terms of the given coefficients in Eq. (11-140). Also, obtain the state equations, i.e., the matrices **A**, **B**, **C**, and **D**.

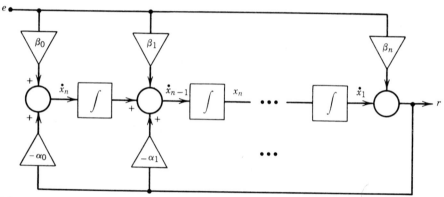

**PROBLEM 11-24.**

**11-25** Consider a zero-input network described by the equation

$$\ddot{r} + b_1\dot{r} + b_2r = 0$$

with the initial conditions $r(0_-)$, $\dot{r}(0_-)$ given. Find the expression for $R(s)$ as a rational function. Then formulate the state equation and draw a simulation diagram.

**11-26** Obtain the state equations, $\phi(t)$, and the simulation diagram by a parallel decomposition (partial fractions) for

$$H(s) = \frac{(s-1)(s-2)}{(s+1)(s+2)(s+3)(s+4)}.$$

**11-27** The diagonal form of **A** obtained by the parallel decomposition (see Eq. (11-168) and following equations) suggests another approach: Let us choose a new state vector $\hat{\mathbf{x}}$, related to **x** by

$$\mathbf{x} = \mathbf{M}\hat{\mathbf{x}}$$

where **M** is a nonsingular, constant matrix yet to be determined. Obtain the state equations for $\hat{\mathbf{x}}$, that is,

$$\dot{\hat{\mathbf{x}}} = \hat{\mathbf{A}}\hat{\mathbf{x}} + \hat{\mathbf{B}}\mathbf{e}$$

$$\mathbf{r} = \hat{\mathbf{C}}\hat{\mathbf{x}} + \hat{\mathbf{D}}\mathbf{e}$$

where $\hat{\mathbf{A}}$ is required to be diagonal.

a) Relate the new matrices $\hat{\mathbf{A}}$, $\hat{\mathbf{B}}$, $\hat{\mathbf{C}}$, and $\hat{\mathbf{D}}$ to **A**, **B**, **C**, **D**, and **M**.

b) Show with an example of your choice that when the eigenvalues of **A** are all simple, matrix **M** may be chosen as

$$\mathbf{M} = [\mathbf{V}_1 \vdots \mathbf{V}_2 \vdots \cdots \vdots \mathbf{V}_n]$$

where $\mathbf{V}_i$ is the column matrix satisfying the equation,

$$(\lambda_i \mathbf{U} - \mathbf{A})\mathbf{V}_i = 0 \qquad i = 1, 2, \ldots, n$$

This $\mathbf{V}_i$ is called the *eigenvector* corresponding to $\lambda_i$.

11-28 Draw a simulation diagram and obtain the state equations for:

a) $H(s) = \dfrac{s^2 + 2s - 3}{(s + 1)^2(s + 2)}$

b) $H(s) = \dfrac{s^4 - 2}{(s + 1)^2(s + 2)^2}$

by parallel decomposition (partial fractions).

11-29 Use the chain decomposition to draw a simulation diagram and state equations for:

a) Problem 11-20(b)
b) Problem 11-28(a)
c) Problem 11-28(b)

# REFERENCES

1. H. M. Power, "Useful relations for partial fraction expansion of proper rational functions and transition matrices," *IEEE Trans. on Education*, vol. E-10, no. 3, September 1967, pp. 179–180.
2. E. R. Davis, "Simulation diagrams for state-space equations," *IEEE Trans. on Education*, vol. E-9, no. 3, September 1966, pp. 173–174.

# STABILITY

The concepts of stability are very important in the analysis and in the design of networks. We shall discuss in this chapter only linear, constant networks, and two of the stability concepts associated with them.

## 12-1  ZERO-INPUT STABILITY AND THE CHARACTERISTIC EQUATION

We say that a network is *zero-input stable*[†] if the zero-input response (in magnitude) approaches zero for $t \to \infty$. This is a rather sensible definition: in the absence of outside inputs, only the initial state excites the network. The response of such a network, if stable, will approach zero as time increases.

From our studies of loop, node, or state variable analysis we recognize that such a zero-input response will be of the following form, obtained from a partial fraction expansion

$$f(t) = \sum_k A_k e^{s_k t} + \sum_q B_q t^n e^{s_q t} \tag{12-1}$$

where the $A_k$ and $B_q$ are constants, and $s_k$ and $s_q$ are the natural frequencies (eigenvalues) of the network. The first sum in Eq. (12-1) accounts for *simple* natural frequencies ($n = 1$); the second sum accounts for *multiple* natural frequencies ($n = 2, 3, \ldots$).

Typically, $f(t)$ is a zero-input loop current for loop analysis [Eq. (3-11)]; a zero-input node voltage for node analysis [Eq. (3-12)]; or a zero-input state variable in such an analysis [Eqs. (11-11) and (11-21)]. In all cases, zero-input stability requires

$$\lim_{t \to \infty} |f(t)| = 0 \tag{12-2}$$

---

[†] Also called *asymptotically stable*.

where the natural frequencies $s_k$, $s_q$ are the roots of the characteristic equation

$$\det \mathbf{Z} = 0 \tag{12-3}$$

in loop analysis, or

$$\det \mathbf{Y} = 0 \tag{12-4}$$

in node analysis, or

$$\det (s\mathbf{U} - \mathbf{A}) = 0 \tag{12-5}$$

in state variable analysis.

The zero-input stability, then, requires that in Eq. (12-1)

$$\lim_{t \to \infty} |e^{s_k t}| = 0 \tag{12-6}$$

and

$$\lim_{t \to \infty} |t^n e^{s_q t}| = 0 \tag{12-7}$$

Since, in general, a natural frequency is a complex number, we rewrite Eq. (12-6) as

$$\lim_{t \to \infty} |e^{(\sigma_k + j\omega_k)t}| = \lim_{t \to \infty} |e^{\sigma_k t}||e^{j\omega_k t}| = 0 \tag{12-8}$$

where, as usual, $s_k = \sigma_k + j\omega_k$. Now, due to Euler's identity, we have

$$|e^{j\omega_k t}| \equiv 1 \tag{12-9}$$

Consequently, Eq. (12-8) will be satisfied if

$$\sigma_k < 0 \tag{12-10}$$

Similarly, for Eq. (12-7) we have

$$\lim_{t \to \infty} |t^n e^{\sigma_q t}| = 0 \tag{12-11}$$

provided

$$\sigma_q < 0 \tag{12-12}$$

This result is established in Eq. (12-11) by using l'Hôpital's rule.

---

### Example 1

Using Eq. (12-12), we rewrite Eq. (12.11) as

$$\lim_{t \to \infty} \left| \frac{t^n}{e^{\sigma_q t}} \right|$$

This is an indeterminate form $\infty/\infty$. Differentiating $n$ times numerator and denominator separately (according to l'Hôpital), we obtain

$$\lim_{t \to \infty} \left| \frac{n!}{(\sigma_q)^n e^{\sigma_q t}} \right| = \lim_{t \to \infty} \left| \frac{k}{e^{\sigma_q t}} \right| = 0 \qquad \square$$

In conclusion: *zero-input stability requires that all the characteristic values (natural frequencies) of the network have negative real parts, $\sigma < 0$; in other words, all the natural frequencies must be inside the left half (lhp) of the s-plane.* See Fig. 12-1.

Specific locations of natural frequencies and their associated waveforms in the time domain are shown in Fig. 12-2; only cases $(a)-(d)$ satisfy Eq. (12-2), and these correspond to the requirement of $\sigma < 0$. Cases $(e)$ and $(f)$ correspond to $\sigma = 0$, and we will discuss them soon.                                **Probs. 12-1**
                                                                                                      **12-2**

## 12-2   HURWITZ POLYNOMIALS. THE ROUTH CRITERION AND THE HURWITZ TEST

An obvious way to verify the location of the eigenvalues would be to actually find them. For first-degree or second-degree characteristic equations this is no problem. Cubic and quartic (degree four) equations may still be solved by algebraic means, although the formulas are forbidding. Numerical and computer methods are available too. However, in most cases, we are not interested in actually finding the roots. Rather, we wish to ascertain whether all of them are in the left-half plane (lhp). Also, we might want to consider variations of one or more network parameters and the effects of such variations on stability. Here, again, it would be helpful if we could make such an investigation without the need of actually finding the roots.

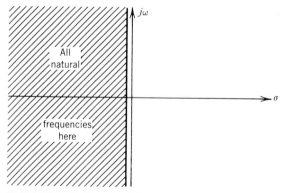

**FIGURE 12-1.** Permitted location of natural frequencies, $\sigma < 0$.

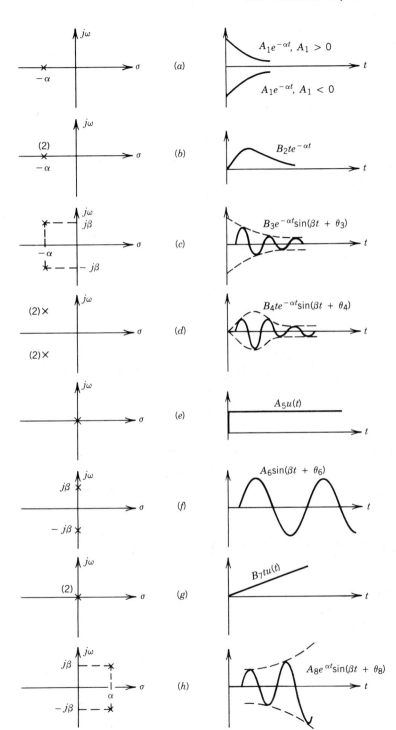

**FIGURE 12-2.** Various waveforms.

## Example 2

Consider the network shown in Fig. 12-3. We are interested in the effect of variations of $k$ on the stability of the network.

The characteristic equation here is

$$2s^2 + (1 - k)s + 2 = 0$$

In this case, we do not need any elaborate methods in order to see that, for $0 < k < 1$, the network is zero-input stable, and for $k > 1$ it is zero-input unstable. In more complicated cases, a similar situation might arise, and we would like to be able to make similar predictions without having to solve high-order equations.

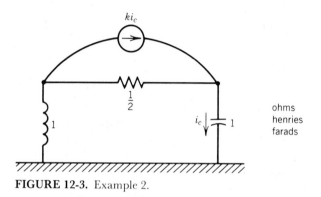

**FIGURE 12-3.** Example 2.  □

First, we need the following definition: a polynomial $Q(s)$ with real coefficients is called a *strict Hurwitz* polynomial if its zeros[†] all lie in the lhp (Re $s = \sigma < 0$). Thus, for zero-input stability, the characteristic polynomial must be strict Hurwitz.

According to this definition, a strict Hurwitz polynomial may have only the following zeros:

1. On the negative real axis, and of any multiplicity. They give rise to factors of the form $(s + \alpha)^k$, where $\alpha > 0$ and $k = 1, 2, \ldots$.

2. Complex conjugate with a negative real part, of any multiplicity. These give rise to factors of the form,

$$(s + \beta + j\gamma)^q(s + \beta - j\gamma)^q = (s^2 + 2\beta s + \beta^2 + \gamma^2)^q$$

where $\beta > 0$, $\gamma > 0$, and $q = 1, 2, \ldots$.

A strict Hurwitz polynomial, $Q_s(s)$, can therefore be written as follows:

$$Q_s(s) = \prod_j (s + \alpha_j) \prod_p (s^2 + 2\beta_p s + \beta_p^2 + \gamma_p^2) \tag{12-13}$$

---

[†] A zero of a polynomial $Q(s)$ is a value of $s$, say $s_1$, for which $Q(s_1) = 0$. If $Q(s)$ is the denominator of a rational function $H(s)$, then $s_1$ is a pole of $H(s)$.

where the subscript "$s$" in $Q$ denotes "strict Hurwitz," and the symbol $\prod$ indicates "the product of ..."

From these remarks, we can make the following observations as *necessary* (but not sufficient) conditions for a polynomial to be strict Hurwitz:

a) All the coefficients of the polynomial must be real and *positive*. This is so because no negative terms can appear in the products of Eq. (12-13).

b) No powers can be missing in the polynomial between the highest and the lowest powers of $s$. The reason is the same as before: in the products of Eq. (12-13), terms can cancel only by subtraction; however, there can be no subtraction without any negative signs.

These two observations are a simple test by inspection: The given polynomial must pass these before any further test. If the polynomial fails the inspection test, there is no need to continue: it is not strict Hurwitz. If, however, it passes this inspection, we cannot draw any definite conclusion as yet, since this test is a necessary, but not sufficient, condition.

---

## Example 3

The polynomial

$$Q(s) = s^5 + 14s^4 + 3s^3 + 2s + 1$$

is *not* strict Hurwitz. The second power $(s^2)$ is missing, and therefore $Q(s)$ has one or more zeros with a positive real part. In fact, two zeros are at $0.3232 \pm j0.4721$.
The polynomial

$$Q(s) = 3s^5 + 4s^4 + 2s^3 + 5s^2 + 3s + 1$$

may, or may not, be strict Hurwitz. Further investigation is needed.
The polynomial

$$Q(s) = s^3 + s^2 + 2s + 24$$

may, or may not, be strict Hurwitz. ☐

---

After the test by inspection, the given polynomial $Q(s)$ must be subjected to a conclusive test which will establish whether $Q(s)$ is strict Hurwitz or not. There are several such tests, and here we shall discuss a few.

## The Routh Criterion

Given a polynomial,

$$Q(s) = b_0 s^n + b_1 s^{n-1} + b_2 s^{n-2} + \cdots + b_{n-1}s + b_n \tag{12-14}$$

that passed the test by inspection, we form an array of numbers related to the coefficients of $Q(s)$, as follows. The first two rows of the array are the given coefficients of $Q(s)$, written as:

$$
\begin{array}{c|cccc}
s^n & b_0 & b_2 & b_4 & b_6 & \cdots \\
s^{n-1} & b_1 & b_3 & b_5 & b_7 & \cdots
\end{array}
$$

i.e., the first row contains the first, third, fifth, ..., coefficients, while the second row contains the second, fourth, sixth, ..., coefficients. The marginal $s^n$ and $s^{n-1}$ are helpful for later purposes. Every subsequent row (the third, the fourth, etc.) will be derived from the two rows immediately preceding it. The third row is formed as follows:

$$
\begin{array}{c|ccc}
s^{n-2} & \dfrac{b_1 b_2 - b_0 b_3}{b_1} & \dfrac{b_1 b_4 - b_0 b_5}{b_1} & \dfrac{b_1 b_6 - b_0 b_7}{b_1}
\end{array}
$$

and for convenience, denote this row as

$$
\begin{array}{c|ccc}
s^{n-2} & c_1 & c_2 & c_3 & \cdots
\end{array}
$$

The fourth row is formed along the same lines, using rows two and three. The result is

$$
\begin{array}{c|cc}
s^{n-3} & \dfrac{c_1 b_3 - b_1 c_2}{c_1} & \dfrac{c_1 b_5 - b_1 c_3}{c_1} & \cdots
\end{array}
$$

and is denoted conveniently as

$$
\begin{array}{c|ccc}
s^{n-3} & d_1 & d_2 & d_3 & \cdots
\end{array}
$$

The fifth row will be formed similarly, using rows three and four. Every row is obtained in this fashion, using its two preceding rows. This process ends when the marginal power is zero, $s^0$, and we have the *Routh array*:

$$
\begin{array}{c|cccc}
s^n & b_0 & b_2 & b_4 & b_6 & \cdots \\
s^{n-1} & b_1 & b_3 & b_5 & b_7 & \cdots \\
s^{n-2} & c_1 & c_2 & c_3 & & \cdots \\
s^{n-3} & d_1 & d_2 & d_3 & & \cdots \\
\vdots & \vdots & \vdots & \vdots & \vdots \\
s^0 & \cdots & \cdots & \cdots & \cdots
\end{array}
$$

---

### Example 4

Form the Routh array for $Q(s) = s^5 + 2s^4 + 4s^3 + 4s^2 + 3s + 8$. The first two rows are

$$
\begin{array}{c|ccc}
s^5 & 1 & 4 & 3 \\
s^4 & 2 & 4 & 8
\end{array}
$$

The third row is formed

$$s^3 \left|\ \frac{8-4}{2}\quad \frac{6-8}{2}\right.$$

that is,

$$s^3 \left|\quad 2\quad -1\right.$$

The fourth row is formed from the second and third rows,

$$s^2 \left|\ \frac{8-(-2)}{2}\quad \frac{16-0}{2}\right.$$

that is,

$$s^2 \left|\quad 5\quad 8\right.$$

Similarly, we continue to get the fifth row,

$$s^1 \left|\ \frac{-5-16}{5}\right.$$

that is,

$$s^1 \left|\ -\frac{21}{5}\right.$$

and finally,

$$s^0 \left|\quad 8\right.$$

The complete Routh array is

$$
\begin{array}{c|ccc}
s^5 & 1 & 4 & 3 \\
s^4 & 2 & 4 & 8 \\
s^3 & 2 & -1 & \\
s^2 & 5 & 8 & \\
s^1 & -\dfrac{21}{5} & & \\
s^0 & 8 & &
\end{array}
$$

□

***Example 5***

The Routh array for $Q(s) = s^3 + s^2 + 2s + 24$ (example 3 above) is

$$
\begin{array}{c|cc}
s^3 & 1 & 2 \\
s^2 & 1 & 24 \\
s^1 & -22 & \\
s^0 & 24 &
\end{array}
$$

□

Finally, *Routh's criterion* is: *the given polynomial* $Q(s)$ *has as many zeros in the right-half plane (rhp) as there are changes of signs going down the first column of the Routh array.* This remarkable criterion gives us the precise conclusion that we need.

Using it, in example 4, we count two changes of sign in the first column: one from $+5$ to $-21/5$ and the other from $-21/5$ to $+8$. Therefore $Q(s) = s^5 + 2s^4 + 4s^2 + 3s + 8$ has two zeros in the rhp. (They are $s = 0.8772 \pm j1.17$.) Similarly, the polynomial in example 5, $Q(s) = s^3 + s^2 + 2s + 24$, has two zeros in the rhp, since there are two changes in the signs of the first column of its Routh array. (In fact, these zeros are $s = 1 \pm j\sqrt{7}$.)

Consequently, *we require for zero-input stability that the first column in the Routh array contain only positive numbers.*

Several comments should be made about the Routh test:

1. At any stage, an entire row may be conveniently multiplied or divided by a positive, nonzero constant without affecting the results of the criterion. Thus, the array in example 4 above can be modified to read:

| | | | | |
|---|---|---|---|---|
| $s^5$ | 1 | 4 | 3 | |
| $s^4$ | ~~2~~ | ~~4~~ | ~~8~~ | divide by 2 |
| $s^4$ | 1 | 2 | 4 | |
| $s^2$ | $\dfrac{5}{2}$ | 4 | | multiply by 2 |
| $s^2$ | 5 | 8 | | |
| $s^1$ | $-\dfrac{21}{5}$ | | | multiply by 5 |
| $s^1$ | $-21$ | | | |
| $s^0$ | 8 | | | |

This is a convenient means to avoid unwieldy numbers.

2. What to do if the first entry of a row is zero while the others are nonzero? The difficulty causes two problems: (a) the sign ($+$ or $-$) of this term is questionable, and (b) the subsequent row will be indeterminate, since it is obtained by a division by this zero.

---

## Example 6

The Routh array for
$$Q(s) = 2s^6 + s^5 + 6s^4 + 3s^3 + 10s^2 + s + 1$$
begins with

| | | | | |
|---|---|---|---|---|
| $s^6$ | 2 | 6 | 10 | 1 |
| $s^5$ | 1 | 3 | 1 | |
| $s^4$ | 0 | 8 | 1 | |

and, in attempting to form the next row, we run into a division by zero. □

---

We can use one of the following ways to remedy this situation:

1. Multiply $Q(s)$ by $(s + a)$, where $a > 0$ is *not* a zero of $Q(s)$. Carry out then the Routh criterion on the augmented polynomial: One of its zeros is known, $s = -a$, and therefore if $Q(s)$ was strict Hurwitz to begin with, so will be $Q(s)(s + a)$.

---

## Example 7

In example 6, $s = -1$ is not a zero of $Q(s)$, as can be verified by a direct substitution. Then

$$Q(s)(s + 1) = 2s^7 + 3s^6 + 7s^5 + 9s^4 + 13s^3 + 11s^2 + 2s + 1.$$

The Routh array for this augmented polynomial is

| | | | | |
|---|---|---|---|---|
| $s^7$ | 2 | 7 | 13 | 2 |
| $s^6$ | 3 | 9 | 11 | 1 |
| $s^5$ | 3 | 17 | 4 | (after multiplication by 3) |
| $s^4$ | $-8$ | 7 | 1 | |

At this stage we know already that $Q(s)$ is not strict Hurwitz because of the minus sign in the first column!  □

---

2. Replace the zero by $\varepsilon \approx 0$. Carry on, and eventually let $\varepsilon \to 0$.

---

## Example 8

Following this approach, we obtain the following array for example 6:

| | | | | |
|---|---|---|---|---|
| $s^6$ | 2 | 6 | 10 | 1 |
| $s^5$ | 1 | 3 | 1 | |
| $s^4$ | $\varepsilon$ | 8 | 1 | (after multiplication by $\varepsilon$) |
| $s^3$ | $3\varepsilon - 8$ | $\varepsilon - 1$ | | |

etc.

It does not matter if $\varepsilon < 0$ or $\varepsilon > 0$: the sign changes in the first column are the same in number.  □

---

What to do if an entire row is full of zeros?

*Example 9*

$$Q(s) = 2s^5 + s^4 + 4s^3 + 2s^2 + 2s + 1$$

The Routh array will begin with

$$
\begin{array}{c|ccc}
s^5 & 2 & 4 & 2 \\
s^4 & 1 & 2 & 1
\end{array}
$$

and the next row will be full of zeros. The use of $\varepsilon$ will not work here.   □

---

This situation occurs whenever *the preceding two rows are proportional to each other.* Here we can resort to either of two ways. One is to form an auxiliary polynomial, using the coefficients of the last nonzero row. This auxiliary polynomial will be all-even or all-odd, and its highest power will be shown in the left margin. *This auxiliary polynomial is a factor of* $Q(s)$! And so we can factor partially the original polynomial.

---

*Example 10*

In example 9, the auxiliary polynomial will be $s^4 + 2s^2 + 1$, using the row

$$
\begin{array}{c|ccc}
s^4 & 1 & 2 & 1
\end{array}
$$

Remember: the auxiliary polynomial is always all-even or all-odd, and its highest power is determined by the marginal notation. Factoring, then, $Q(s)$, we obtain

$$Q(s) = (s^4 + 2s^2 + 1)(2s + 1)$$

and the zeros of $Q(s)$ are found from

$$2s + 1 = 0 \qquad s = -\tfrac{1}{2}$$
$$(s^4 + 2s^2 + 1) = (s^2 + 1)^2 = 0 \qquad s = \pm j, \pm j$$

Therefore, this $Q(s)$ is not strict Hurwitz, because of the zeros on the $j\omega$ axis.   □

---

The second way is to form the auxiliary polynomial, as before, and replace the row of zeros by the coefficients of the *derivative* of the auxiliary polynomial.

---

*Example 11*

Consider

$$Q(s) = s^6 + 9s^4 + 23s^2 + 15$$

If we try to set up the Routh array, we wind up with

$$
\begin{array}{c|cccc}
s^6 & 1 & 9 & 23 & 15 \\
s^5 & 0 & 0 & 0 &
\end{array}
$$

The auxiliary polynomial here is $Q(s)$ itself (and therefore no immediate factorization is possible). The derivative is

$$6s^5 + 36s^3 + 46s$$

making the second row

$$
\begin{array}{c|ccc}
s^5 & 6 & 36 & 46
\end{array}
$$

Proceeding now in the usual manner, we find that $Q(s)$ is not strict Hurwitz.   ☐

Finally, let us mention (again, without proof) that when a row full of zeros occurs in the Routh array, the zeros of the auxiliary polynomial have a symmetry with respect to the origin, as shown in Fig. 12-4.

In conclusion, the Routh criterion is a powerful tool in determining the location (lhp or rhp) of zeros of a polynomial. In certain cases, it leads also to a partial (or complete) factorization of the polynomial.[†]                    ***Prob. 12-3***

### The Hurwitz Test

The Hurwitz test is an alternate method to determine the location (lhp or rhp) of the zeros of a polynomial; in addition, as we shall see, its mathematical development is related to a certain type of electrical networks.

The Hurwitz test starts with the first two rows of the Routh array. We form the auxiliary polynomial corresponding to each row: one all-even, the other all-odd. Denote them by $Q_E(s)$ and $Q_o(s)$, respectively. Of course,

$$Q(s) = Q_E(s) + Q_o(s) \tag{12-15}$$

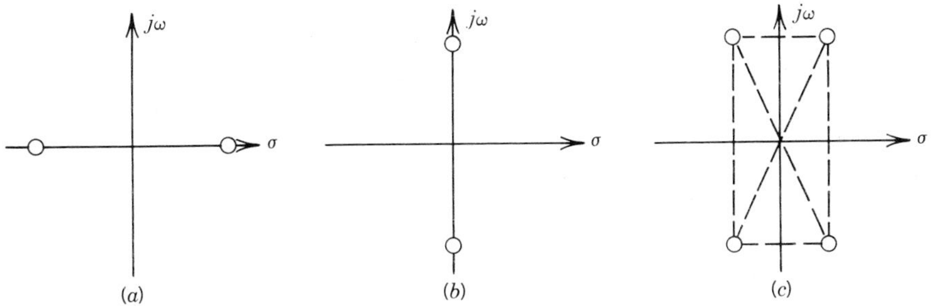

**FIGURE 12-4.** Quadrant symmetry.

[†] See also Appendix E.

Let us assume that $Q_E(s)$ corresponds to the first row of the Routh array and that $Q_o(s)$ corresponds to the second one. Thus, $Q_E(s)$ is exactly one degree higher than $Q_o(s)$. (The same is true if $Q_o(s)$ corresponds to the first row and $Q_E(s)$ to the second.) Next, divide $Q_E(s)$ by $Q_o(s)$ once. The result is

$$\frac{Q_E(s)}{Q_o(s)} = c_0 s + \frac{Q_1(s)}{Q_o(s)} \tag{12-16}$$

where $c_0 > 0$ and $Q_1(s)$ is one degree lower than $Q_o(s)$. Invert this remainder and divide once again,

$$\frac{Q_o(s)}{Q_1(s)} = c_1 s + \frac{Q_2(s)}{Q_1(s)} \tag{12-17}$$

Repeat this operation, each time inverting the remainder and dividing once. This process is finite, since the degree of every $Q_j(s)$, $j = 1, 2, 3, \ldots$ is reduced after each step. The final result is the continued fraction

$$\frac{Q_E(s)}{Q_o(s)} \text{ or } \frac{Q_o(s)}{Q_E(s)} = c_0 s + \cfrac{1}{c_1 s + \cfrac{1}{c_2 s + \cfrac{1}{\ddots + \cfrac{1}{c_n s}}}} \tag{12-18}$$

The Hurwitz test is: *all the zeros of $Q(s)$ are in the lhp if, and only if, all the $c$'s in Eq. (12-18) are positive and the process does not end prematurely.*

---

### Example 12

Consider

$$Q(s) = s^3 + 2s^2 + 2s + 2$$

Here,

$$Q_o(s) = s^3 + 2s \qquad Q_E(s) = 2s^2 + 2$$

The "divide once, invert, divide" procedure is

$$\frac{s^3 + 2s}{2s^2 + 2} = \tfrac{1}{2} s + \frac{s}{2s^2 + 2}$$

$$\frac{2s^2 + 2}{s} = 2s + \frac{2}{s}$$

$$\frac{s}{2} = \tfrac{1}{2} s$$

or, as in Eq. 12-18

$$\frac{s^3 + 2s}{2s^2 + 2} = \tfrac{1}{2}s + \cfrac{1}{2s + \cfrac{1}{\frac{1}{2}s}}$$

All the $c$'s, here 1/2, 2, and 1/2, are positive. Therefore $Q(s)$ is strict Hurwitz.   ☐

---

### Example 13

Let

$$Q(s) = s^3 + 2s^2 + 10s + 22$$

The Hurwitz test here is

$$\frac{s^3 + 10s}{2s^2 + 22} = \tfrac{1}{2}s + \cfrac{1}{-2s + \cfrac{1}{-22s}}$$

and since all the $c$'s, 1/2, $-2$, $-1/22$, are *not* positive, $Q(s)$ is not strictly Hurwitz (actually, we could have stopped after obtaining $-2s$).   ☐

---

### Example 14

Apply the Hurwitz test to

$$Q(s) = 2s^5 + s^4 + 4s^3 + 2s^2 + 2s + 1$$

of example 9. We have here:

$$\frac{Q_o(s)}{Q_E(s)} = \frac{2s^5 + 4s^3 + 2s}{s^4 + 2s^2 + 1} = 2s + 0$$

and the continued fraction expansion ("divide once, invert, divide") ends prematurely. What happened? $Q_o(s)$ and $Q_E(s)$ have a common factor which, by cancellation, causes this premature termination. *This common factor is the last divisor;* here, it is $s^4 + 2s^2 + 1 = (s^2 + 1)^2$. Naturally, if this factor is common to $Q_E(s)$ and $Q_o(s)$, it is also a factor of $Q(s)$. This is, then, the auxiliary polynomial found by the Routh criterion. Thus a premature termination of the Hurwitz test enables us to find a factor of $Q(s)$. The last divisor, prior to the termination, is this factor. This process is equivalent to obtaining the auxiliary polynomial in the Routh test when a row full of zeros is encountered.   ☐

---

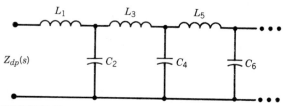

**FIGURE 12-5.** An $LC$ ladder network.

The relation between the Hurwitz test and certain electrical networks is very interesting: consider an $LC$ ladder network, shown in Fig. 12-5. Its driving-point impedance can be written as a continued fraction [see Eq. (7-8)]:

$$Z_{dp}(s) = L_1 s + \cfrac{1}{C_2 s + \cfrac{1}{L_3 s + \cfrac{1}{C_4 s} + \ddots}} \tag{12-19}$$

With all $L$'s and $C$'s positive, this is identical with the process of the Hurwitz test. We have here a close correspondence between lossless $(LC)$ ladder networks and the Hurwitz test.

☐ **Probs. 12-4**
**12-5**
**12-6**
**12-9**

## *12-3 A GRAPHICAL METHOD: THE ENCIRCLEMENT THEOREM

Consider again $Q(s)$, as in Eq. (12-14),

$$Q(s) = b_0 s^n + b_1 s^{n-1} + \cdots + b_{n-1} s + b_n \tag{12-14}$$

It has exactly $n$ zeros (simple or multiple). Write $Q(s)$ in its factored form as

$$Q(s) = b_0 (s - p_1)(s - p_2) \cdots (s - p_n) \tag{12-20}$$

We want to know how many, if any, of these are in the rhp.

Let us plot $Q(s)$ for different values of $s$. If $s$ is, in general, a complex number, say $s = s_0 = |s_0| e^{j\phi_0}$, we obtain $Q(s_0)$, a complex number written also in its polar form,

$$Q(s_0) = |Q(s_0)| e^{j\theta_0} \tag{12-21}$$

See Fig. 12-6. Similarly, for $s = s_1 = |s_1| e^{j\phi_1}$, we get $Q(s_1) = |Q(s_1)| e^{j\theta_1}$. As $s$ varies, $s = s_0$, $s = s_1$, etc., its locus is the curve $C$; the locus of the corresponding points $Q(s_0)$, $Q(s_1)$, etc., is the curve $\Gamma$. Let us consider $C$ to be a closed curve that does not intersect itself, as shown in Fig. 12-7a. There are three possibilities for the corresponding curve $\Gamma$: (a) It does not encircle the origin of the $Q$-plane, or (b) It

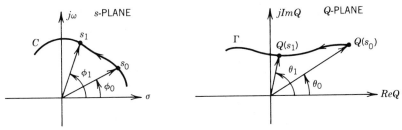

**FIGURE 12-6.** Mapping of the *s*-plane onto the *Q*-plane.

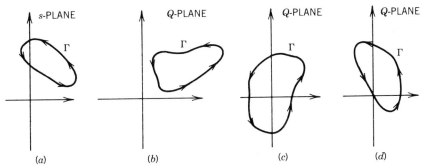

**FIGURE 12-7.** Possible mappings of a curve *C*.

encircles the origin, or (c) It passes through the origin. These are shown in Fig. 12-7b, c, and d, respectively. In the first case, when $\Gamma$ does not encircle the origin, the net change in $\theta$—the argument of $Q$—is zero as $C$ is traversed once. In the second case, the net change in $\theta$ is $2k\pi$, where $k$ is the number of times that $\Gamma$ encircles the origin. We repeat these statements as follows:

$$\Delta\theta = 0 \qquad \text{(origin not encircled)}, \tag{12-22}$$

$$\Delta\theta = 2k\pi \qquad \text{(origin enclosed, } k = 1, 2, \dots) \tag{12-23}$$

where $\Delta\theta$ is the net change in $\theta$.

To evaluate $\Delta\theta$, consider again Eq. (12-20) for a particular value of $s$, say $s = s_0$:

$$Q(s_0) = b_0(s_0 - p_1)(s_0 - p_2) \cdots (s_0 - p_n) \tag{12-24}$$

A typical term on the right-hand side of Eq. (12-24), $(s_0 - p_j)$, can be interpreted in the *s*-plane as a complex number, represented by a directed line from the point $p_j$ to the point $s_0$ (see Fig. 12-8). Therefore, we have

$$|Q(s_0)| = b_0|(s_0 - p_1)| \; |(s_0 - p_2)| \cdots |(s_0 - p_n)| \tag{12-25}$$

that is, the magnitude of $Q(s_0)$ is equal to the product of the magnitudes of the terms $(s_0 - p_j), j = 1, 2, \dots, n$, while

$$\theta_0 = \text{Arg } Q(s_0) = \text{Arg }(s_0 - p_1) + \text{Arg }(s_0 - p_2) + \cdots + \text{Arg }(s_0 - p_n) \tag{12-26}$$

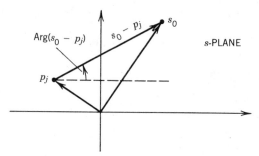

**FIGURE 12-8.** Geometry of $(s_0 - p_j)$.

the angle of $Q(s_0)$ is the sum of the angles of $(s_0 - p_j), j = 1, 2, \ldots, n$. The net change in $\theta$ as $C$ is traced once is then

$$\Delta\theta = \sum_{j=1}^{n} \Delta \operatorname{Arg} (s - p_j) \qquad (12\text{-}27)$$

To evaluate $\Delta\theta$ (which, according to Eq. (12-22, 23) is either 0 or $2k\pi$), we must distinguish between two cases:

1. All the $p_j$'s are outside the curve $C$. In this case, as $s$ traces $C$ once around we have

$$\Delta \operatorname{Arg} (s - p_j) = 0 \qquad j = 1, 2, \ldots, n \qquad (12\text{-}28)$$

which means that

$$\Delta\theta = 0 \qquad (12\text{-}22)$$

2. Some of the $p_j$'s say $p_1, p_2, \ldots, p_k$, are inside the curve $C$. Now, as $s$ traces $C$ once around,

$$\Delta \operatorname{Arg} (s - p_j) = 2\pi \qquad j = 1, 2, \ldots k \qquad (12\text{-}29)$$

and therefore,

$$\Delta\theta = 2k\pi \qquad (12\text{-}23)$$

A third case, shown in Fig. 12-7d, happens when one (or more) of the $p_j$ are *on* the curve $C$. Then, obviously, $Q(p_j) = 0$ and $\Gamma$ passes through the origin. In summary: *if $Q(s)$ has $k$ zeros inside a closed curve $C$ of the s-plane, the net change in the argument of $Q(s)$ is $2k\pi$ as $s$ traces $C$ once around.* This important result from the theory of complex variables is known as the *encirclement theorem*, or as the *principle of the argument*.

How can we apply this theorem to determine the locations of the $p_j$'s? Naturally, we would like the curve $C$ to enclose the entire right-half plane; that is, $C$ will consist of a large semicircle of radius $R(R \to \infty)$ and the $j\omega$ axis, as shown in Fig. 12-9. Let us compute the contributions to $\Delta\theta$ when $s$ traces this $C$. On the semicircle we have

$$s = R e^{j\phi} \qquad -\frac{\pi}{2} \le \phi \le \frac{\pi}{2} \qquad (12\text{-}30)$$

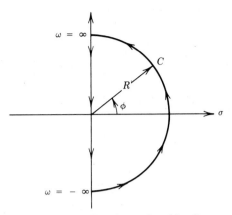

**FIGURE 12-9.** The rhp enclosed by $C$.

As $R \to \infty$, thus enclosing the entire right-half plane, $s \to \infty$ and the highest power in Eq. (12-14) dominates. That is,[†]

$$Q(s) \doteq b_0 s^n = b_0 R^n e^{jn\phi} \tag{12-31}$$

Therefore, as $s$ traces the semicircle from $\phi = -\pi/2$ to $\phi = +\pi/2$, the argument of $Q(s)$, $\theta$, goes from $-n(\pi/2)$ to $+n(\pi/2)$. That is

$$\Delta\theta = n\pi \qquad -\frac{\pi}{2} \le \phi \le \frac{\pi}{2} \tag{12-32}$$

To complete the curve $C$, we must vary $s$ along the $j\omega$ axis, from $\omega = +\infty$ to $\omega = -\infty$. If we note the net change of the argument of $Q(s)$ here, then by adding to it $n\pi$, we shall have the total $\Delta\theta$. That is,

$$\Delta\theta = \Delta\theta(\omega) + n\pi, \tag{12-33}$$

where $\Delta\theta(\omega)$ denotes the net change in $\theta$ as $s$ varies along the $j\omega$ axis. But plots of $Q(s)$ for $s = j\omega$ are quite familiar to us from chapter 7! (See section 7-4.) These plots are simplified if we recall that $Q(s)$ is real when $s$ is real and

$$Q^*(j\omega) = Q(-j\omega) \tag{12-34}$$

As a result, $\theta(\omega)$, the argument of $Q(j\omega)$, is an odd function and we need plot it only from $\omega = 0$ to $\omega = \infty$. Also, the plot of $Q(j\omega)$ is symmetric about the horizontal axis.

---

### *Example 15*

Consider

$$Q(s) = s^2 + s + 1 \qquad (n = 2)$$

[†] $\doteq$ means "approximately equals."

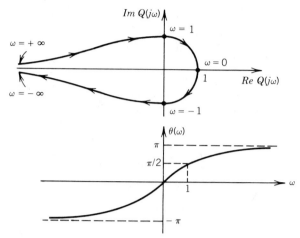

**FIGURE 12-10.** Example 15.

We have

$$Q(j\omega) = 1 - \omega^2 + j\omega = Re\, Q(j\omega) + jIm\, Q(j\omega)$$

The plots of $Q(j\omega)$ and of $\theta(\omega)$ are shown in Fig. 12-10. We have

$$\theta(\omega) = \tan^{-1}\frac{\omega}{1 - \omega^2}$$

and, as $s$ varies from $+\infty$ to $-\infty$ along the $j\omega$ axis, $\theta(\omega)$ varies from $\pi$ to $-\pi$. Therefore $\Delta\theta(\omega) = -\pi - \pi = -2\pi$. To this, add $n\pi = 2\pi$ according to Eq. (12-33), and therefore the net change is $\Delta\theta = 0$. Hence, $Q(s)$ has no zeros inside the right-half plane. In addition, since the locus of $Q(s)$, $\Gamma$, does not pass through the origin, there are no zeros of $Q(s)$ on $C$, i.e., *on the $j\omega$ axis*. Hence, all the zeros are inside the left-half plane, and $Q(s)$ is a strict Hurwitz polynomial. □

*Example 16*

Consider

$$Q(s) = s^3 + s^2 + 2s + 24 \qquad (n = 3)$$

as in example 5 of the previous section. We have

$$Q(j\omega) = 24 - \omega^2 + j(2\omega - \omega^3) = Re\, Q(\omega) + jIm\, Q(\omega)$$

$$\therefore \quad \theta(\omega) = \tan^{-1}\frac{2\omega - \omega^3}{24 - \omega^2}$$

The plots of $Q(j\omega)$ and $\theta(\omega)$ are shown in Fig. 12-11. Here, as $s$ traces the $j\omega$ axis from

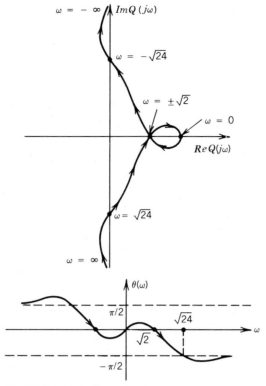

**FIGURE 12-11.** Example 16.

$+\infty$ to $-\infty$, $\theta(\omega)$ varies from $-\pi/2$ to $\pi/2$. That is, $\Delta\theta(\omega) = \pi/2 - (-\pi/2) = \pi$. According to Eq. (12-33), we get the total $\Delta\theta = \pi + 3\pi = 4\pi$, and so $k = 4\pi/2\pi = 2$, i.e., two encirclements. Therefore $Q(s)$ has two zeros inside the rhp.

*Probs. 12-10*
*12-11*
*12-12*
*12-13*

## 12-4   ZERO-STATE (BIBO) STABILITY

Another type of stability is concerned with input-output relations. More precisely, we consider a zero-state $[\mathbf{x}(0_-) = \mathbf{0}]$ and define this type of stability as follows: For every bounded input, the output is bounded. This bounded-input, bounded-output (BIBO) stability seems very natural to adopt: if we apply an input which is bounded, i.e.,

$$|e(t)| \leq A_1 \qquad \text{for all } t \qquad (12\text{-}35)$$

where $A_1$ is a finite constant, then we expect from a BIBO stable network an output which is bounded, that is,

$$|r(t)| \leq A_2 \qquad \text{for all } t \tag{12-36}$$

where $A_2$ is another finite constant.

Let us explore some consequences in terms of the impulse response $h(t)$. First, let us recall (Chapter 1) that for a causal and time-invariant system we must have

$$h(t) = 0 \qquad t < 0 \tag{12-37}$$

Now, let the input $e(t)$ be bounded. The zero-state response is obtained by convolution

$$
\begin{aligned}
r(t) &= e(t) * h(t) \\
&= \int_0^\infty e(t - x)h(x) \, dx
\end{aligned}
\tag{12-38}
$$

It is a known property of an integral that its absolute value never exceeds the integral of the absolute value of the integrand, that is,

$$|r(t)| = \left| \int_0^\infty e(t - x)h(x) \, dx \right| \leq \int_0^\infty |e(t - x)||h(x)| \, dx \tag{12-39}$$

Using Eq. (12-35), we obtain from Eq. (12-39)

$$|r(t)| \leq A_1 \int_0^\infty |h(x)| \, dx \tag{12-40}$$

and if the output is to be bounded we must have, from Eqs. (12-36) and (12-40)

$$\int_0^\infty |h(t)| \, dt \leq A_3 \tag{12-41}$$

where $A_3$ is a finite constant, $A_3 = A_2/A_1$. Note that $x$ can be replaced by $t$ in Eq. (12-41), since either one is a dummy variable. Geometrically, Eq. (12-41) means that the total area under $|h(t)|$ must be finite.

## 12-5 BIBO STABILITY AND THE NETWORK FUNCTION $H(s)$

For linear, constant networks we have developed a relationship between the zero-state output $R(s)$ and the input $E(s)$. It is given by

$$R(s) = H(s)E(s) \tag{12-42}$$

where $H(s)$ is the appropriate network function. We make the following observations:

a) The poles of $R(s)$ are those of $H(s)$ *and* of $E(s)$. There may be a rare case of a pole-zero cancellation in forming the product $H(s)E(s)$.

b) Since $e(t)$ is bounded by assumption, the poles of $E(s)$ are entirely in the left-half plane (1hp).

c) $E(s)$ may have also poles *on* the $j\omega$ axis, but these must be *simple*. A simple pole at $s = 0$ corresponds to a step function, Fig. 12-2e, while a pair of simple poles at $s = \pm j\beta$ correspond to a bounded sinusoidal waveform, Fig. 12-2f. Multiple poles on the $j\omega$ axis yield unbounded waveforms such as $t \sin \beta t$. Poles in the right-half plane (rhp) yield unbounded waveforms, Fig. 12-2h.

So much for the inputs. In order to guarantee a bounded output, restrictions must be placed on $H(s)$. In view of our discussion so far, it is rather simple to show that the output $r(t)$ will be bounded provided that:

1. In $H(s)$, the degree of the numerator is less than or, at the most, equal to the degree of the denominator; and

2. All the poles of $H(s)$ are *inside* the lhp, that is, the denominator of $H(s)$ must be a strict Hurwitz polynomial.

Let us illustrate with a few examples.

---

## Example 17

Consider the network function

$$H(s) = \frac{s - 1}{s + 2}$$

for a single-input-single-output network. It satisfies the two restrictions mentioned before, and every bounded input will result in a bounded output. Hence the network is BIBO stable. ☐

---

## Example 18

For a simple $LC$ series circuit, the characteristic equation is

$$LCs^2 + 1 = 0$$

with the network function being $Y_{dp}(s) = Cs/(LCs^2 + 1)$, and having simple poles at $s = +j\omega_0$, $s = -j\omega_0$, $\omega_0^2 = 1/LC$. However, for the *bounded* input

$$e(t) = \cos \omega_0 t$$

the output is *unbounded* (show this!) and therefore this network is not BIBO stable. The denominator of the network function is not strict Hurwitz. ☐

*Example 19*

Let

$$H(s) = \frac{s^4 + 4s^3 + 6s^2 + 4s + 1}{s^3 + 2s^2 + 2s + 2}$$

Even though the denominator is a strict Hurwitz polynomial, the network is BIBO unstable because the numerator is of higher degree than the denominator. Thus, for example, the *bounded* input $e(t) = u(t)$ will result in an output containing an *unbounded* impulse.                           ☐

Note that nothing has been said so far about the zeros of $H(s)$. Generally, no restrictions are placed on the location of these zeros. However, if $H(s)$ is a driving-point function

$$H(s) = Z_{dp}(s) \text{ or } Y_{dp}(s) \tag{12-43}$$

then its reciprocal is also a network function. For this reciprocal, poles are the zeros of $H(s)$. Therefore, poles *and* zeros of driving-point functions are restricted to the lhp for BIBO stability. This is not true for transfer functions, since the reciprocal of such a function is *not* a network function in general.

For the multiple-input, multiple-output case, where we have a matrix network function $\mathbf{H}(s)$, the same restrictions apply to every element $H_{jk}(s)$ of $\mathbf{H}(s)$.

*Prob. 12-14*

## 12-6   CONCLUSION

It is important to realize that a network may be zero-input *unstable* even though it is zero-state BIBO *stable*, as illustrated in the following example.

*Example 20*

Let

$$\mathbf{A} = \begin{bmatrix} 1 & -1 \\ -6 & 0 \end{bmatrix} \quad \mathbf{B} = \begin{bmatrix} 2 \\ -1 \end{bmatrix} \quad \mathbf{C} = [-\tfrac{2}{3} \quad -\tfrac{1}{3}] \quad \mathbf{D} = 0$$

The characteristic equation is

$$\det (s\mathbf{U} - \mathbf{A}) = s^2 - s - 6 = (s - 3)(s + 2) = 0$$

Since one eigenvalue is in the rhp ($s = 3$), we have clearly zero-input instability. On the other hand, the calculation of $H(s)$ according to Eq. (11-108) yields $H(s) = 1/(s + 2)$. The pole of $H(s)$ is inside the l hp ($s = -2$), indicating BIBO stability. What happened? The detailed calculation of $H(s)$ gives (verify!)

$$H(s) = \frac{s - 3}{(s - 3)(s + 2)}$$

and a cancellation of the factor $(s - 3)$ occurs. We see therefore that even if a network is BIBO stable it may be possible for some internal current or voltage to grow without bounds. This unstable mode $(s = 3)$ is *not observable* at the output.

☐

Finally, to stress these points again, let us remember that a network function describes only the zero-state response. Therefore, we cannot obtain the zero-input response from $H(s)$ alone.

## PROBLEMS

12-1 Obtain $\mathbf{\Phi}(s)$, the Laplace transform of $\phi(t)$, for problem 11-2 using signal flow graph methods:

a) Set the input to zero.

b) Assume initial conditions $\mathbf{x}(0_-)$ and use them as sources for the signal flow graph.

c) Use Mason's formula to obtain $\mathbf{\Phi}(s)$, the overall gain between $\mathbf{x}(0_-)$ and $\mathbf{X}(s)$.

12-2 Consider example 18 in section 12-5. Let the excitation be

$$e(t) = K \sin \omega_0 t \qquad \omega_0^2 = 1/LC$$

Discuss the boundedness of the response in terms of the poles of $I(s)$, and in terms of exciting a natural frequency, that is, resonance.

12-3 Apply the Routh test to the following polynomials:

a) $Q(s) = 12s^5 + 12s^4 + 16s^3 + 10s^2 + 5s + 1$

b) $Q(s) = s^4 + 4s^3 + 6s^2 + 12s + 9$

c) $Q(s) = s^5 + 2s^4 + 2s^3 + 4s^2 + 9s + 18$

d) $Q(s) = 1.2s^4 + 3.84s^3 + 16.5s^2 + 24.4s + 11.3$

e) $Q(s) = s^4 + s^3 + s^2 + 2s + 2$

f) $Q(s) = s^7 + 4s^6 + 21s^5 + 32s^4 + 12s^3 + 11s^2 + 8s + 4$

g) $Q(s) = 12s^4 + 12as^3 + 10s^2 + 6as + 1 \qquad a > 0$

h) $Q(s) = s^5 + s^4 + 3s^3 + 2s^2 + 2s$

i) $Q(s) = s^4 + 2s^3 + s^2 + 3s + 3$

j) $Q(s) = s^5 + 1.5s^4 + 3.5s^3 + 3.5s^2 + 3s + 1$

**12-4** Apply the Hurwitz test to each of the polynomials in problem 12-3. Whenever $Q(s)$ is a strict Hurwitz polynomial, obtain an associated $LC$ ladder network.

**12-5** For what value(s) of $K$ is the polynomial

$$Q(s) = 6s^5 + 2s^4 + Ks^3 + 3s^2 + 2s + 10$$

strict Hurwitz?

**12-6** For the equation:

$$s^5 + 2s^4 + 3s^3 + 8s^2 + 4s + 6 = 0$$

find out how many roots are in the right-half plane and how many are in the left-half plane. If there are any roots in the rhp, can you predict where they might be (*on the* $+\sigma$ axis, *off* it)? Explain.

**12-7** Repeat problem 12-6 for

$$2s^4 + 2s^3 + 4s^2 + 5s + 2 = 0$$

**12-8** Check for BIBO stability:

a) $H(s) = \dfrac{s^2 - 6}{s^7 + 3s^6 + 8s^5 + 15s^4 + 17s^3 + 12s^2 + 4s}$

b) $H(s) = \dfrac{s^3 + s - 10}{s^8 + 7s^6 + 17s^4 + 17s^2 + 6}$

**12-9** The characteristic equation of a network is

$$s^3 + Ks^2 + 2s + 4 = 0$$

Investigate the zero-input stability of the network in terms of $K$.

**12-10** Apply the encirclement theorem to the following polynomials:

a) $Q(s) = 2s^3 + 4s^2 + s + 1$
b) $Q(s) = s^3 + s^2 + s + 1$
c) $Q(s) = s^3 + Ks^2 + 2s + 4$ (see problem 12-9)

**12-11** Extend the encirclement theorem to a rational function,

$$H(s) = K \frac{(s - z_1)(s - z_2) \cdots (s - z_m)}{(s - p_1)(s - p_2) \cdots (s - p_n)}$$

and prove that

$$\Delta\theta = 2\pi(k - m)$$

where, as before, $k$ is the number of zeros of $H(s)$ and $m$ is the number of poles of $H(s)$ enclosed by $C$.

**12-12** Apply the result of problem 12-11 to

$$H(s) = 3\frac{s - 2}{s^2 + s + 1}$$

and use $C$ as the contour enclosing the entire rhp.

*Hint:* Since $H(s) \to 0$ as $s \to \infty$, there is not contribution to $\Delta\theta$ along the semicircular portion of $C$. Therefore,

$$\Delta\theta = \Delta\theta(\omega) = 2\pi(k - m)$$

12-13  Repeat Prob. 12-12 for

$$H(s) = \frac{s^3 + s^2 + s + 1}{2s^3 + 4s^2 + s + 1}$$

12-14  For each of the given rational functions, check which can qualify as: 1) network functions, and 2) driving-point functions. Give reasons.

a)  $\dfrac{s(s + 1)(s - 2)}{(s^2 + 2)(s^2 + s + 1)}$

b)  $\dfrac{(s + j2)(s + 1)}{s^2 + 2s + 3}$

c)  $\dfrac{s^2 + 1}{s^2(s + 1)}$

d)  $\dfrac{s^2 + s + 1}{s^2(s + 1)}$

e)  $\dfrac{s}{(s^2 + 1)^2(s^2 + 4)}$

f)  $\dfrac{s}{s^2 - s + 10}$

g)  $\dfrac{s + 2}{s^2 + s + 10}$

# 13

# PERIODIC WAVEFORMS AND THE FOURIER SERIES

Periodic waveforms are very common in power distribution, electronic circuits, communication networks, and computers. A few such waveforms are shown in Fig. 13-1. Two major characteristics of a periodic function are: 1) its period, $T$, the time defined by

$$f(t \pm T) = f(t) \qquad -\infty < t < +\infty \qquad (13\text{-}1)$$

that is, the shortest time needed for the function to repeat itself; 2) its cycle, $f_1(t)$, the portion of $f(t)$ during one period. These are shown also in Fig. 13-1. We are interested in finding the zero-state response of a circuit when the input is such a periodic waveform. The Laplace transform may be used, of course, as will be shown in an example. However, we'll see that the Laplace transform is quite complicated in such problems. Furthermore, we will be interested only in the steady-state part of the response, and, in particular, in the *frequency contents* of that response. These ideas are closely related to the frequency response (magnitude and phase) studied in Chapter 9.

For these reasons, and other which follow in our discussion, we proceed to study the Fourier series and, in the next chapter, the Fourier integral.

## 13-1  THE LAPLACE TRANSFORM REVISITED

In order to review the use of the Laplace transform here, consider the following example.

---

*Example 1*

A square wave current source, shown in Fig. 13-2a is applied at $t = 0$ to the $RL$ network, Fig. 13-2b, with the initial condition being zero. The zero-state response

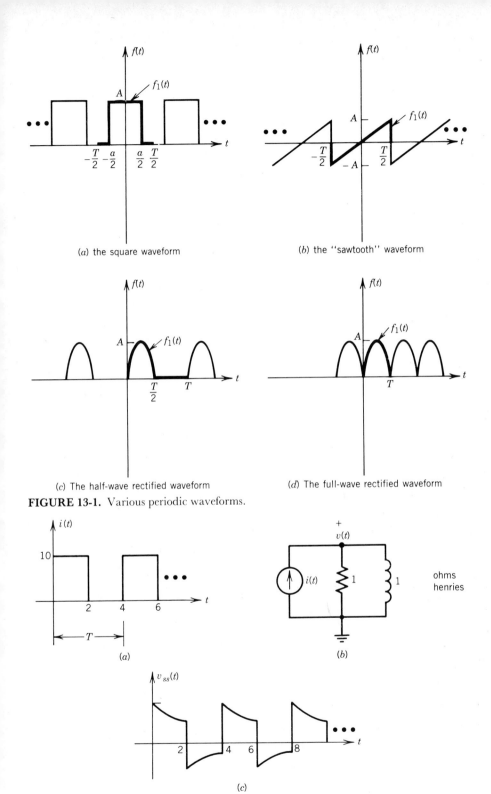

(a) the square waveform

(b) the "sawtooth" waveform

(c) The half-wave rectified waveform

(d) The full-wave rectified waveform

**FIGURE 13-1.** Various periodic waveforms.

ohms
henries

(a)

(b)

(c)

**FIGURE 13-2.** Example 1.

$V(s)$ is, as usual,

$$V(s) = Z(s)I(s) = \frac{s}{s+1} I(s)$$

The Laplace transform of $i(t)$ can be found as outlined in Chapter 2, Example 15 (with Figure 2-18). Specifically,

$$I(s) = \frac{I_1(s)}{1 - e^{-Ts}}$$

where $T$ is the period (in seconds) of $i(t)$, and $I_1(s)$ is the Laplace transform of the first cycle, over the first period, of $i(t)$. We have here

$$I_1(s) = \int_{0-}^{2} 10e^{-st}\, dt + \int_{2}^{4} 0e^{-st}\, dt = \frac{10(1 - e^{-2s})}{s}$$

$$\therefore I(s) = \frac{10(1 - e^{-2s})}{s(1 - e^{-4s})}$$

$$\therefore V(s) = \frac{10(1 - e^{-2s})}{(s+1)(1 - e^{-4s})}$$

Expand $V(s)$ by partial fractions into

$$V(s) = \frac{K_1}{s+1} + \frac{V_1(s)}{1 - e^{-4s}}$$

Why? The reasons are fairly obvious: 1) the first term, with the characteristic value (pole) at $s = -1$, yields the time-domain term $K_1 e^{-t}u(t)$ which vanishes for $t \to \infty$. This is $v_{tr}(t)$, the *transient part* of $v(t)$; this will always be the case when all the characteristic values, the poles of $H(s)$, are in the left-half plane, that is when the network is BIBO stable (see Chapter 12). 2) More interesting is the second term. It represents a periodic voltage, as evidenced by the denominator $(1 - e^{-4s})$, of the same period ($T = 4$) as the input; the Laplace transform of its first cycle is $V_1(s)$. This periodic voltage contains the *steady-state part* of $v(t)$, the part that remains nonzero as $t \to \infty$.

To calculate $v(t)$, we use the usual method for $K_1$

$$K_1 = V(s)(s+1)]_{s=-1} = 10 \frac{1 - e^2}{1 - e^4} = 1.19$$

$$\therefore v_{tr}(t) = 1.19\, e^{-t}u(t)$$

To find $V_1(s)$, we recombine the partial fractions and equate the result to $V(s)$. Specifically, we get

$$10(1 - e^{-2s}) = K_1(1 - e^{-4s}) + (s+1) V_1(s)$$

or,

$$V_1(s) = \frac{1}{s+1} [10(1 - e^{-2s}) - 1.19(1 - e^{-4s})]$$

$$= \frac{10}{s+1} - \frac{10}{s+1} e^{-2s} - \left( \frac{1.19}{s+1} - \frac{1.19}{s+1} e^{-4s} \right)$$

Consequently, the first cycle is

$$\mathcal{L}^{-1} V_1(s) = v_1(t) = 10[e^{-t}u(t) - e^{-(t-2)}u(t-2)] - 1.19[e^{-t}u(t) - e^{-(t-4)}u(t-4)]$$

The steady-state response, $v_{ss}(t)$, is shown in Fig. 13-2c.

As we remarked earlier, this method is straightforward. However, it can become very involved with more complicated waveforms and networks. It also yields the transient response which is of little interest to us. We can also argue, with justification, that this method, using the Laplace transform, is valid only for a "pseudo-periodic" function, $i(t)$ being periodic for $t > 0$ but $i(t) = 0$ for $t < 0$, as is assumed for all the functions in our study of the Laplace transform. A true periodic function is periodic over the entire range of time, $-\infty < t < \infty$. The fact that the correct steady-state response is obtained via the Laplace transform rests on waiting for a long time after $t = 0$ to obtain the steady state; by then it really does not matter if the input is pseudo-periodic (starting at $t = 0$) or truly periodic. Finally, the Laplace transform method gives no information about the frequency response of the output—a very important aspect. The Fourier series and integral will provide us with the steady-state response and will focus on the frequency response. □

## 13-2   THE FOURIER SERIES

A periodic function $f(t)$ satisfies Eq. (13-1), repeated here

$$f(t \pm T) = f(t) \qquad -\infty < t < \infty \qquad (13\text{-}1a)$$

The *frequency* of $f(t)$ is its number of cycles per second (cps), and is measured in hertz (Hz)

$$f = \frac{1}{T} \qquad (13\text{-}1b)$$

The French mathematician Fourier proved that, subject to certain conditions (discussed later), such a periodic function can be written as a sum of pure sinusoids in the *fundamental frequency* $\omega_0$

$$\omega_0 = \frac{2\pi}{T} = 2\pi f \qquad (\text{rad/s}) \qquad (13\text{-}2)$$

and in its integral multiples, $n\omega_0$, called *harmonics*. Specifically,

$$f(t) = \frac{a_0}{2} + a_1 \cos \omega_0 t + a_2 \cos 2\omega_0 t + \cdots + a_n \cos n\omega_0 t + \cdots$$

$$+ b_1 \sin \omega_0 t + b_2 \sin 2\omega_0 t + \cdots + b_n \sin n\omega_0 t + \cdots$$

(13-3)

Here the first term, $a_0/2$, is a constant. The reason for the division by 2 is just a convenience, and will become obvious soon. We note that both sines and cosines may appear, and that the series is, in general, infinite. In engineering, where approximations are often done, we will stop after a finite number of terms (a *truncated series*) for a given problem.

---

### Example 2

The Fourier series for the "sawtooth" wave in Figure 13-3a is given by $f(t) = 2 \sin t - \sin 2t + \frac{2}{3} \sin 3t + \cdots$ that is, $a_0/2 = 0$, $a_n = 0$, $b_1 = 2$, $b_2 = -1$, $b_3 = 2/3, \ldots$ . In Fig. 13-3b we show the plots of the three partial sums $f_a(t) = 2 \sin t$, $f_b(t) = 2 \sin t - \sin 2t$, and $f_c(t) = 2 \sin t - \sin 2t + \frac{2}{3} \sin 3t$. We see here how the approximation improves with more terms.

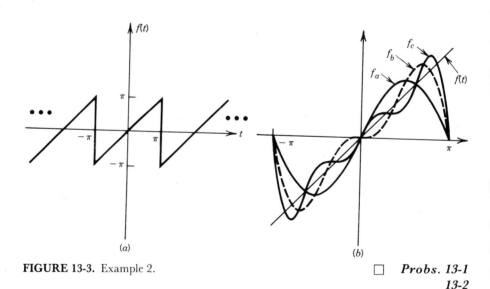

(a)  (b)

**FIGURE 13-3.** Example 2.  □ **Probs. 13-1**

**13-2**

---

The main problem that we have is to calculate the Fourier coefficients, the $a$'s and the $b$'s in Eq. (13-3), for a given $f(t)$.

## 13-3   EVALUATION OF THE FOURIER COEFFICIENTS

In order to calculate the $a$'s and the $b$'s in Eq. (13-3), we will use the following relationships

$$\int^T \sin k\omega_0 t \, dt = 0 \qquad \text{for all integer } k \tag{13-4}$$

$$\int^T \cos k\omega_0 t \, dt = 0 \qquad k \neq 0 \tag{13-5}$$

$$\int^T (\sin k\omega_0 t)(\cos p\omega_0 t) \, dt = 0 \qquad \text{for all } k, p \tag{13-6}$$

$$\int^T (\sin k\omega_0 t)(\sin p\omega_0 t) \, dt = \begin{cases} 0 & k \neq p \\ \dfrac{T}{2} & k = p \end{cases} \tag{13-7}$$

$$\int^T (\cos k\omega_0 t)(\cos p\omega_0 t) \, dt = \begin{cases} 0 & k \neq p \\ \dfrac{T}{2} & k = p \end{cases} \tag{13-8}$$

Here the integral $\int^T$ means the integral over one period, that is, from $t_0$ to $(t_0 + T)$, where $t_0$ is any point of our choice. These relationships are easy to prove by using known trigonometric identities such as

$$\cos x \cos y = \tfrac{1}{2} \left[ \cos(x + y) + \cos(x - y) \right] \tag{13-9}$$

**Prob. 13-3**

Functions which satisfy equations like (13-6), (13-7), and (13-8) are called *orthogonal functions*; the sinusoidal functions are orthogonal over the period $T$.

To calculate $a_0/2$ in Eq. (13-3), we integrate both sides of the equation over one period. We get

$$\int^T f(t) \, dt = \int^T \frac{a_0}{2} \, dt + \int^T a_1 \cos \omega_0 t \, dt + \cdots + \int^T b_1 \sin \omega_0 t \, dt + \cdots \tag{13-10}$$

All the integrals on the right, except the first one, vanish due to Eqs. (13-4) and (13-5). Therefore,

$$\int^T f(t) \, dt = \frac{a_0}{2} T \tag{13-11}$$

or,

$$\frac{a_0}{2} = \frac{1}{T} \int^T f(t) \, dt \tag{13-12}$$

This is the net area under one cycle, divided by $T$. Therefore, the constant term is the *average value* of $f(t)$, sometimes called the *dc term*, and, in many cases, we will be able to calculate it by inspection.

To calculate $a_1$ we multiply by $\cos \omega_0 t$ both sides of Eq. (13-3), then integrate over $T$.

$$\int^T f(t) \cos \omega_0 t \, dt = \int^T \frac{a_0}{2} \cos \omega_0 t \, dt + \int^T a_1 (\cos \omega_0 t)^2 \, dt \qquad (13\text{-}13)$$

$$+ \int^T a_2 \cos 2\omega_0 t \cos \omega_0 t \, dt + \cdots$$

$$+ \int^T b_1 \sin \omega_0 t \cos \omega_0 t \, dt + \cdots$$

Because of the orthogonality relations, Eqs. (13-4) through (13-8), the only non-zero integral on the right side in Eq. (13-13) is the second one. Thus

$$\int^T f(t) \cos \omega_0 t \, dt = a_1 \frac{T}{2} \qquad (13\text{-}14)$$

or

$$a_1 = \frac{2}{T} \int^T f(t) \cos \omega_0 t \, dt \qquad (13\text{-}15)$$

The pattern is clear now. To calculate $a_2$, we multiply Eq. (13-3) by $\cos 2\omega_0 t$, then integrate over $T$. All the integrals, except the one with $a_2$, will vanish. The result is

$$a_2 = \frac{2}{T} \int^T f(t) \cos 2\omega_0 t \, dt \qquad (13\text{-}16)$$

In general,

$$a_k = \frac{2}{T} \int^T f(t) \cos k\omega_0 t \, dt \qquad k = 0, 1, 2, \ldots \qquad (13\text{-}17\text{a})$$

similarly,

$$b_k = \frac{2}{T} \int^T f(t) \sin k\omega_0 t \, dt \qquad k = 1, 2, \ldots \qquad (13\text{-}17\text{b})$$

In Eq. (13-17) we have the desired results for computing the Fourier coefficients. We notice that Eq. (13-17a) is good also for $k = 0$, thus explaining the need to divide $a_0$ by 2 in order to obtain Eq. (13-12).

---

## Example 3

Let us verify the Fourier series for the "sawtooth" waveform, given in example 2. Here we have

$$f(t) = t \qquad -\pi < t < \pi$$

and $T = 2\pi$. Consequently, $\omega_0 = 2\pi/T = 1$. Using Eq. (13-12), we see *by inspection* that $a_0/2 = 0$, since the total (net) area under $f(t)$ over one period is zero. Use now Eq. (13-17a):

$$a_k = \frac{2}{2\pi} \int_{-\pi}^{\pi} t \cos kt \, dt = \frac{1}{\pi} \left[ \frac{\cos kt}{k^2} + \frac{t \sin kt}{k} \right]_{-\pi}^{\pi}$$

where the integration was done by parts or by looking it up in tables. Finally,

$$a_k = \frac{1}{\pi} \left[ \frac{\cos k\pi}{k^2} + \frac{\pi \sin k\pi}{k} - \frac{\cos(-k\pi)}{k^2} + \frac{\pi \sin(-k\pi)}{k} \right] = 0$$

Next, use Eq. (13-17b) to find $b_k$:

$$b_k = \frac{2}{2\pi} \int_{-\pi}^{\pi} t \sin kt \, dt = \frac{1}{\pi} \left[ \frac{\sin kt}{k^2} - \frac{t \cos kt}{k} \right]_{-\pi}^{\pi}$$

for $k = 1$, we get

$$b_1 = \frac{1}{\pi} \left[ \frac{\pi}{1} + \frac{\pi}{1} \right] = 2$$

for $k = 2$,

$$b_2 = \frac{1}{\pi} \left[ \frac{-\pi}{2} + \frac{-\pi}{2} \right] = -1$$

for $k = 3$,

$$b_3 = \frac{1}{\pi} \left[ \frac{\pi}{3} + \frac{\pi}{3} \right] = \frac{2}{3}$$

in agreement with Example 2.                                                    □

---

When $f(t)$ has different expressions over its period, the integrals in Eq. (13-17) must be evaluated as sums of several integrals, each over an appropriate time interval with its appropriate expression for $f(t)$. For example, in Fig. 13-1a, we have the following expressions of $f(t)$

$$f(t) = 0 \qquad -T/2 < t < -a/2$$
$$f(t) = A \qquad -a/2 < t < a/2$$
$$f(t) = 0 \qquad a/2 < t < T/2$$

**Probs. 13-4
through
13-9**

When $f(t)$ satisfies certain *symmetry* conditions, there are further simplifications for its Fourier series and coefficients. We will discuss two such conditions.

a) if $f(t)$ is an *even* function, satisfying

$$f(t) = f(-t) \tag{13-18}$$

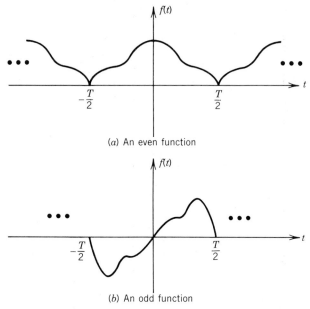

(a) An even function

(b) An odd function

**FIGURE 13-4.** Even and odd functions.

that is, $f(t)$ is reflected (as in a mirror) about the vertical axis. See Fig. 13-4a. For such a function, it can be shown that $b_k \equiv 0$; in other words, the Fourier series contains only cosine terms (and possibly a constant, $a_0/2$).

b) if $f(t)$ is an *odd* function, satisfying

$$f(t) = -f(-t) \tag{13-19}$$

that is, $f(t)$ is reflected about the vertical and the horizontal axes, as shown in Fig. 13-4b. For such a function all $a_k \equiv 0$, and the Fourier series contains only sine terms.

These results are also plausible intuitively: the cosine function itself is even, while the sine is an odd function. It stands to reason, then, that an even $f(t)$ contains only cosines, while an odd $f(t)$ contains only sines. **Probs. 13-10**

**13-11**

## 13-4 CONVERGENCE OF THE FOURIER SERIES

It was mentioned earlier that if the periodic function $f(t)$ satisfies certain conditions, it can be expanded in a Fourier series. These conditions are called the *Dirichlet conditions*:

1. $f(t)$ may have, at most, a finite number of discontinuities ("jumps") over one period.

2. $f(t)$ may have, at most, a finite number of maxima and minima over one period.

3. $f(t)$ must be absolutely integrable over one period,

$$\int^T |f(t)|\, dt = M \qquad M < \infty \qquad (13\text{-}20)$$

Most periodic engineering functions satisfy the Dirichlet conditions.

At a point of discontinuity, $t = t_0$ in Fig. 13-5, the Fourier series converges to the average value

$$f(t_0) = \tfrac{1}{2}[f(t_{0-}) + f(t_{0+})] \qquad (13\text{-}21)$$

where $t_{0-}$ is the point just before $t_0$ and $t_{0+}$ is just after $t_0$.

Consider again the expressions for the Fourier coefficients, Eq. (13-17). As $k$ increases, both $\cos k\omega_0 t$ and $\sin k\omega_0 t$ will alternate signs faster and faster, thus reducing the value of the integrals. In other words, we expect $a_k$ and $b_k$ to approach zero as $k$ increases. This is, in fact, the case:

1. When $f(t)$ has a finite number of discontinuities, $f'(t)$, its first derivative, contains the impulse function. Then $a_k$ and $b_k$ decrease in magnitude as $1/k$. An example is the square wave, Fig. 13-1a and problem 13-4.

2. When $f(t)$ is continuous, and $f'(t)$ has discontinuities so that $f''(t)$ contains impulses, then $a_k$ and $b_k$ decrease in magnitude as $1/k^2$. An example is given in Problem 13-21.

3. When $f(t)$ and $f'(t)$ are continuous but $f''(t)$ has discontinuities so that $f'''(t)$ has impulses, then $a_k$ and $b_k$ decrease as $1/k^3$. As an example, consider a periodic parabolic function, $t^2$.

4. In general, when $f(t), f'(t), \ldots f^{(n-1)}(t)$ are continuous, but the $n^{\text{th}}$ derivative, $f^{(n)}(t)$, has discontinuities, $a_k$ and $b_k$ decrease as $1/k^{n+1}$.    ***Prob. 13-12***

Let us consider the error in approximating a periodic $f(t)$ by a *finite* Fourier series. Specifically, let

$$S_n(t) = \frac{a_0}{2} + \sum_{k=1}^{n} (a_k \cos k\omega_0 t + b_k \sin k\omega_0 t) \qquad (13\text{-}22)$$

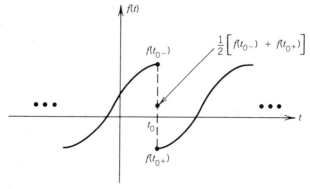

**FIGURE 13-5.** A point of discontinuity.

be the *finite* series of $(2n + 1)$ terms. The error of such an approximation is

$$\varepsilon_n(t) = f(t) - S_n(t) \tag{13-23}$$

and, in general, it changes in magnitude and sign over the period. A commonly used measure of this error, in many engineering applications, is the average of the squared $\varepsilon_n(t)$, called the *mean square error*,

$$E_n = \frac{1}{T} \int_0^T [\varepsilon_n(t)]^2 \, dt \tag{13-24}$$

The reasons for its usefulness are: (1) we are not interested in instantaneous positive or negative errors, so we square $\varepsilon_n(t)$ to get $[\varepsilon_n(t)]^2$, a positive quantity; (2) we average this quantity over a period, to get a *constant*, $E_n$, dependent on $n$, the number of terms taken. In our specific case, we have

$$E_n = \frac{1}{T} \int_0^T \left[ f(t) - \frac{a_0}{2} - \sum_{k=1}^n a_k \cos k\omega_0 t - \sum_{k=1}^n b_k \sin k\omega_0 t \right]^2 dt \tag{13-25}$$

We now wish to *minimize* $E_n$, a reasonable step for an approximation. If we consider $E_n$ as a function of the $a$'s and $b$'s,

$$E_n = E_n(a_0, a_1, a_2, \ldots a_n, b_1, b_2, \ldots b_n) \tag{13-26}$$

then $E_n$ will be minimum if

$$\frac{\partial E_n}{\partial a_0} = 0, \frac{\partial E_n}{\partial a_1} = 0, \ldots, \frac{\partial E_n}{\partial a_n} = 0; \frac{\partial E_n}{\partial b_1} = 0, \ldots, \frac{\partial E_n}{\partial b_n} = 0 \tag{13-27}$$

Take the partial derivative of $E_n$, Eq. (13-25), with respect to $a_0$ and equate it to zero:

$$\frac{\partial E_n}{\partial a_0} = \frac{a_0}{2} - \frac{1}{T} \int_0^T f(t) \, dt = 0 \tag{13-28}$$

Therefore

$$\frac{a_0}{2} = \frac{1}{T} \int_0^T f(t) \, dt \tag{13-29}$$

Similarly setting $\partial E_n / \partial a_k = 0$ from Eq. (13-25) and using the orthogonality properties of the sine and cosine, we get

$$\frac{\partial E_n}{\partial a_k} = a_k - \frac{2}{T} \int_0^T f(t) \cos k\omega_0 t \, dt \tag{13-30}$$

that is,

$$a_k = \frac{2}{T} \int_0^T f(t) \cos k\omega_0 t \, dt \tag{13-31}$$

Similar results are obtained by setting $\partial E_n / \partial b_k = 0$, namely,

$$\frac{\partial E_n}{\partial b_k} = b_k - \frac{2}{T} \int_0^T f(t) \sin k\omega_0 t \, dt = 0 \tag{13-32}$$

that is,

$$b_k = \frac{2}{T} \int^T f(t) \sin k\omega_0 t \, dt \qquad (13\text{-}33)$$

The important conclusion that we draw is: of all possible finite trigonometric sums of the form

$$f(t) = \alpha_0 + \alpha_1 \cos \omega_0 t + \alpha_2 \cos 2\omega_0 t + \cdots + \alpha_n \cos n\omega_0 t$$
$$+ \beta_1 \sin \omega_0 t + \beta_2 \sin 2\omega_0 t + \cdots + \beta_n \sin n\omega_0 t \qquad (13\text{-}34)$$

which approximate a periodic $f(t)$, the Fourier series does it with *the least* mean-square error, that is,

$$\alpha_0 = \frac{a_0}{2}$$

$$\alpha_1 = a_1 \qquad \beta_1 = b_1$$
$$\vdots \qquad\qquad \vdots \qquad\qquad (13\text{-}35)$$
$$\alpha_n = a_n \qquad \beta_n = b_n$$

## 13-5   THE COSINE FORM OF THE FOURIER SERIES

In many applications, it is more convenient to rewrite the two terms of the $k^{\text{th}}$ harmonic, $(a_k \cos k\omega_0 t + b_k \sin k\omega_0 t)$, as an equivalent single cosine term. To do so, we divide and multiply by $(a_k^2 + b_k^2)^{1/2}$, as follows:

$$a_k \cos k\omega_0 t + b_k \sin k\omega_0 t$$

$$= \sqrt{a_k^2 + b_k^2} \left[ \frac{a_k}{\sqrt{a_k^2 + b_k^2}} \cos k\omega_0 t + \frac{b_k}{\sqrt{a_k^2 + b_k^2}} \sin k\omega_0 t \right] \qquad (13\text{-}36)$$

Next we refer to Fig. 13-6, where the right-angle triangle shows these relations. From this figure we see that

$$\frac{a_k}{\sqrt{a_k^2 + b_k^2}} \cos k\omega_0 t + \frac{b_k}{\sqrt{a_k^2 + b_k^2}} \sin k\omega_0 t = \cos \theta_k \cos k\omega_0 t$$

$$+ \sin \theta_k \sin k\omega_0 t = \cos(k\omega_0 t - \theta_k) \qquad (13\text{-}37)$$

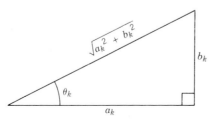

**FIGURE 13-6.** Relations among $a_k$, $b_k$, and $\theta_k$.

Therefore the $k^{th}$ harmonic is

$$\sqrt{a_k^2 + b_k^2} \cos(k\omega_0 t - \theta_k) = A_k \cos(k\omega_0 t - \theta_k) \tag{13-38}$$

with

$$A_k = \sqrt{a_k^2 + b_k^2} \qquad \text{the amplitude}$$
$$\theta_k = \tan^{-1} b_k/a_k \qquad \text{the phase angle}$$

The cosine Fourier series is then

$$f(t) = A_0 + \sum_{k=1}^{\infty} A_k \cos(k\omega_0 t - \theta_k) \tag{13-39}$$

with $A_0 = a_0/2$, the dc (average) term.

---

### Example 4

A certain periodic current ($\omega_0 = 2$) is given by

$$i(t) = 4 + 3.1 \cos 2t - 1.1 \cos 4t - 0.3 \cos 6t + \cdots$$
$$- 2.7 \sin 2t - 1.4 \sin 4t + 0.4 \sin 6t + \cdots$$

The dc term is $A_0 = 4$. The first harmonic has an amplitude of

$$A_1 = \sqrt{(3.1)^2 + (-2.7)^2} = 4.11$$

and a phase angle of

$$\theta_1 = \tan^{-1} \frac{-2.7}{3.1} = -41° \qquad \text{(note the quadrant of } \theta_1!)$$

For the second harmonic, we get

$$A_2 = \sqrt{(-1.1)^2 + (-1.4)^2} = 1.78$$

$$\theta_2 = \tan^{-1} \frac{-1.4}{-1.1} = -128.16° \qquad \text{(note the quadrant)}$$

For the third harmonic, we get

$$A_3 = \sqrt{(-0.3)^2 + (0.4)^2} = 0.5$$

$$\theta_3 = \tan^{-1} \frac{0.4}{-0.3} = 126.87° \qquad \text{(note the quadrant)}$$

Therefore, the cosine Fourier series for $i(t)$ is

$$i(t) = 4 + 4.11 \cos(2t + 41°) + 1.78 \cos(4t + 128.16°)$$
$$+ 0.5 \cos(6t - 126.87°) + \cdots$$

☐ **Probs. 13-13**
**13-14**

## 13-6   FOURIER SERIES, SUPERPOSITION, AND PHASORS

One of the major advantages of using the Fourier series is found in the simple calculation of the steady-state response. The idea is straightforward: if the input (current or voltage source) is periodic and the network is linear, superposition and phasors are used to calculate the steady-state response. Two examples will illustrate this approach.

---

### *Example 5*

Let a periodic current source, $i(t)$, be the input to a parallel $RC$ circuit, as shown in Fig. 13-7$a$. We wish to calculate the steady-state response $v(t)$. The cosine Fourier series for $i(t)$ is given as

$$i(t) = 10 + 6\cos(100t + 45°) + 3\cos(200t - 10°) + 2.1\cos(300t + 35°) + \cdots$$

and is shown in the equivalent circuit in Fig. 13-7$b$. Here

$$i_0(t) = 10$$
$$i_1(t) = 6\cos(100t + 45°)$$
$$i_2(t) = 3\cos(200t - 10°)$$
$$i_3(t) = 2.1\cos(300t + 35°)$$

Let us use superposition, with one source at a time. Since each source is purely sinusoidal, phasor analysis yields

$$\mathbf{V} = \mathbf{ZI}$$

for each source applied separately. Here $\mathbf{V}$ is the phasor voltage, $\mathbf{I}$ the phasor current, and $\mathbf{Z}$ the driving-point sinusoidal impedance of the $RC$ circuit. See Fig. 13-7$c$.

For any single frequency, $\omega$, we have

$$\mathbf{Z} = \frac{1}{\mathbf{Y}} = \frac{1}{2 + (j\omega)(0.02)}$$

With the dc source $i_0(t)$ ($\omega = 0$) alone we have $\mathbf{I}_0 = 10$ and

$$\mathbf{V}_0 = (\tfrac{1}{2})(10) = 5 \qquad \therefore v_0(t) = 5$$

With the first harmonic source $i_1(t)$, $\omega = 100$, we have $\mathbf{I}_1 = 6\underline{|45°}$ and

$$\mathbf{V}_1 = \left(\frac{1}{2 + j2}\right) 6\underline{|45°} = \frac{6\underline{|45°}}{2.828\underline{|45°}} = 2.121\underline{|0°}$$

$$\therefore v_1(t) = 2.121\cos 100t$$

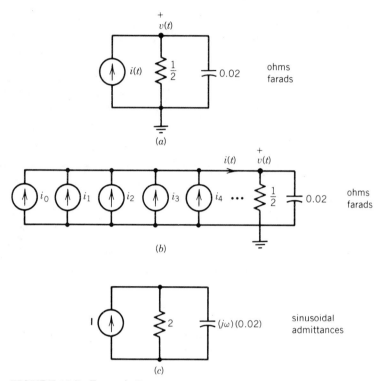

**FIGURE 13-7.** Example 5.

With the second harmonic source $i_2(t)$, $\omega = 200$, we have $\mathbf{I}_2 = 3\lfloor -10°$ and

$$\mathbf{V}_2 = \left(\frac{1}{2 + j4}\right) 3\lfloor -10° = \frac{3\lfloor -10°}{4.472\lfloor 63.4°} = 0.671\lfloor -73.4°$$

$$\therefore v_2(t) = 0.671 \cos(200t - 73.4°)$$

With the third harmonic source, $i_3(t)$, $\omega = 300$, we have $\mathbf{I}_3 = 2.1\lfloor 35°$ and

$$\mathbf{V}_3 = \left(\frac{1}{2 + j6}\right) 2.1\lfloor 35° = \frac{2.1\lfloor 35°}{6.325\lfloor 71.56°} = 0.332\lfloor -36.56°$$

$$\therefore v_3(t) = 0.332 \cos(300t - 36.56°)$$

Consequently, by superposition, the steady-state output voltage $v(t)$ is the sum of the individual outputs,

$$v(t) = 5 + 2.121 \cos 100t + 0.671 \cos(200t - 73.4°)$$
$$+ 0.332 \cos(300t - 36.56°) + \cdots$$

It is periodic (with the same period as the input), and given by its Fourier cosine series.   □

## Example 6

Let the half-wave rectified sine wave, Fig. 13-1$c$ and problems 13-5 and 13-13, have an amplitude of 156 V and a period $T = 1/60$ s. Such a waveform, $v_r(t)$, is the output of a half-wave rectifier circuit connected to the common household outlet, as shown in Fig. 13-8. It is to be used for a resistive load, for example, a car battery that needs charging. Quite obviously, the variations of $v_r(t)$ are not very desirable, even though $v_r(t)$ does not change polarity. To smooth out $v_r(t)$, an $LC$ filter is introduced, as shown. We know that (see problem 13-13)

$$v_r(t) = \frac{156}{\pi} + 78\cos(\omega_0 t - 90°) - \frac{312}{\pi} \sum_{k=2,4,6,...} \frac{1}{k^2 - 1} \cos k\omega_0 t$$

where $\omega_0 = 2\pi(60) = 377$ rad/s.

Let us calculate a few terms of the output voltage $v_R(t)$, using superposition and phasors. For any one frequency, $\omega_k$, the phasor relation is given by

$$\mathbf{V}_R = \mathbf{V}_r \frac{\mathbf{Z}_{RC}}{\mathbf{Z}_{RC} + \mathbf{Z}_L}$$

a voltage divider equation. Specifically, here

$$\mathbf{V}_R = \mathbf{V}_r \frac{\dfrac{1}{(j\omega)10^{-5} + 10^{-3}}}{\dfrac{1}{(j\omega)10^{-5} + 10^{-3}} + (j\omega)(2)} = \mathbf{V}_r \frac{1}{1 - (2)(10^{-5})\omega^2 + j(2)10^{-3}\omega}$$

The individual calculations are:

For dc ($\omega = 0$):

$$\mathbf{V}_{R_0} = \frac{156}{\pi} \qquad \therefore v_{R_0}(t) = \frac{156}{\pi}$$

which is obvious since, at dc, the inductor is a short circuit and the capacitor an open circuit.

For the first harmonic ($\omega = \omega_0 = 377$):

$$\mathbf{V}_{R_1} = 78\underline{|-90°} \frac{1}{-1.84 + j0.754} = \frac{78\underline{|-90°}}{1.99\underline{|157.7°}} = 39.2\underline{|112.3°}$$

$$\therefore v_{R_1}(t) = 39.2\cos(377t + 112.3°)$$

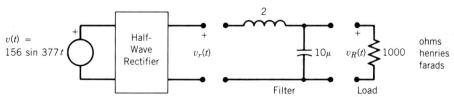

FIGURE 13-8. Example 6.

*For the second harmonic, $k = 2$ (to be subtracted):*

$$\mathbf{V}_{R_2} = 33.1\ \underline{|0°}\ \frac{1}{-10.37 + j1.51} = \frac{33.1\ \underline{|0°}}{10.48\ \underline{|171.7°}} = 3.16\ \underline{|-171.7°}$$

$$\therefore v_{R_2}(t) = 3.16\ \cos(754t - 171.7°)$$

and so on for the higher harmonics. Therefore

$$v_R(t) = \frac{156}{\pi} + 39.2\ \cos(377t + 112.3°) - 3.16\ \cos(754t - 171.7°) + \cdots$$

showing a marked decrease in the amplitudes of the harmonics, and therefore a smoother waveform. A measure of comparison for the smoothness (flatness) of such waveforms is the *ripple factor* of any harmonic, defined as

$$\%\ \text{ripple factor} = \frac{\text{amplitude of harmonic}}{\text{dc term}} \times 100$$

For the rectified wave, $v_r(t)$, the ripple factors for the first and second harmonics are, respectively,

$$\frac{78}{156/\pi} \times 100 = 157\%$$

and

$$\frac{(312)\pi}{(3\pi)156} \times 100 = 67\%$$

For the output voltage across the resistor, these ripple factors are improved

$$\frac{39.2}{156/\pi} \times 100 = 79\%$$

and

$$\frac{3.16}{156/\pi} \times 100 = 6.4\% \qquad \square\quad \textbf{Prob. 13-15}$$

## 13-7 THE EXPONENTIAL FOURIER SERIES

A third form of the Fourier series is equally useful. It expresses the sine and cosine as

$$\sin k\omega_0 t = \frac{1}{2j}(e^{jk\omega_0 t} - e^{-jk\omega_0 t}) \qquad (13\text{-}40)$$

and

$$\cos k\omega_0 t = \tfrac{1}{2}(e^{jk\omega_0 t} + e^{-jk\omega_0 t}) \qquad (13\text{-}41)$$

as derived from Euler's identity $e^{jx} = \cos x + j \sin x$.

If we use Eqs. (13-40) and (13-41) in the Fourier series, Eq. (13-3), we get

$$f(t) = \frac{a_0}{2} + \sum_{k=1}^{\infty} a_k \frac{e^{jk\omega_0 t} + e^{-jk\omega_0 t}}{2} + \sum_{k=1}^{\infty} b_k \frac{e^{jk\omega_0 t} - e^{-jk\omega_0 t}}{2j} \qquad (13\text{-}42)$$

Let us rearrange these terms, collecting positive exponentials and negative exponentials

$$f(t) = \frac{a_0}{2} + \sum_{k=1}^{\infty} \left( \frac{a_k - jb_k}{2} \right) e^{jk\omega_0 t} + \sum_{k=1}^{\infty} \left( \frac{a_k + jb_k}{2} \right) e^{-jk\omega_0 t} \qquad (13\text{-}43)$$

where we use $1/j = -j$. Define now a new coefficient, $c_k$, as

$$c_k = \frac{a_k - jb_k}{2} \qquad c_0 = \frac{a_0}{2} \qquad (13\text{-}44)$$

which, in general, is complex. Then Eq. (13-43) can be written compactly as

$$f(t) = c_0 + \sum_{k=1}^{\infty} c_k e^{jk\omega_0 t} + \sum_{k=1}^{\infty} c_k^* e^{-jk\omega_0 t} \qquad (13\text{-}45)$$

with $c_k^* = \frac{1}{2}(a_k + jb_k)$, the complex conjugate of $c_k$. In both summations, $k$ goes from $+1$ to $+\infty$. As a final simplification in notation, let us introduce

$$c_{-k} = c_k^* \qquad (13\text{-}46)$$

Then Eq. (13-45) reads

$$f(t) = \sum_{k=-\infty}^{k=+\infty} c_k e^{jk\omega_0 t} \qquad (13\text{-}47)$$

Notice that here the index $k$ goes from $-\infty$ to $+\infty$; for negative integers, from $-\infty$ to $-1$, we get the terms in the second summation of Eq. (13-45); then, when $k = 0$, we obtain the dc term, $c_0$; finally, as $k$ goes through all positive integers, $+1$ to $+\infty$, we get the first summation in Eq. (13-45). Equation (13-47) is a very compact form of the exponential form of the Fourier series.

The explicit form of the coefficient $c_k$ is obtained from Eqs. (13-44) and (13-17):

$$c_k = \tfrac{1}{2}(a_k - jb_k) = \frac{1}{2} \left[ \frac{2}{T} \int_0^T f(t) \cos k\omega_0 t \, dt - j\frac{2}{T} \int_0^T f(t) \sin k\omega_0 t \, dt \right]$$

$$= \frac{1}{T} \int_0^T f(t) (\cos k\omega_0 t - j \sin k\omega_0 t) \, dt$$

$$= \frac{1}{T} \int_0^T f(t) e^{-jk\omega_0 t} \, dt \qquad (13\text{-}48)$$

where, in the last step, we used again Euler's identity. Equation (13-48) is valid for all $k$ (negative, zero, positive).

***

### *Example 7*

Let us evaluate $c_k$ for the square waveform in Fig. 13-1a, assuming $f(t) = A$, $-\dfrac{T}{4} < t < \dfrac{T}{4}$. We write here

$$
c_k = \frac{1}{T} \int^T f(t) e^{-jk\omega_0 t}\, dt = \frac{1}{T} \int_{-T/4}^{T/4} A e^{-jk\omega_0 t}\, dt
$$

$$
= \frac{A}{T} \left. \frac{e^{-jk\omega_0 t}}{-jk\omega_0} \right]_{-T/4}^{T/4}
$$

$$
= \frac{A}{T} \frac{e^{jk\omega_0 T/4} - e^{-jk\omega_0 T/4}}{jk\omega_0}
$$

$$
= \frac{A}{T} \frac{e^{jk\pi/2} - e^{-jk\pi/2}}{jk\omega_0}
$$

$$
= \frac{A}{2} \frac{1}{\dfrac{k\omega_0 T}{4}} \frac{e^{jk\pi/2} - e^{-jk\pi/2}}{2j}
$$

$$
= \frac{A}{2} \frac{\sin k\pi/2}{k\pi/2}
$$

As expected, $c_k$ is real in this case: $f(t)$ is an even function, with $b_k \equiv 0$, so $c_k = \dfrac{a_k}{2}$ is real. The first few coefficients are therefore

$$
c_0 = \frac{A}{2} \left( \frac{\sin k\pi/2}{k\pi/2} \right)_{k=0} = \frac{A}{2} \left( \frac{(\pi/2)\cos k\pi/2}{\pi/2} \right)_{k=0} = \frac{A}{2}
$$

Here, l'Hôpital's rule was used to evaluate the indeterminate form $0/0$.

$$
c_1 = \frac{A}{2} \frac{\sin \pi/2}{\pi/2} = \frac{A}{\pi}
$$

$$
c_2 = \frac{A}{2} \frac{\sin \pi}{\pi} = 0
$$

$$
c_3 = \frac{A}{2} \frac{\sin 3\pi/2}{3\pi/2} = -\frac{A}{3\pi}
$$

$$
\vdots
$$

etc.                                                    □  **Probs. 13-16**
                                                              **13-17**

## 13-8   POWER RELATIONSHIPS

Let us calculate the power in a network when the voltage and current are periodic and expressed in their Fourier series. Specifically, let us use the cosine Fourier series

$$v(t) = V_0 + V_1 \cos(\omega_0 t + \alpha_1) + V_2 \cos(2\omega_0 t + \alpha_2) + \cdots \quad (13\text{-}49\text{a})$$

and

$$i(t) = I_0 + I_1 \cos(\omega_0 t + \beta_1) + I_2 \cos(2\omega_0 t + \beta_2) + \cdots \quad (13\text{-}49\text{b})$$

where we have made use of an earlier observation that both functions have the same period (and therefore the same $\omega_0$). The instantaneous power is $p(t) = v(t)i(t)$, but of more practical interest is the *average power*

$$P_{\text{avg}} = \frac{1}{T} \int_0^T p(t)\, dt = \frac{1}{T} \int_0^T v(t)i(t)\, dt \quad (13\text{-}50)$$

Evaluating this integral with the expressions for $v(t)$ and $i(t)$ as given, and with the orthogonality relations of the cosine function, we obtain

$$P_{\text{avg}} = V_0 I_0 + \frac{V_1 I_1}{2} \cos(\alpha_1 - \beta_1) + \frac{V_2 I_2}{2} \cos(\alpha_2 - \beta_2) + \cdots \quad (13\text{-}51)$$

For each sinusoidal harmonic, we define the *effective values* of voltage and current as

$$V_{\text{eff}_k} = \frac{V_k}{\sqrt{2}} \qquad I_{\text{eff}_k} = \frac{I_k}{\sqrt{2}} \quad (13\text{-}52)$$

and, consequently,

$$P_{\text{avg}} = V_0 I_0 + \sum_{k=1}^{\infty} V_{\text{eff}_k} I_{\text{eff}_k} \cos\phi_k$$

$$= P_{\text{avg}_0} + \sum_{k=1}^{\infty} P_{\text{avg}_k} \quad (13\text{-}53)$$

where $P_{\text{avg}_k}$ is the average power of the $k^{\text{th}}$ harmonic

$$P_{\text{avg}_k} = V_{\text{eff}_k} I_{\text{eff}_k} \cos\phi_k \quad (13\text{-}54)$$

and $\cos\phi_k = \cos(\alpha_k - \beta_k)$ is the power factor of the $k^{\text{th}}$ harmonic.

The result obtained in Eq. (13-53) is important: the total average power is the sum of the average powers of all harmonics. *Different* harmonics of voltage and current, for example, $V_3 \cos(3\omega_0 t + \alpha_3)$ and $I_4 \cos(4\omega_0 t + \beta_4)$, *do not* contribute to the average power.

---

### Example 8

If

$$v(t) = 10 + 6\cos(t + 45°) + 1.8\cos(2t - 10°)$$

$$i(t) = 3 + 1.4\cos(t + 20°) + 0.5\cos 2t$$

Then

$$P_{avg} = (10)(3) + \left(\frac{6}{\sqrt{2}}\right)\left(\frac{1.4}{\sqrt{2}}\right)\cos 25° + \left(\frac{1.8}{\sqrt{2}}\right)\left(\frac{0.5}{\sqrt{2}}\right)\cos(-10°) = 34.25W$$

□ **Probs. 13-18**<br>**13-19**<br>**13-20**

## 13-9  DISCRETE SPECTRA

As we have seen, a periodic $f(t)$ is specified completely by its Fourier series. That is, from the given $f(t)$ we can get its, $a_k$, $b_k$ [Eq. (13-17)] or its $A_k$, $\theta_k$ [Eq. (13-38)], or its $c_k$ [Eq. (13-48)]. Conversely, given $a_k$ and $b_k$, or $A_k$ and $\theta_k$, or $c_k$, we can get $f(t)$. This idea is shown in Table 13-1 below, and we recognize here the familiar features of a transform:

### TABLE 13-1  Frequency Transform

| Time domain | Frequency domain $(\omega_0, 2\omega_0, \ldots)$ |
|---|---|
| $f(t)$ | $a_k, b_k$<br>OR: $A_k, \theta_k$<br>OR: $c_k$ |

a time-domain function $f(t)$ is transformed into the frequency domain. Conversely, from the frequency domain description, we can get back (an inverse transform!) the time domain function. The formal derivation of the Fourier transform will be discussed in the next chapter. Another feature of a transform, perhaps the most important one, is also evident here: the ease of calculations in the frequency domain. The various amplitudes and phase angles of the harmonics are easily displayed and calculated; in the time domain, on the other hand, very little can be done (or understood) with, say, the fact that $f(t)$ is a periodic square wave. The insight into the frequency domain and the ease of calculations offered by the Fourier series are, indeed, the reasons for this transform.

In order to highlight further these aspects in the frequency domain, we resort to a common feature in engineering, namely, graphical displays. Consider the cosine form of the Fourier series, repeated here

$$f(t) = A_0 + \sum_{k=1}^{\infty} A_k \cos(k\omega_0 t - \theta_k) \tag{13-55}$$

The amplitudes $A_0, A_1, \ldots$ and the angles $\theta_0, \theta_1, \ldots$ of the various harmonics can be plotted versus $k\omega_0$. These plots are called the *discrete amplitude spectrum* and the *discrete phase spectrum* for $f(t)$, and they provide the graphical information on $A_k$ and $\theta_k$ for Eq. (13-55). In plotting $A_k$, we treat amplitudes as positive quantities. If a particular harmonic has a negative sign in front of it, we subtract 180° from the angle, since

$$-A_k \cos(k\omega_0 t - \theta_k) = +A_k \cos(k\omega_0 t - \theta_k - 180°) \tag{13-56}$$

### Example 9

For the half-wave rectified sine (amplitude $= A$, period $= T$) in Fig. 13-1$c$ we have

$$f(t) = \frac{A}{\pi} + \frac{A}{2} \cos(\omega_0 t - 90°) + \frac{2A}{\pi} \sum_{k=2,4,6...} \frac{1}{k^2 - 1} \cos(k\omega_0 t - 180°)$$

The discrete amplitude and phase spectra are shown in Fig. 13-9. Values of amplitude and phase exist only for $k\omega_0 = 0, \omega_0, 2\omega_0, \ldots.$ Therefore we call these plots *discrete*.

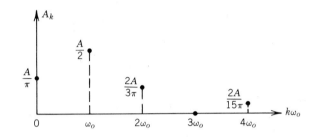

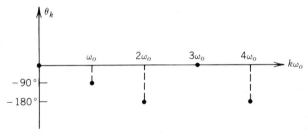

**FIGURE 13-9.**  Example 9.

The same ideas apply to the exponential form of the Fourier series. Since

$$c_k = \frac{a_k - jb_k}{2} = |c_k|e^{j\theta_k} \qquad (13\text{-}57)$$

we obtain

$$|c_k| = \tfrac{1}{2}\sqrt{a_k^2 + b_k^2} = \tfrac{1}{2}A_k \qquad (13\text{-}58)$$

that is, the discrete magnitude spectrum, $|c_k|$, is a scaled version (by $\tfrac{1}{2}$) of the amplitude spectrum $A_k$. Also, in plotting $|c_k|$ versus $k\omega_0$, we remember that

$$c_{-k} = c_k^* \qquad (13\text{-}59a)$$

and therefore

$$|c_{-k}| = |c_k^*| = |c_k| \qquad (13\text{-}59b)$$

so the plot extends from $-\infty$ to $+\infty$, and is *even*.

## Example 10

Consider the square wave of Fig. 13-1$a$ with

$$
f(t) = \begin{cases} A & -\dfrac{a}{2} < t < \dfrac{a}{2} \\[2ex] 0 & -\dfrac{T}{2} < t < -\dfrac{a}{2}; \dfrac{a}{2} < t < \dfrac{T}{2} \end{cases}
$$

This is an even function ($b_k = 0$), and we calculate $c_k$ as follows:

$$
\begin{aligned}
c_k &= \frac{1}{T} \int^T f(t) e^{-jk\omega_0 t}\, dt = \frac{A}{T} \int_{-a/2}^{a/2} e^{-jk\omega_0 t}\, dt \\[2ex]
&= \frac{A}{T} \frac{1}{jk\omega_0} (e^{jk\omega_0 a/2} - e^{-jk\omega_0 a/2}) \\[2ex]
&= \frac{Aa}{T} \frac{\sin(k\omega_0 a/2)}{(k\omega_0/2)}
\end{aligned}
$$

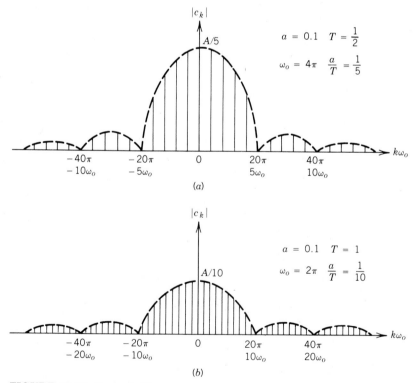

FIGURE 13-10.  Example 10.

and since $\omega_0 = 2\pi/T$ we get

$$c_k = \frac{Aa}{T} \frac{\sin(k\pi a/T)}{(k\pi a/T)}$$

As expected, $c_k$ is real, since $c_k = \frac{1}{2}(a_k - jb_k)$ and $b_k = 0$. Therefore the phase $\theta_k$ is zero. The amplitude spectrum, $|c_k|$ plotted vs. $k\omega_0$, is shown in Fig. 13-10 for two cases:

a)  $a = 0.1$, $T = 1/2$     $\therefore$ $\omega_0 = 4\pi$

Here $a/T = 1/5$ and the discrete values of $|c_k|$ are:

$$k = 0 \qquad |c_0| = \frac{A}{5}\left(\frac{\sin k\pi/5}{k\pi/5}\right)_{k=0} = \frac{A}{5}$$

where l'Hôpital's rule was used for the indeterminate form $0/0$. Alternatively, (and more easily!) we could have calculated $c_0 = a_0/2$ as the average value of $f(t)$, $(1/T)(Aa) = A/5$.

$$k = 1 \qquad |c_1| = \frac{A}{5}\frac{\sin \pi/5}{\pi/5} = \frac{A}{5}(0.935)$$

$$k = 2 \qquad |c_2| = \frac{A}{5}\frac{\sin 2\pi/5}{2\pi/5} = \frac{A}{5}(0.757)$$

$$k = 3 \qquad |c_3| = \frac{A}{5}\frac{\sin 3\pi/5}{3\pi/5} = \frac{A}{5}(0.505)$$

$$k = 4 \qquad |c_4| = \frac{A}{5}\frac{\sin 4\pi/5}{4\pi/5} = \frac{A}{5}(0.234)$$

$$k = 5 \qquad |c_5| = \frac{A}{5}\frac{\sin \pi}{\pi} = 0$$

$$k = 6 \qquad |c_6| = \frac{A}{5}\left|\frac{\sin 6\pi/5}{6\pi/5}\right| = \frac{A}{5}(0.156)$$

$$k = 7 \qquad |c_7| = \frac{A}{5}\left|\frac{\sin 7\pi/5}{7\pi/5}\right| = \frac{A}{5}(0.216)$$

$$k = 8 \qquad |c_8| = \frac{A}{5}\left|\frac{\sin 8\pi/5}{8\pi/5}\right| = \frac{A}{5}(0.189)$$

$$k = 9 \qquad |c_9| = \frac{A}{5}\left|\frac{\sin 9\pi/5}{9\pi/5}\right| = \frac{A}{5}(0.104)$$

$$k = 10 \qquad |c_{10}| = \frac{A}{5}\frac{\sin 2\pi}{2\pi} = 0$$

In fact, $c_k = 0$ whenever $\sin(k\pi/5) = 0$, that is, for $k = \pm 5, \pm 10, \pm 15 \ldots$ as shown in Fig. 13.10a.

b) $a = 0.1$, $T = 1$ $\quad \therefore \omega_0 = 2\pi$

Here $a/T = 1/10$ and the discrete values of $|c_k|$ are:

$$k = 0 \qquad |c_0| = \frac{A}{10}$$

$$k = 1 \qquad |c_1| = \frac{A}{10}\frac{\sin \pi/10}{\pi/10} = \frac{A}{10}(0.984)$$

$$k = 2 \qquad |c_2| = \frac{A}{10}\frac{\sin \pi/5}{\pi/5} = \frac{A}{10}(0.935)$$

$$k = 3 \qquad |c_3| = \frac{A}{10}\frac{\sin 3\pi/10}{3\pi/10} = \frac{A}{10}(0.858)$$

$$k = 4 \qquad |c_4| = \frac{A}{10}\frac{\sin 4\pi/10}{4\pi/10} = \frac{A}{10}(0.757)$$

$$k = 5 \qquad |c_5| = \frac{A}{10}\frac{\sin \pi/2}{\pi/2} = \frac{A}{10}(0.637)$$

$$k = 6 \qquad |c_6| = \frac{A}{10}\frac{\sin 3\pi/5}{3\pi/5} = \frac{A}{10}(0.505)$$

$$k = 7 \qquad |c_7| = \frac{A}{10}\frac{\sin 7\pi/10}{7\pi/10} = \frac{A}{10}(0.368)$$

$$k = 8 \qquad |c_8| = \frac{A}{10}\frac{\sin 4\pi/5}{4\pi/5} = \frac{A}{10}(0.234)$$

$$k = 9 \qquad |c_9| = \frac{A}{10}\frac{\sin 9\pi/10}{9\pi/10} = \frac{A}{10}(0.109)$$

$$k = 10 \qquad |c_{10}| = \frac{A}{10}\frac{\sin \pi}{\pi} = 0$$

And here $|c_k| = 0$ whenever $\sin (k\pi/10) = 0$, that is, for $k = \pm 10, \pm 20, \pm 30, \dots$ as shown in Fig. 13-10b.

In both parts of this figure, the plots are discrete; the dashed line joining the discrete points is used simply because it represents the plot of $|\sin x/x|$, a standard curve; it will appear in more detail in the next chapter.

Another important observation should be made here. As the period $T$ increases, $\omega_0$ decreases ($\omega_0 = 2\pi/T$), and more and more harmonics are present in the Fourier series. This makes sense also in an intuitive way: in order to approximate $f(t) = 0$ between $-T/2$ and $-a/2$ and $a/2$ and $T/2$, more and more sines and cosines of many harmonics are needed as $T$ gets larger, so that they can cancel each other ($+$ and $-$ values) to give zero. $\qquad\qquad$ ☐ **Probs. 13-21**

**13-22**

**13-23**

## PROBLEMS

13-1 a) Plot $\sin^2 t$ versus $t$ and find its period $T$.

b) Use Euler's identity

$$\sin t = \frac{1}{2j} (e^{jt} - e^{-jt}) \qquad \cos t = \tfrac{1}{2}(e^{jt} + e^{-jt})$$

to find the *finite* Fourier series for $\sin^2 t$.

13-2 Repeat problem 13-1 for $\cos^3 t$.

13-3 Prove Eqs. (13-4) through (13-8).

13-4 Obtain the expressions for the Fourier coefficients of the square wave in Fig. 13-1a, with

$$f(t) = \begin{cases} 0 & -4 < t < -2 \\ 10 & -2 < t < 2 \\ 0 & 2 < t < 4 \end{cases}$$

Write the Fourier series up to and including the sixth harmonic terms.

13-5 Repeat problem 13-4 for the half-wave rectified sine wave in Fig. 13-1c, of amplitude $A$ and period $T$.

13-6 Repeat Problem 13-4 for the full-wave rectified sine wave in Fig. 13-1d, of amplitude $A$ and period $T = 1$ s.

13-7 Calculate the Fourier coefficients for the function $f(t)$ shown below, up to and including the terms of the sixth harmonic. Plot successively: (a) the dc term plus the first harmonic; (b) add to (a) the second harmonic; (c) add to (b) the third harmonic; (d) continue till you've added the sixth harmonic. Your plots should show better and better approximations to $f(t)$. *Note*: in general, it is quite difficult to estimate in advance the error of a Fourier series. Luckily, we don't have to do it, since we know the exact values of $f(t)$—they are given. For a particular truncation of a series, then, we can calculate the *actual* error.

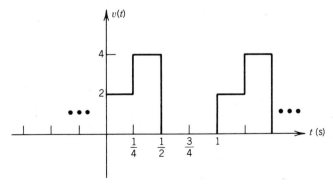

**PROBLEM 13-7.**

13-8 Any period $a \leq T \leq b$ can be scaled properly to a standard interval $0 \leq T_s \leq 2\pi$. Let $t$ be the unscaled time variable, defined on $[a, b]$. Let $\tau$ be the new (scaled) time variable, defined on $[0, 2\pi]$. Derive the relationship between $t$ and $\tau$. *Hint*: assume $t = m\tau + n$ and find $m$, $n$.

13-9 Since the expressions for the Fourier coefficients, Eq. (13-17), are integrals, we can approximate them numerically as sums. This is particularly useful when: (a) the periodic function $f(t)$ is given graphically (from an oscilloscope, for example), or (b) a computer program is to be written to implement Eq. (13-17). Specifically, let $T = 2\pi$ as in problem 13-8. Divide $\tau$ into $2N$ equal divisions

$$\tau_r = \frac{r\pi}{N} \qquad r = 0, 1, 2, \ldots, 2N - 1$$

Then the Fourier series coefficients are

$$a_k = \frac{1}{N} \sum_{\tau=0}^{2N-1} f(\tau) \cos \frac{k\pi}{N} \tau \qquad k = 0, 1, 2, \ldots, N$$

$$b_k = \frac{1}{N} \sum_{\tau=0}^{2N-1} f(\tau) \sin \frac{k\pi}{N} \tau \qquad k = 1, 2, \ldots, N - 1$$

Implement this method (with a short computer program) on (a) The half-wave rectified sine wave, problem 13-5, (b) The waveform of problem 13-7, and compare results.

13-10 Expand $f(t) = |t|$, $-\pi < t < \pi$ ($T = 2\pi$) in its Fourier series, making use of the fact that it is an even function. Develop the Fourier coefficients through the fourth harmonic, and plot the partial sums of successive terms.

13-11 Let $f(t) = 1-t$, $0 < t < 1$, and periodic, $T = 1$. Expand it in its Fourier series. *Hint*: consider first another function $g(t) = f(t) - \frac{1}{2}$. Plot $g(t)$, verify its symmetry property, then expand $g(t)$. Finally go back to $f(t)$.

13-12 a) Prove *Parseval's theorem* which is stated as

$$\frac{1}{T} \int^T [f(t)]^2 \, dt = \frac{a_0^2}{4} + \frac{1}{2} \sum_{k=1}^{\infty} (a_k^2 + b_k^2)$$

*Hint*: Multiply Eq. (13-3) by $f(t)$, and integrate over a period, using the orthogonality properties of the sine and cosine functions.

b) Give a circuit interpretation to Parseval's theorem, letting $f(t) = i(t)$, a periodic current flowing through a 1-ohm resistor.

13-13 Write the Fourier cosine series for the half-wave rectified sine wave of Fig. 13-1c.

13-14 Write the Fourier cosine series for the staircase waveform of problem 13-7.

13-15 Rework Example 6 in the text, with a full-wave rectified voltage, i.e.,

$$v_r(t) = 156 \sin 377t \qquad 0 < t < \tfrac{1}{120}\,\text{s}$$

and the same $LC$ filter and resistive load. Sketch also the output voltage.

13-16 Repeat problem 13-7 for the exponential form of the Fourier series.

13-17 Repeat problem 13-10 for the exponential form of the Fourier series.

13-18 Derive the expression for the average power when $v(t)$ and $i(t)$ are given in their exponential Fourier series

$$v(t) = \sum_{-\infty}^{\infty} c_{v_k} e^{jk\omega_0 t}$$

$$i(t) = \sum_{-\infty}^{\infty} c_{i_k} e^{jk\omega_0 t}$$

and obtain

$$P_{\text{avg}} = \sum_{-\infty}^{\infty} (c_{v_k})(c_{i-k}) = \sum_{-\infty}^{\infty} c_{v_k} c_{i_k}^*$$

13-19 Use the result of problem 13-18 to derive Eq. (13-53) in the text, i.e., show that

$$P_{\text{avg}} = \sum_{-\infty}^{\infty} c_{v_k} c_{i_k}^* = V_0 I_0 + \sum_{k=1}^{\infty} V_{\text{eff}_k} I_{\text{eff}_k} \cos \phi_k$$

13-20 Use the results of problem 13-18 and of problem 13-12 to prove that the effective (rms) value of a periodic current (or voltage), defined by

$$F_{\text{eff}} = \sqrt{\frac{1}{T} \int^{T} [f(t)]^2 \, dt}$$

is the square root of the sum of the squares of the effective values of the harmonics,

$$F_{\text{eff}}^2 = F_0^2 + F_{1\text{eff}}^2 + F_{2\text{eff}}^2 + \cdots$$

13-21 Calculate and plot the discrete magnitude spectrum $|c_k|$ of the triangle waveform shown, for: 1) $a/T = 1/5$ and 2), $a/T = 1/10$.

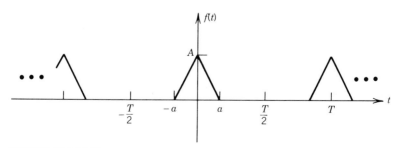

**PROBLEM 13-21.**

13-22 Compare the discrete magnitude and phase spectra of a periodic $f(t)$ with those of the same function shifted by $t_0$ units on the time axis. *Hint*: write the exponential form of the Fourier series for $f(t)$, and then substitute $(t - t_0)$ in it to account for the shift.

13-23 Rework example 1 in this chapter using Fourier series, with $i(t) = 10$, $0 < t < 2$s and $i(t) = 0$, $2 < t < 8$s, with a period $T = 8$s. (*Suggestion*: do it via the exponential Fourier series).

# 14

---

# THE FOURIER
# TRANSFORM

In this chapter we extend the methods of Fourier series, studied previously, to nonperiodic functions. These functions are common in many network problems, and the extension is done by letting the period $T$ approach infinity. The Fourier series then becomes a Fourier integral.

## 14-1 THE SQUARE PULSE REVISITED

Let us refer back to the square pulse in the previous chapter, shown in Fig. 14-1a. As derived there, the discrete Fourier coefficients of the exponential form are given by

$$c_k = \frac{Aa}{T} \frac{\sin (k\pi a/T)}{k\pi a/T} = \frac{Aa}{T} \frac{\sin (k\omega_0 a/2)}{k\omega_0 a/2} \tag{14-1}$$

Two plots of $|c_k|$ versus $k\omega_0$ are shown in Fig. 14-1b, c. Let us repeat several important observations:

1. As $T$ increases (from, say, $T = 5a$ to $T = 10a$ in this example), the fundamental frequency $\omega_0$ decreases, since $\omega_0 = 2\pi/T$.

2. The discrete points $k\omega_0$ become closer to each other. In other words, there are more and more harmonics, and their separation $(k + 1)\omega_0 - k\omega_0 = \omega_0$ is smaller.

3. The *envelope* (dashed line) of the discrete magnitude spectrum is a continuous function given by

$$(c_k) = \frac{Aa}{T} \frac{\sin (\omega a/2)}{\omega a/2} \tag{14-2}$$

where $(c_k)$ means the envelope (continuous) plot. This plot is the famous curve $\sin x/x$, $(x = \omega a/2)$, shown in Fig. 14-2. The plots in Fig. 14-1 are for the magnitude $|c_k|$, hence only positive lobes.

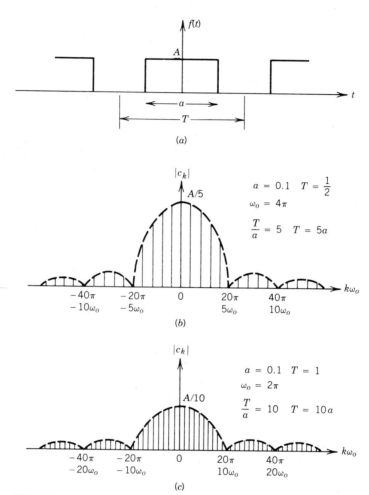

FIGURE 14-1. The square pulse.

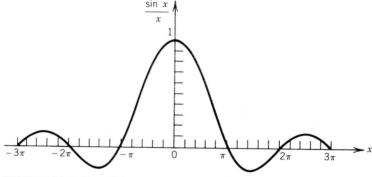

FIGURE 14-2. Plot of $\sin x/x$.

4. The amplitude spectrum is zero whenever $\sin k\pi a/T$ is zero, that is, for $ka/T = \pm 1, \pm 2, \pm 3, \ldots$ (integers). If $a/T = 1/5$ as in Fig. 14-1b, these zeros occur for $k = \pm 5, \pm 10, \ldots$, i.e., at $\pm 5\omega_0, \pm 10\omega_0, \ldots$. For $a/T = 1/10$, they occur at $\pm 10\omega_0, \pm 20\omega_0, \ldots$. For $a/T = 1/N$, they occur at $\pm N\omega_0, \pm 2N\omega_0, \ldots$, and as $T$ increases ($N$ increases) there are more and more harmonics between these points.

5. As $T \to \infty$, the function becomes *nonperiodic*, and is shown in Fig. 14-3.

## 14-2   THE FOURIER INTEGRAL AND TRANSFORM

We are now ready to take the limit as $T \to \infty$ of the exponential Fourier series

$$f(t) = \sum_{k=-\infty}^{+\infty} c_k e^{jk\omega_0 t} \qquad \omega_0 = \frac{2\pi}{T} \qquad (14\text{-}3)$$

and

$$c_k = \frac{1}{T} \int_{-T/2}^{T/2} f(t) e^{-jk\omega_0 t} \, dt \qquad (14\text{-}4)$$

From Eq. (14-4) we have

$$c_k T = \int_{-T/2}^{T/2} f(t) e^{-jk\omega_0 t} \, dt \qquad (14\text{-}5)$$

As $T \to \infty$, $k\omega_0$ becomes the continuous variable $\omega$ while $\omega_0$ itself becomes $d\omega$. Thus, the integral in Eq. (14-5) is

$$\int_{-\infty}^{\infty} f(t) e^{-j\omega t} \, dt = F(\omega) \qquad (14\text{-}6)$$

Eq. (14-3) is written as

$$f(t) = \frac{1}{2\pi} \sum_{k=-\infty}^{+\infty} c_k T e^{jk\omega_0 t} \omega_0 \qquad (14\text{-}7)$$

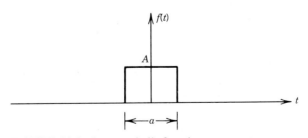

**FIGURE 14-3.** A nonperiodic function.

by multiplying and dividing it by $2\pi = \omega_0 T$. Now, as $T \to \infty$ and $\omega_0 \to d\omega$, the summation in Eq. (14-7) becomes an integral and we have, with Eq. (14-6),

$$f(t) = \frac{1}{2\pi} \int_{-\infty}^{\infty} F(\omega) e^{j\omega t} \, d\omega \qquad (14\text{-}8)$$

Equation (14-6) is the Fourier transform of $f(t)$, written symbolically as

$$\mathscr{F} f(t) = F(\omega) \qquad (14\text{-}9)$$

while Eq. (14-8) is the inverse Fourier transform of $F(\omega)$, written symbolically as

$$\mathscr{F}^{-1} F(\omega) = f(t) \qquad (14\text{-}10)$$

The frequency-domain function $F(\omega)$, Eq. (14-6), is seen to be the extension of Eq. (14-4) for a nonperiodic function, while $f(t)$, the inverse of $F(\omega)$, is the extended version of Eq. (14-3) for a nonperiodic function.

This derivation, while not rigorous, is quite satisfactory for our engineering applications. For example, we did not dwell on the conditions for the validity (existence) of Eqs. (14-6) and (14-8). As in the case of the Fourier series, the *sufficient* (but not necessary) conditions for the Fourier transform are:

1. $f(t)$ is absolutely integrable, that is,

$$\int_{-\infty}^{\infty} |f(t)| \, dt = M < \infty \qquad (14\text{-}11)$$

2. $f(t)$ may have, at most, a finite number of maxima and minima in any finite interval.

3. $f(t)$ may have, at most, a finite number of (finite) discontinuities in any finite interval.

These conditions, named (as in the Fourier series) after Dirichlet, are only sufficient. We will see how several important functions, which do not satisfy them, still have Fourier transforms, with the help of impulse functions.

Since $F(\omega)$, as an extension of $c_k$, is complex in general, we write $F(\omega)$ in its rectangular or polar forms

$$F(\omega) = R(\omega) + jX(\omega) \qquad (14\text{-}12\text{a})$$

$$F(\omega) = |F(\omega)| e^{j\phi(\omega)} \qquad (14\text{-}12\text{b})$$

We call $|F(\omega)|$ the *continuous amplitude spectrum* of $f(t)$; $\phi(\omega)$ is the *continuous phase spectrum* of $f(t)$. These are the extensions of the discrete spectra for a periodic function to continuous spectra for a nonperiodic function.

---

### Example 1

Find the Fourier transform, and plot the continuous spectra, of $f(t)$ as shown in Fig. 14-3, with $A = \frac{1}{2}$, $a = 2$. The sufficiency of Eq. (14-11) is established,

$M = Aa = 1$, and we write

$$F(\omega) = \int_{-\infty}^{\infty} f(t) e^{-j\omega t}\, dt = \int_{-1}^{1} \frac{1}{2} e^{-j\omega t}\, dt = \frac{\sin \omega}{\omega}$$

and we have (again) the function shown in Fig. 14-2. It is not a surprise—we have been expecting it since the end of the previous chapter!

The continuous amplitude spectrum, $|F(\omega)|$, will consist of the same plot as Fig. 14-2, but with the negative lobes ($\pm \pi < \omega < \pm 2\pi, \pm 3\pi < \omega < \pm 4\pi, \ldots$) plotted positive. The continuous phase plot, $\phi(\omega)$, will be

$$\phi(\omega) = 0° \qquad -\pi < \omega < \pi \qquad 2\pi < \omega < 3\pi, \ldots$$

$$= 180° \qquad -2\pi < \omega < -\pi \qquad \pi < \omega < 2\pi, \ldots$$

☐ **Prob. 14-1**

## Example 2

Let us find the Fourier Transform of $f(t) = e^{-at}u(t)$, $a > 0$, as shown in Fig. 14-4a. We write

$$F(\omega) = \mathscr{F} e^{-at} u(t) = \int_{0}^{\infty} e^{-at} e^{-j\omega t}\, dt = \frac{1}{a + j\omega}$$

Its magnitude spectrum is

$$|F(\omega)| = \frac{1}{\sqrt{a^2 + \omega^2}}$$

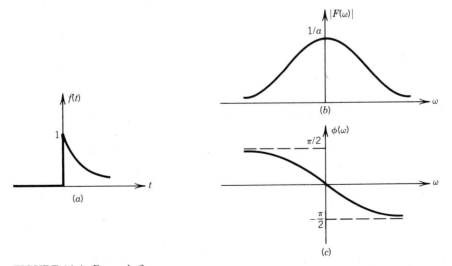

**FIGURE 14-4.** Example 2.

and its phase spectrum is

$$\phi(\omega) = -\tan^{-1}(\omega/a)$$

They are plotted in Fig. 14-4b, c.

☐ **Probs. 14-2**

## Example 3

To illustrate the use of the impulse function in Fourier transforms, consider the direct evaluation of

$$\mathscr{F}1 = \int_{-\infty}^{\infty} 1e^{-j\omega t} \, dt = ?$$

the integral, quite obviously, fails to converge. Also, $f(t) = 1, -\infty < t < \infty$, fails to satisfy the first Dirichlet condition, Eq. (14-11).

Nevertheless, let us try an inverse transform. Specifically, let us calculate

$$\mathscr{F}^{-1}2\pi\delta(\omega) = \frac{1}{2\pi}\int_{-\infty}^{\infty} 2\pi\delta(\omega)e^{j\omega t} \, d\omega = e^{j\omega t}\Big]_{\omega=0} = 1$$

where we have used the sampling property of the impulse function. Therefore we have

$$\mathscr{F}1 = 2\pi\delta(\omega)$$

as shown in Fig. 14-5. Compare also with problem 14-2.

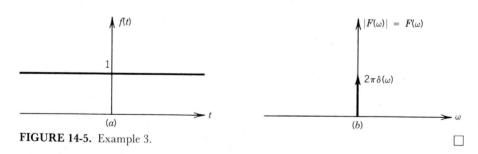

FIGURE 14-5. Example 3.

☐

## 14-3 PROPERTIES OF THE FOURIER TRANSFORM

The Fourier transform possesses many properties which are similar to the Laplace transform. Let us derive some of them, for later use.

### Linearity

$$\mathscr{F}[a_1 f_1(t) + a_2 f_2(t)] = a_1 F_1(\omega) + a_2 F_2(\omega) \qquad (14\text{-}13)$$

The proof, as in the Laplace transform, rests on Eq. (14-6) and on the basic property of an integral of a sum.

## Differentiation

If we differentiate Eq. (14-8) with respect to $t$, we get

$$\frac{d}{dt} f(t) = \frac{1}{2\pi} \int_{-\infty}^{\infty} \frac{d}{dt} [F(\omega) e^{j\omega t}] \, d\omega$$

$$= \frac{1}{2\pi} \int_{-\infty}^{\infty} [F(\omega) \cdot j\omega] e^{j\omega t} \, d\omega \qquad (14\text{-}14)$$

Here, we interchanged the order of differentiation (with respect to $t$) and integration (with respect to $\omega$), recognizing that $t$ and $\omega$ are independent. In Eq. (14-14) we have

$$\mathscr{F} \frac{df(t)}{dt} = j\omega F(\omega) \qquad (14\text{-}15)$$

similar to the Laplace transform, where multiplication by $s$ corresponded to differentiation. (Question: why are there no initial conditions, $f(0_-)$, in Eq. (14-15)?)

## Time Shifting

Let us calculate the Fourier transform of $f(t - a)$, a time-shifted function. We write

$$\mathscr{F} f(t - a) = \int_{-\infty}^{\infty} f(t - a) e^{-j\omega t} \, dt \qquad (14\text{-}16)$$

and substitute $t - a = y$, $dt = dy$. The result is

$$\mathscr{F} f(t - a) = \int_{-\infty}^{\infty} f(y) e^{-j\omega(y + a)} \, dy = e^{-j\omega a} F(\omega) \qquad (14\text{-}17)$$

where $F(\omega) = \mathscr{F} f(t)$, the Fourier transform of the original, nonshifted function. Again, the similarity of the Laplace transform is apparent.

*Probs. 14-3*
*14-4*
*14-5*

## Frequency Shifting

If

$$\mathscr{F} f(t) = F(\omega)$$

then

$$\mathscr{F} f(t) e^{j\omega_0 t} = F(\omega - \omega_0) \qquad (14\text{-}18)$$

Other properties are summarized in Table 14-1.

To illustrate the Fourier transform of an integrated function, let us first recognize that, since time differentiation tranforms to multiplication by $j\omega$, time

## TABLE 14-1  The Fourier Transform

### Properties and Operations

| $f(t)$ | $F(\omega)$ |
|---|---|
| Linearity $a_1 f_1(t) + a_2 f_2(t)$ | $a_1 F_1(\omega) + a_2 F_2(\omega)$ |
| Time differentiation $\dfrac{d^n f(t)}{dt^n}$ | $(j\omega)^n F(\omega)$ |
| Time integration $\displaystyle\int_{-\infty}^{t} f(x)\,dx$ | $\dfrac{1}{j\omega} F(\omega) + \pi F(0)\delta(\omega)$ |
| Time shifting $f(t - a)$ | $F(\omega)e^{-j\omega a}$ |
| Modulation $\quad f(t)e^{j\omega_0 t}$ (frequency shifting) | $F(\omega - \omega_0)$ |
| Time scaling $f(at)$ | $\dfrac{1}{\lvert a\rvert} F\!\left(\dfrac{\omega}{a}\right)$ |
| Time convolution $\displaystyle\int_{-\infty}^{\infty} h(\lambda)x(t - \lambda)\,d\lambda$ | $H(\omega)X(\omega)$ |
| Frequency convolution $\quad h(t)x(t)$ | $\dfrac{1}{2\pi} \displaystyle\int_{-\infty}^{\infty} H(\omega - \lambda)X(\lambda)\,d\lambda$ |
| Symmetry $F(t)$ | $2\pi f(-\omega)$ |
| Frequency differentiation $\quad (-jt)^n f(t)$ | $\dfrac{d^n F(\omega)}{d\omega^n}$ |

### Some Elementary Functions

| $f(t)$ | $F(\omega)$ |
|---|---|
| $\delta(t)$ | $1$ |
| $1$ | $2\pi\delta(\omega)$ |
| $e^{-at}u(t) \quad a > 0$ | $\dfrac{1}{a + j\omega}$ |
| $e^{-a\lvert t\rvert}$ | $\dfrac{2a}{a^2 + \omega^2}$ |
| $e^{j\omega_0 t}$ | $2\pi\delta(\omega - \omega_0)$ |
| $\cos\omega_0 t$ | $\pi[\delta(\omega + \omega_0) + \delta(\omega - \omega_0)]$ |
| $\sin\omega_0 t$ | $j\pi[\delta(\omega + \omega_0) - \delta(\omega - \omega_0)]$ |
| $u(t)$ | $\pi\delta(\omega) + \dfrac{1}{j\omega}$ |
| $\mathrm{sgn}(t)$ | $\dfrac{2}{j\omega}$ |

integration would require *division* by $j\omega$, as was the case in the Laplace transform. This is true, but not complete. The Fourier transform of integration is

$$\mathscr{F}\left[\int_{-\infty}^{t} f(x)\, dx\right] = \frac{1}{j\omega} F(\omega) + \pi F(0)\delta(\omega) \qquad (14\text{-}19)$$

where the additional term with the impulse accounts for the constant (dc) term of integration. Here

$$F(0) = F(\omega)]_{\omega=0} = \int_{-\infty}^{\infty} f(t)\, dt \qquad (14\text{-}20)$$

If $F(0) = 0$, then integration in the time domain transforms into just division by $j\omega$ in the frequency domain.

---

## Example 4

We will obtain the Fourier transform of $u(t)$, the unit step function, first by a limit process, and then by integration.

Consider the function $f(t)$ defined as

$$f(t) = \begin{cases} e^{-at}u(t) & t > 0 \\ -e^{at}u(-t) & t < 0 \end{cases}$$

and shown in Fig. 14-6$a$. We have its Fourier transform

$$F(\omega) = \frac{1}{j\omega + a} - \frac{1}{-j\omega + a} = \frac{-2j\omega}{\omega^2 + a^2}$$

Now let $a \to 0$. The function $f(t)$ becomes

$$\lim_{a\to 0} f(t) = \begin{cases} 1 & t > 0 \\ -1 & t < 0 \end{cases}$$

and it is commonly called sgn($t$), read as "signum t" or "sign t",

$$\text{sgn}(t) = \frac{t}{|t|} = \begin{cases} 1 & t > 0 \\ -1 & t < 0 \end{cases}$$

and shown in Fig. 14-6$b$. Its Fourier transform is therefore

$$\mathscr{F}\,\text{sgn}(t) = \lim_{a\to 0} \frac{-2j\omega}{\omega^2 + a^2} = \frac{2}{j\omega}$$

The unit step function $u(t)$ can be written as

$$u(t) = \frac{1}{2} + \frac{1}{2}\,\text{sgn}(t)$$

and therefore

$$\mathscr{F}\,u(t) = \mathscr{F}\,\frac{1}{2} + \frac{1}{2}\,\mathscr{F}\,\text{sgn}(t) = \pi\delta(\omega) + \frac{1}{j\omega}$$

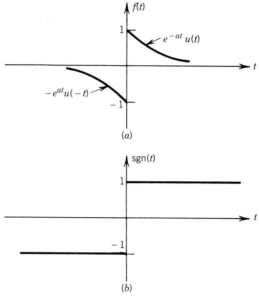

FIGURE 14-6. Example 4.

Alternately, since

$$u(t) = \int_{-\infty}^{t} \delta(x)\,dx$$

and $\mathscr{F}\delta(t) = 1 = F(\omega)$, $F(0) = 1$, we have from Table 14-1

$$\mathscr{F}u(t) = \frac{1}{j\omega} \cdot 1 + (\pi)(1)\delta(\omega) = \frac{1}{j\omega} + \pi\delta(\omega)$$

as before.

☐  ***Probs. 14-6***
***14-7***

## 14-4  APPLICATIONS TO CIRCUITS

Of the numerous applications of the Fourier transform, we consider several examples.

### Example 5

In the network shown in Fig. 14-7a, the applied voltage is $v(t) = 10e^{-t}u(t)$. The differential equation for $i(t)$ is KVL

$$1\frac{di}{dt} + 2i = 10e^{-t}u(t)$$

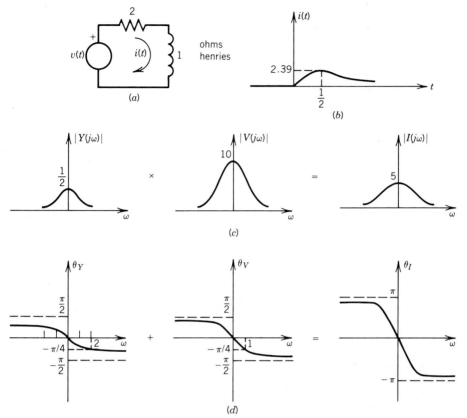

**FIGURE 14-7.** Example 5.

and its Fourier transform yields

$$j\omega I(\omega) + 2I(\omega) = \frac{10}{1 + j\omega}$$

or

$$I(\omega) = \frac{10}{(1 + j\omega)(2 + j\omega)}$$

Expanding in partial fractions, we get

$$I(\omega) = \frac{10}{1 + j\omega} - \frac{10}{2 + j\omega}$$

inverting, we get

$$i(t) = \begin{cases} 10e^{-t} - 10e^{-2t} & t > 0 \\ 0 & t < 0 \end{cases}$$

This response is shown in Fig. 14-7*b*. To emphasize the network function approach, write $I(\omega)$ as[†]

$$I(j\omega) = \frac{1}{2 + j\omega} \cdot \frac{10}{1 + j\omega} = Y(j\omega) \cdot V(j\omega)$$

in the form of

$$R(j\omega) = H(j\omega) \cdot E(j\omega)$$

where $R(j\omega)$ is the Fourier transform of the response, $H(j\omega)$ is the appropriate network function, and $E(j\omega)$ is the Fourier transform of the input.

Obviously, the magnitude spectrum of $I(j\omega)$ is

$$|I(j\omega)| = |Y(j\omega)| \cdot |V(j\omega)|$$

and the phase is

$$\theta_I = \theta_Y + \theta_V$$

These results are shown in Fig. 14-7*c* and *d*.

This entire example looks familiar from our previous study of network functions (Chap. 7) and frequency response (Chap. 9). Indeed, this is the case. Here, we solved this problem from the frequency spectrum point of view.  □  **Probs. 14-8**
**14-9**
**14-10**

---

## Example 6

A *filter* is a linear network with certain specific amplitude spectrum characteristics. For example, a *distortionless* filter is one whose response $g(t)$ has the same form as the input $f(t)$

$$g(t) = Kf(t - t_0)$$

as shown in Fig. 14-8*a*. We allow a time delay in $g(t)$, but, within a constant multiplier, the waveform of $g(t)$ is the same as $f(t)$. In other words, the frequency spectrum of $F(\omega)$ and $G(\omega)$ are, according to the time-shifting theorem,

$$G(\omega) = Ke^{-j\omega t_0}F(\omega)$$

So the network function of the filter is

$$H(\omega) = Ke^{-j\omega t_0}$$

with a constant amplitude and a linear phase

$$|H(\omega)| = K \qquad \theta_H = -\omega t_0$$

---

[†] Throughout the Fourier transform, the notation $I(\omega)$, $F(\omega)$, etc., is simpler than $I(j\omega)$, $F(j\omega)$. Here we use the latter, in order to stress the relation to our previous study of network functions.

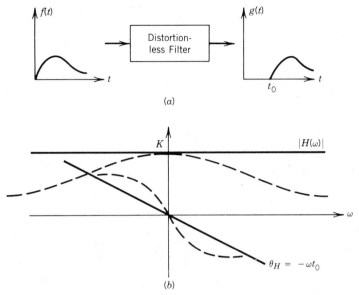

**FIGURE 14-8.**  Example 6.

as shown in Fig. 14-8$b$.

An all-resistive filter is distortionless (with $t_0 = 0$), but otherwise not very interesting. The *ideal* characteristics of amplitude and phase are often approximated by real filters as shown in dashed lines. Consequently, some distortion is introduced. □

## Example 7

An ideal *low pass* filter is one whose amplitude is given by

$$|H(\omega)| = \begin{cases} K & |\omega| < \omega_c \\ 0 & |\omega| > \omega_c \end{cases}$$

where $\omega_c$ is the *cutoff frequency*. The *bandwidth* of this filter is $0 < \omega < \omega_c$.

The network function of this filter is

$$H(\omega) = \begin{cases} Ke^{-j\omega t_0} & |\omega| < \omega_c \\ 0 & |\omega| > \omega_c \end{cases}$$

See Fig. 14-9. Let us find the impulse response $h(t)$ of this filter. With $f(t) = \delta(t)$ as input, we have $F(\omega) = 1$, and therefore

$$H(\omega) = \begin{cases} Ke^{-j\omega t_0} \cdot 1 & -\omega_c < \omega < \omega_c \\ 0 & \omega > \omega_c \end{cases}$$

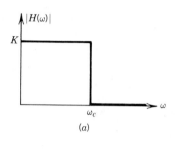

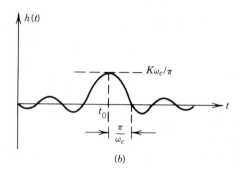

(a)

(b)

**FIGURE 14-9.** Example 7.

Inverting, we get

$$h(t) = \frac{1}{2\pi} \int_{-\omega_c}^{\omega_c} K e^{-j\omega t_0} \cdot e^{j\omega t} \, d\omega$$

Evaluation of such integrals was done previously (see Example 1.) The result is

$$h(t) = K \frac{\sin \omega_c(t - t_0)}{\pi(t - t_0)}$$

and its plot is shown in Fig. 14-9*b*                                                 ☐ *Prob. 14-11*

---

### *Example 8*

Since input–output Fourier transforms are related by

$$R(\omega) = H(\omega)E(\omega)$$

let us calculate the response to an input $e^{j\omega_0 t}$. Since

$$E(\omega) = \mathscr{F} e^{j\omega_0 t} = 2\pi\delta(\omega - \omega_0)$$

we get

$$R(\omega) = H(\omega) \cdot 2\pi\delta(\omega - \omega_0)$$

and inversion yields

$$r(t) = \frac{1}{2\pi} \int_{-\infty}^{\infty} R(\omega)e^{j\omega t} \, d\omega = H(\omega_0)e^{j\omega_0 t}$$

We have proved here (again!) the well-known result: a constant, linear network responds to a sinusoid $(e^{j\omega_0 t})$ at a frequency $\omega_0$ with a scaled version, multiplied by $H(\omega_0)$, of the *same* sinusoid. Among the applications of this principle is the use of phasors and the frequency response techniques of Chapter 9 and here.

☐ ***Probs. 14-12***
***14-13***

## 14-5   PARSEVAL'S THEOREM AND THE ENERGY SPECTRUM

Let us derive Parseval's theorem for the Fourier transform, as we did in the previous chapter for the Fourier series. We write the frequency convolution relationship

$$\mathscr{F}\{f_1(t) \cdot f_2(t)\} = F_1(\omega) * F_2(\omega) \qquad (14\text{-}21)$$

$$\therefore \int_{-\infty}^{\infty} f_1(t) f_2(t) e^{-j\omega t}\, dt = \frac{1}{2\pi} \int_{-\infty}^{\infty} F_1(\lambda) F_2(\omega - \lambda)\, d\lambda \qquad (14\text{-}22)$$

Now let $\omega = 0$ and $f_1(t) = f_2(t) = f(t)$

$$\int_{-\infty}^{\infty} [f(t)]^2\, dt = \frac{1}{2\pi} \int_{-\infty}^{\infty} F(\omega) F(-\omega)\, d\omega \qquad (14\text{-}23)$$

where the dummy variable of integration $\lambda$ was replaced by $\omega$. We have also (problem 14-14) that $F(-\omega) = F^*(\omega)$ if $f(t)$ is real. Consequently,

$$\int_{-\infty}^{\infty} [f(t)]^2\, dt = \frac{1}{2\pi} \int_{-\infty}^{\infty} |F(\omega)|^2\, d\omega \qquad (14\text{-}24)$$

which is Parseval's theorem.

If $f(t)$ is a voltage source across a $1\,\Omega$ resistor, the integral on the left-hand side is the total energy in the signal $f(t)$. It is equal to the sum (integration), over all time, of the power, that is, of the energy per unit time $|f(t)|^2$. According to Parseval's theorem, the same result is obtained if we calculate the energy per unit frequency, $|F(\omega)|^2/2\pi$, and sum (integrate) over all frequencies. Therefore, the quantity $|F(\omega)|^2$ is often call the *energy spectral density* of $f(t)$.

Parseval's theorem and its consequences provide very strong ties among time-domain and frequency-domain calculations.

---

*Example 9*

Find the energy spectral density of

$$f(t) = 10e^{-t}u(t)$$

and confirm it by Parseval's theorem.

In the time domain, we have here

$$\int_{0}^{\infty} [10e^{-t}]^2\, dt = 50\,\text{J}$$

In the frequency domain, we have

$$\frac{1}{2\pi} \int_{-\infty}^{\infty} \left| \frac{10}{j\omega + 1} \right|^2 d\omega = \frac{1}{\pi} \int_{0}^{\infty} \frac{100}{\omega^2 + 1} d\omega = \frac{100}{\pi} \tan^{-1}\omega \Big]_{0}^{\infty} = 50 \, J$$

It is interesting, and extremely powerful, that we can calculate the energy spectrum directly from $F(\omega)$, without having to invert it into $f(t)$! ☐ **Probs. 14-14**

**14-15**

**14-16**

**14-17**

## 14-6 COMPARISON BETWEEN THE FOURIER AND THE LAPLACE TRANSFORMS

Similarities between the Laplace transform and the Fourier transform are apparent. To review and summarize, let us consider first the direct transforms

$$F_L(s) = \mathscr{L}f(t)u(t) = \int_{0}^{\infty} f(t)e^{-st} \, dt \tag{14-25}$$

$$F_F(\omega) = \mathscr{F}f(t) = \int_{-\infty}^{\infty} f(t)e^{-j\omega t} \, dt \tag{14-26}$$

Here we have added the subscripts $L$ ("Laplace") and $F$ ("Fourier") in order to distinguish clearly between the two letters "$F$".

### From $F_L(s)$ to $F_F(\omega)$

If $f(t)$ is *causal*, that is, if $f(t) = 0$ for $t < 0$, and all the poles of $F_L(s)$ are *inside* the left-half plane, $(\sigma < 0)$, then Eq. (14-26) is a special case of Eq. (14-25) with $s = j\omega$

$$F_F(\omega) = F_L(s)]_{s=j\omega} \tag{14-27}$$

The reason for the causality of $f(t)$ is the lower limit on the integral in Eq. (14-25), $t = 0$, by comparison to $t = -\infty$ in Eq. (14-26). Obviously, if $f(t)$ is causal, the lower limit in Eq. (14-26) becomes also $t = 0$. The restriction on the poles of $F_L(s)$ guarantees that $f(t)$ is absolutely integrable, Eq. (14-11) and, therefore, that the Fourier integral converges.

---

*Example 10*

If $F_L(s) = K/(s + a)$, $a > 0$, then $f(t) = Ke^{-at}u(t)$ and

$$F_F(\omega) = \frac{K}{s + a}\bigg]_{s=j\omega} = \frac{K}{j\omega + a} \qquad ☐$$

---

If the poles of $F_L(s)$ are inside the left-half plane ($\sigma < 0$) and *on* the $j\omega$ axis ($\sigma = 0$), where they are simple, then we write $F_L(s)$ as

$$F_L(s) = \hat{F}_L(s) + \sum_{p=1}^{n} \frac{K_p}{s - j\omega_p} \qquad (14\text{-}28)$$

where $\hat{F}_L(s)$ has only poles inside the lhp. ($\sigma < 0$), and the summation exhibits all the *simple* poles on the $j\omega$ axis. Therefore

$$F_F(\omega) = F_L(s)]_{s=j\omega} + \pi \sum_{p=1}^{n} K_p \delta(\omega - \omega_p) \qquad (14\text{-}29)$$

The first term is in accordance with Eq. (14-27), and the second one contains the Fourier transform of the function $K_p e^{j\omega_p t} u(t)$. This, in turn, is obtained from Table 14-1—but be careful: there the time function is $e^{j\omega_p t}$ for $-\infty < t < +\infty$, while here we have a causal function $e^{j\omega_p t} u(t)$. Hence we see that, instead of $2\pi$, we multiply only by $\pi$.

---

## Example 11

The unit step function $u(t)$ has for its Laplace transform

$$F_L(s) = \frac{1}{s}$$

Its Fourier transform is therefore

$$F_F(\omega) = \frac{1}{s}\bigg]_{s=j\omega} + \pi\delta(\omega - 0) = \frac{1}{j\omega} + \pi\delta(\omega) \qquad \square$$

---

If $F_L(s)$ has poles in the right-half plane ($\sigma > 0$), there is no Fourier transform for $f(t)$. The reason, again, is that the Fourier integral fails to converge.

---

## Example 12

Consider the function

$$f(t) = e^{2t} u(t)$$

If we set up its Fourier integral

$$\int_0^{\infty} e^{2t} e^{-j\omega t} \, dt = \int_0^{\infty} e^{(2 - j\omega)t} \, dt = ?$$

we recognize that it does not converge (at the upper limit); hence the Fourier transform does not exist. The Laplace transform, on the other hand, is

$$\mathcal{L} e^{2t} u(t) = \int_0^{\infty} e^{2t} e^{-(\sigma + j\omega)t} \, dt = \int_0^{\infty} e^{(2-\sigma)t} e^{-j\omega t} \, dt$$

and the integral *converges* for $\sigma > 2$. Therefore, the Laplace transform exists, with a abcissa of convergence $\sigma_c = 2$.

$$F_L(s) = \frac{1}{s - 2} \qquad \sigma > 2$$

as derived in Chapter 2. In other words, the Laplace transform has a "built-in" convergence factor, $e^{-\sigma t}$, which the Fourier transform lacks. This is one reason for using the Laplace transform.  □

The three cases discussed here are summarized in Fig. 14-10.

## *From $F_F(\omega)$ to $F_L(s)$

From the first two cases discussed in this section, we know that $F_F(\omega)$ exists when $F_L(s)$ has a denominator which is a strict or a modified Hurwitz polynomial ($\sigma \leq 0$). Then

$$F_L(s) = F_F(\omega)]_{\omega = s/j} \tag{14-30}$$

which is, in fact, the reverse process of Eq. (14-27) and (14-29).

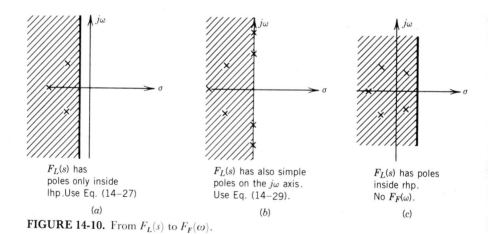

$F_L(s)$ has poles only inside lhp. Use Eq. (14–27)

(a)

$F_L(s)$ has also simple poles on the $j\omega$ axis. Use Eq. (14–29).

(b)

$F_L(s)$ has poles inside rhp. No $F_F(\omega)$.

(c)

**FIGURE 14-10.** From $F_L(s)$ to $F_F(\omega)$.

## Example 13

Given

$$F_F(\omega) = \frac{1}{j\omega + j2} + \pi\delta(\omega + 2)$$

we write from Eqs. (14-29) and (14-30)

$$F_L(s) = \frac{1}{s + j2}$$

corresponding to $f(t) = e^{-j2t}u(t)$.  ☐

---

Next, let us consider the Fourier transform of the function $f(t)e^{-\sigma t}u(t)$, with $\sigma > 0$.

$$\mathcal{F}\{f(t)e^{-\sigma t}u(t)\} = \int_0^\infty f(t)e^{-\sigma t}e^{-j\omega t}\, dt$$

$$= \int_0^\infty f(t)e^{-(\sigma + j\omega)t}\, dt = F_L(s) \qquad s = \sigma + j\omega \qquad (14\text{-}31)$$

that is, the Laplace transform $F_L(s)$ of $f(t)$, with $s = \sigma + j\omega$, $\sigma > 0$, is the Fourier transform of the function $f(t)e^{-\sigma t}u(t)$. What has happened here? we simply made sure, from the start, that a convergence factor $e^{-\sigma t}$ is present to ensure the existence of the Fourier transform.

Now we use the frequency convolution, from Table 14-1: the function $f(t)e^{-\sigma t}u(t)$ can be considered the product of $f(t)$ and $e^{-\sigma t}u(t)$. We know that

$$\mathcal{F}f(t) = F_F(\omega) \qquad (14\text{-}32a)$$

and

$$\mathcal{F}e^{-\sigma t}u(t) = \frac{1}{\sigma + j\omega} \qquad (14\text{-}32b)$$

The frequency convolution theorem states then

$$F_L(s)]_{s=\sigma + j\omega} = \frac{1}{2\pi}F_F(\omega) * \frac{1}{\sigma + j\omega} = \frac{1}{2\pi}\int_{-\infty}^\infty \frac{F_F(\lambda)}{\sigma + j(\omega - \lambda)}\, d\lambda$$

$$= \frac{1}{2\pi}\int_{-\infty}^\infty \frac{F_F(\lambda)}{s - j\lambda}\, d\lambda \qquad \sigma > 0 \qquad \textbf{\textit{Prob. 14-19}}$$

(14-33)

While Eq. (14-33) establishes the formal relationship for obtaining $F_L(s)$ from $F_F(\omega)$, it is very tedious to evaluate this integral directly.

Let us conclude our discussion by relating the inverses of the two transforms. Since $F_L(s)$ is the Fourier transform of $f(t)e^{-\sigma t}u(t)$, we have from Eq. (14-8),

$$f(t)e^{-\sigma t}u(t) = \mathcal{F}^{-1}F_L(s) = \frac{1}{2\pi}\int_{-\infty}^\infty F_L(\sigma + j\omega)e^{j\omega t}\, d\omega \qquad (14\text{-}34)$$

Multiplying both sides by $e^{\sigma t}$, we get

$$f(t)u(t) = \frac{1}{2\pi}\int_{-\infty}^\infty F_L(\sigma + j\omega)e^{(\sigma + j\omega)t}\, d\omega \qquad (14\text{-}35)$$

where $e^{\sigma t}$ is allowed as a *constant* multiplier inside the integral which is with respect to $\omega$. Now we introduce the complex variable $s$,

$$s = \sigma + j\omega \qquad \frac{1}{j}\,ds = d\omega$$

and

$$\omega = -\infty \to s = \sigma - j\infty$$
$$\omega = +\infty \to s = \sigma + j\infty$$

So finally

$$f(t)u(t) = \frac{1}{2\pi j}\int_{\sigma-j\infty}^{\sigma+j\infty} F_L(s)e^{st}\,ds \qquad (14\text{-}36)$$

which is the explicit formula for inverting the Laplace transform $F_L(s)$, as listed in Chapter 2.

The integral in Eq. (14-36) is a *line integral* in the complex plane, along the line shown in Fig. 14-11. Here $\sigma > \sigma_c$, where $\sigma_c$ is the abcissa of convergence for the particular $F_L(s)$. Again, the actual evaluation of this line integral requires a fair amount of knowledge of complex variables and contour integration. Fortunately for us, the *algebraic* methods developed earlier (partial fractions) yield the desired answers in a simpler, but *totally equivalent* way.

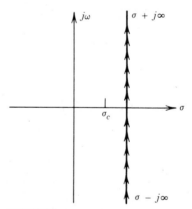

**FIGURE 14-11.** Path of integration.

## PROBLEMS

14-1 Calculate the Fourier transform, and plot the continuous spectra for $f(t)$ as shown. Compare with Example 1 in the text.

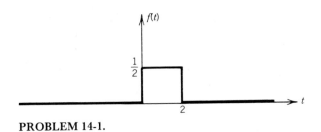

**PROBLEM 14-1.**

14-2 a) Calculate the Fourier transform and plot the continuous spectra for $f(t)$ as shown.

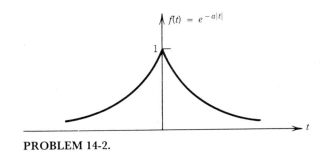

**PROBLEM 14-2.**

b) In part (a), let $a \to 0$. What becomes of $f(t)$? What becomes of $F(\omega)$? *Hint*: calculate the total area under $F(\omega)$ first.

14-3 Using the inversion formula, Eq. (14-8), find
$$\mathscr{F}^{-1}2\pi\delta(\omega - \omega_0)$$

14-4 Using the result of problem 14-3 and linearity, find the Fourier transform of $\cos \omega_0 t$, and of $\sin \omega_0 t$. Plot, for each, the magnitude and phase spectra.

14-5 Prove the frequency shifting property, that is
$$\mathscr{F}^{-1}F(\omega - \omega_0) = f(t)e^{j\omega_0 t}$$

14-6 Calculate the Fourier transform of $g(t)$, the triangular pulse shown, and confirm the result by relating $g(t)$ to $f(t)$ in Problem 14-1.

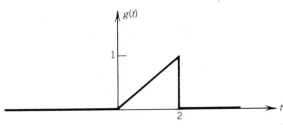

**PROBLEM 14-6.**

14-7 a)  Prove the symmetry property of the Fourier transform, namely, if

$$\mathscr{F}f(t) = F(\omega)$$

then

$$\mathscr{F}F(t) = 2\pi f(-\omega)$$

Here $F(t)$ is the function $F(\omega)$ with $\omega$ replaced by $t$. Similarly, $f(-\omega)$ is $f(t)$ with $t$ replaced by $-\omega$. Hint: just look at Eqs. (14-6), (14-8)!

b)  From the known square pulse

$$f(t) = \begin{cases} 1 & -\dfrac{a}{2} < t < \dfrac{a}{2} \\ 0 & \text{otherwise} \end{cases}$$

and it transform

$$F(\omega) = a\,\frac{\sin(\omega a/2)}{(\omega a/2)}$$

as in Fig. 14-2 and 14-3, obtain by symmetry the time function $F(t)$ and its Fourier transform. Plot all four functions.

14-8  Let $g(t)$ of problem 14-6 be a current source applied to a parallel $RC$ circuit, with $R = 1\,\Omega$, $C = \frac{1}{2}\mathrm{F}$. The voltage across this circuit is $v(t)$. Calculate and plot the amplitude and phase spectra of $v(t)$.

14-9  Repeat problem 14-8 with $i(t) = f(t)$ of Problem 14-1.

14-10  Repeat problem 14-8 with $i(t) = f(t)$ of Problem 14-2.

14-11  Prove that an ideal low-pass filter acts as an ideal delay line (time-shift) to an input which has a low frequency spectrum, that is $F(\omega) = 0$ for $|\omega| > \omega_c$.

14-12  Calculate the response $g(t)$ of an ideal low-pass filter to a periodic input

$$f(t) = \sum_{k=-\infty}^{+\infty} c_k e^{jk\omega_0 t}$$

by considering first the response to $e^{jk\omega_0 t}$ for $k\omega_0 < \omega_c$ and for $k\omega_0 > \omega_c$. Then use superposition. Your final answer should make sense intuitively—state it in words.

14-13  An ideal *high-pass* filter has

$$|H(\omega)| = \begin{cases} 0 & |\omega| < \omega_c \\ K & |\omega| > \omega_c \end{cases}$$

and $\theta_H = -\omega t_0$. Prove that its impulse response is

$$h(t) = K\delta(t - t_0) - K\,\frac{\sin \omega_c(t - t_0)}{\pi(t - t_0)}$$

and plot $h(t)$.

14-14 a) Prove that if $f(t)$ is real, then its Fourier transform satisfies

$$F(-\omega) = F^*(\omega)$$

*Hint*: write Eq. (14-6) using Euler's identity $e^{-j\omega t} = \cos \omega t - j \sin \omega t$, then separate $F(\omega)$ into its rectangular form

$$F(\omega) = R(\omega) + jX(\omega)$$

b) Verify this result with several of the transforms in this chapter.

14-15 Calculate the energy spectral density function for:

1) the impulse function $\delta(t)$
2) the square pulse of problem 14-1
3) the time function of problem 14-2
4) the triangular pulse of problem 14-6

14-16 a) Prove that

$$\int_{-\infty}^{\infty} f_1(t) f_2(t) \, dt = \frac{1}{2\pi} \int_{-\infty}^{\infty} F_1(\omega) F_2^*(\omega) \, d\omega$$

and give an interpretation of energy to this relationship by letting $f_1(t) = i(t)$ and $f_2(t) = v(t)$ for a given one-port network.

14-17 The input voltage to an ideal low-pass filter $(\omega_c = 20 \text{ rad/s}, K = 1)$ is $f(t) = 100e^{-10t}u(t)$ volts

a) Plot $|F(\omega)|^2$
b) Plot $|G(\omega)|^2$ for the output voltage $g(t)$
c) Calculate, using Parseval's theorem, the percentage of the total input energy which is available at the output.

14-18 From your tables of the Laplace transform, obtain the Fourier transform of

a) $f(t) = e^{-at} \cos \beta t \, u(t) \qquad a > 0$
b) $f(t) = e^{-at} \sin \beta t \, u(t) \qquad a > 0$
c) $f(t) = \cos \beta t \, u(t)$
d) $f(t) = \sin \beta t \, u(t)$
e) $f(t) = t^n e^{-at} u(t) \qquad a > 0, n = 1, 2, \ldots$
f) $f(t) = e^{-a|t|} \qquad -\infty < t < \infty \qquad a > 0$

Plot $f(t)$ in each case.

14-19 a) Prove that if $f(t)$ is causal then $F_L(s)$ in Eq. (14-33) can be written as

$$F_L(s) = \frac{1}{\pi} R(\omega) * \frac{1}{\sigma + j\omega}$$

where $R(\omega)$ is the real part of $F_F(\omega)$

$$F_F(\omega) = R(\omega) + jX(\omega)$$

*Hint*: Show first that for a causal $f(t)$

$$f(t)u(t) = 2f_{ev}(t)u(t)$$

where $f_{ev}(t)$ is the even part of $f(t)$ and $\mathscr{F}f_{ev}(t) = R(\omega)$

b) Use the result of part (a) to find the Laplace transform of $\cos \beta t\, u(t)$.

# 15

# TIME-VARYING NETWORKS

In the previous chapters we dealt with lumped, linear, and constant (time invariant) networks. In this chapter we discuss briefly lumped, linear, and *time-varying* networks, together with the application of the Laplace transform, the impulse response, convolution, network functions, and state variables for this type of networks. It must be emphasized that there are no general methods which can be applied successfully to the whole class of time-varying networks. Rather, certain types of problems (sometimes quite restricted) can be attacked only by certain techniques.

Here we can give only a brief outline of these topics. While a detailed discussion is beyond our scope, we shall try to exhibit some of their main features. Frequent comparisons will be made with constant networks.

## 15-1 APPLICATIONS OF THE LAPLACE TRANSFORM

Lumped, linear and time-varying networks are typically described by differential equations whose coefficients are functions of time. (In constant networks, these coefficients are constants.) If these coefficients are polynomials in $t$, we can use the relation

$$\mathscr{L}tf(t) = -\frac{dF(s)}{ds} \tag{15-1a}$$

and, generally

$$\mathscr{L}t^n f(t) = (-1)^n \frac{d^n F(s)}{ds^n} \tag{15-1b}$$

(see Eqs. (2-26), (2-27)). Accordingly, when we transform the differential equation with coefficients which are polynomials in $t$, we obtain a *differential* equation in $s$.[†] This new differential equation may, or *may not*, be easier to solve than the original one.

---

[†] This is in contrast with constant coefficients, where the transformed equation is *algebraic*.

### Example 1

Consider the so-called Bessel's equation of index zero,

$$t\frac{d^2y}{dt^2} + \frac{dy}{dt} + ty = 0 \tag{15-2}$$

with the given initial conditions $y(0) = 1$, $y'(0) = 0$. Laplace transforming Eq. (15-2), we obtain

$$-\frac{d}{ds}[s^2Y(s) - sy(0) - y'(0)] + sY(s) - y(0) - \frac{d}{ds}Y(s) = 0 \tag{15-3}$$

which reduces to

$$(s^2 + 1)\frac{dY(s)}{ds} + sY(s) = 0 \tag{15-4}$$

a first-order differential equation in $Y(s)$. This equation is solved easily by separating variables,

$$\frac{dY(s)}{Y(s)} = -\frac{s}{s^2 + 1}ds \tag{15-5}$$

and integrating

$$Y(s) = \frac{K}{\sqrt{s^2 + 1}} = K(s^2 + 1)^{-1/2} \tag{15-6}$$

where $K$ is an arbitrary constant. The expansion of $(s^2 + 1)^{-1/2}$ by the binomial series yields

$$Y(s) = \frac{K}{s}\left(1 + \frac{1}{s^2}\right)^{-1/2} = \frac{K}{s}\left(1 - \frac{1}{2}\frac{1}{s^2} + \frac{1\cdot 3}{2^2 2!}\frac{1}{s^4} - \cdots\right) \tag{15-7}$$

Inverting, term by term, we get

$$y(t) = K\left(1 - \frac{1}{2^2}t^2 + \frac{1}{2^2\cdot 4^2}t^4 - \frac{1}{2^2\cdot 4^2\cdot 6^2}t^6 + \cdots\right) \tag{15-8}$$

To satisfy the given initial conditions, we must have $K = 1$.[†]

The function in Eq. (15-8) is defined as the Bessel function $J_0(t)$, that is,

$$J_0(t) = 1 - \frac{1}{2^2}t^2 + \frac{1}{2^2\cdot 4^2}t^4 - \cdots \tag{15-9}$$

---

[†] Alternately, we can apply the initial value theorem directly to $Y(s)$ in Eq. (15-6) and obtain the same result, $K = 1$.

This function has been tabulated and studied extensively. It is the solution to Eq. (15-2), and we found its Laplace transform

$$\mathscr{L} J_0(t) = \frac{1}{\sqrt{s^2 + 1}} \qquad \square \quad (15\text{-}10)$$

---

***Example 2***

The first-order differential equation,

$$(t^2 + 1)\frac{dy}{dt} + ty = 0 \qquad y(0) = 0 \qquad (15\text{-}11)$$

when Laplace transformed, yields

$$s\frac{d^2 Y(s)}{ds^2} + \frac{dY(s)}{ds} + sY(s) = 0 \qquad (15\text{-}12)$$

a *second-order* differential equation which is not easier to solve. (The original equation may be solved directly by separation of variables, as in the previous example.)  $\square$

---

Other transforms have been devised and applied to specific types of problems. Problem 15-5 deals with one such transform, and the interested reader is referred to the additional reading listed at the end of this chapter. As mentioned earlier, a single, universal transform method for time-varying networks does not exist.

***Probs. 15-1***
***through***
***15-5***

## 15-2  IMPULSE RESPONSE AND CONVOLUTION

As long as the time-varying network is *linear*, we can apply the principle of superposition to it. We used such an approach to develop the convolution integral for constant (time invariant) networks in Chapter 8. Let us here obtain similar results for a time varying linear network.

First, consider the impulse response. Recall that it is the zero-state response to a unit impulse input. For a constant network we have[†]

$$S\{\delta(t)\} = h(t)$$

and $$(15\text{-}13a)$$

$$S\{(t - \xi)\} = h(t - \xi)$$

---

[†] We use the symbolic notation (see Chapter 1. Fig. 1-1) of $S\{\ \}$ to mean "the network operates on the input $\{\ \}$." Later we shall use specific expressions for $S$.

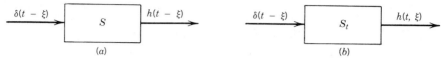

**FIGURE 15-1.** Impulse response of: (a) A constant network; (b) A time-varying network.

See Fig. 15-1a. Thus, in a constant, time invariant network, the response to the impulse $\delta(t)$ is $h(t)$, and the response to a *shifted* impulse $\delta(t - \xi)$ is an identical *shifted* $h(t - \xi)$. In other words, the impulse response of a constant network depends only on the *difference* between the instant of observation $t$ and the instant of application $\xi$ of the impulse.

In a time-varying network this is not true: Since the parameters of the network are time-varying, the response to the impulse $\delta(t - \xi)$ is of the general form $h(t, \xi)$. That is, this response is a function of *both* the instant of application of the impulse, $\xi$, and of the instant of observation, $t$

$$S_t\{\delta(t - \xi)\} = h(t, \xi) \tag{15-13b}$$

See Fig. 15-1b, where the subscript "$t$" in $S_t$ designates a time-varying network. This is further illustrated in Fig. 15-2 where $h$ is plotted as a function of $t$ and $\xi$. In both cases $h \equiv 0$ for $t < \xi$, since we are dealing with non-anticipating (causal) networks. In a constant network, Fig. 15-2a, the shape of $h$ is the same—it is merely shifted. For a time-varying network, Fig. 15-2b, the shape of $h$ is different, being dependent on $t$ and $\xi$.

How to find this impulse response? As in the case of constant networks, the analysis of a linear time-varying network leads, in general, to a linear differential equation relating the response $r(t)$ to the excitation $e(t)$:

$$d_0(t)\frac{d^n r}{dt^n} + d_1(t)\frac{d^{n-1}r}{dt^{n-1}} + \cdots + d_{n-1}(t)\frac{dr}{dt} + d_n(t)r$$

$$= c_0(t)\frac{d^m e}{dt^m} + c_1(t)\frac{d^{m-1}e}{dt^{m-1}} + \cdots + c_{m-1}(t)\frac{de}{dt} + c_m(t)e \tag{15-14a}$$

where the coefficients $c_k$ and $d_k$ are time-varying. We can write Eq. (15-14a) as

$$\left\{\sum_{k=0}^{n} d_k(t)p^k\right\}r(t) = \left\{\sum_{k=0}^{m} c_k(t)p^k\right\}e(t) \tag{15-14b}$$

where $p = d/dt$, or more compactly,

$$L(p; t)\{r(t)\} = K(p; t)\{e(t)\} \tag{15-14c}$$

where

$$L(p; t) = \sum_{k=0}^{n} d_k(t)p^k$$

and $\qquad\qquad\qquad\qquad\qquad\qquad\qquad\qquad\qquad\qquad\qquad\qquad$ (15-14d)

$$K(p; t) = \sum_{k=0}^{m} c_k(t)p^k$$

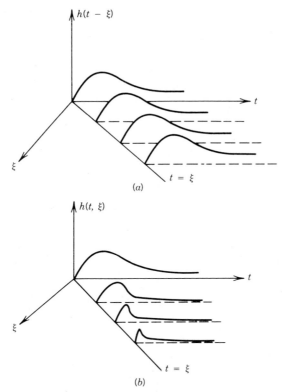

$h(t - \xi)$

$t$

$\xi$

$t = \xi$

(a)

$h(t, \xi)$

$t$

$\xi$

$t = \xi$

(b)

**FIGURE 15-2.** Typical impulse response of: (a) A Constant network; (b) A time-varying network.

are linear differential operators. In the case of constant networks, $L(p)$ and $K(p)$ are constant differential operators independent of $t$.

Now, if the excitation is $\delta(t - \xi)$, a unit impulse applied at time $\xi$, the response $h(t, \xi)$ is then the solution of the differential equation

$$L(p; t)\{h(t, \xi)\} = K(p; t)\{\delta(t - \xi)\} \tag{15-15}$$

Since the network is non-anticipating, there is no response before the application of the excitation; that is,

$$h(t, \xi) = 0 \qquad \text{for } t < \xi \tag{15-16}$$

which can be conveniently written also as $h(t, \xi)u(t - \xi)$. Even though Eq. (15-15) is very basic, the general solution to it is not known, although several *specific* types have been studied and listed.

Since the network is linear, superposition holds. As in the derivation of the convolution for constant networks (Chapter 8), we reason as follows: the input $e$, zero for all $t < 0$, can be written as a sum of impulses,

$$e(t) = \int_0^t e(\xi)\delta(t - \xi)\, d\xi \tag{15-17}$$

[Compare with Eq. (8-5)]. The response to $\delta(t - \xi)$ is $h(t, \xi)u(t - \xi)$. Hence, by superposition, the zero-state response $r(t)$ is

$$r(t) = \int_0^t e(\xi)h(t, \xi)u(t - \xi) \, d\xi \qquad (15\text{-}18a)$$

Since $u(t - \xi)$ is unity between the limits of the integral $0 \le \xi \le t$, Eq. (15-18a) can be simplified to

$$r(t) = \int_0^t e(\xi)h(t, \xi) \, d\xi \qquad (15\text{-}18b)$$

This is the convolution integral for a time-varying network. In the case of a *constant* network we have, by definition,

$$h(t, \xi) = h(t - \xi, 0) = h(t - \xi) \qquad (15\text{-}19a)$$

and the convolution integral assumes the form,

$$r(t) = \int_0^t e(\xi)h(t - \xi) \, d\xi \qquad (15\text{-}19b)$$

as in Eq. (8-6).                                           **Prob. 15-6**

## *15-3   NETWORK FUNCTIONS

Recall (Chapter 7) that the network function for a constant network is defined as

$$H(s) = \frac{R(s)}{E(s)} \qquad (15\text{-}20)$$

the ratio of the (transform) zero-state response to the (transform) excitation. Also we have

$$H(s) = \mathscr{L}h(t) = \mathscr{L}h(t, 0) = \int_0^\infty h(t, 0)e^{-st} \, dt \qquad (15\text{-}21)$$

where we write $h(t, 0)$ to emphasize that it is the response observed at the instant $t$ when the impulse was applied $t$ seconds earlier at $\xi = 0$. The zero-state response is

$$r(t) = \mathscr{L}^{-1}H(s)E(s) \qquad (15\text{-}22)$$

By analogy, the network function for a time-varying network is defined as[1,2]

$$H(s, t) = \int_0^\infty h(t, t - \xi)e^{-s\xi} \, d\xi \qquad (15\text{-}23)$$

Notice that, unlike $H(s)$, the function $H(s, t)$ as defined in Eq. (15-23) is a function of $s$ involving $t$ as a parameter. Also, by analogy to Eq. (15-22), we have for time-varying networks,

$$r(t) = \mathscr{L}^{-1}H(s, t)E(s) \qquad (15\text{-}24)$$

The inverse Laplace is taken with respect to $s$, treating $t$ again as if it were a constant parameter.

In order to find $H(s, t)$ we can use Eq. (15-23), provided the impulse response is known. We illustrate with an example.

---

## Example 3

A time-varying network is described by the differential equation

$$t^2 \frac{d^3r}{dt^3} + 6t \frac{d^2r}{dt^2} + 6 \frac{dr}{dt} = e(t) \tag{15-25a}$$

Its impulse response is[3]

$$h(t, \xi) = \frac{1}{2} \left( \frac{\xi}{t} - 1 \right)^2 u(t - \xi) \tag{15-25b}$$

According to Eq. (15-23), we need first $h(t, t - \xi)$. Replace therefore every $\xi$ in $h(t, \xi)$ by $t - \xi$ to get

$$h(t, t - \xi) = \frac{1}{2} \left( \frac{t - \xi}{t} - 1 \right)^2 u(\xi) = \frac{1}{2} \frac{\xi^2}{t^2} u(\xi) \tag{15-25c}$$

Then we obtain $H(s, t)$ from Eq. (15-23) as

$$H(s, t) = \frac{1}{2} \int_0^\infty \frac{\xi^2}{t^2} e^{-s\xi} u(\xi) \, d\xi = \frac{1}{2t^2} \mathscr{L}\xi^2 = \frac{1}{2t^2} \frac{2}{s^3} = \frac{1}{t^2 s^3}. \tag{15-25d}$$

The zero-state response of this network to an input $e^{-t}u(t)$ is found by using Eq. (15-24),

$$r(t) = \mathscr{L}^{-1} \frac{1}{t^2 s^3} \frac{1}{(s + 1)} = \frac{1}{t^2} \mathscr{L}^{-1} \frac{1}{s^3(s + 1)}$$

$$= \left( \frac{1}{2} - \frac{1}{t} + \frac{1}{t^2} - \frac{e^{-t}}{t^2} \right) u(t) \qquad \Box \tag{15-25e}$$

---

When $h(t, \xi)$ is not known, we can determine the network function $H(s, t)$ directly from the differential equation, Eq. (15-14), as follows: Just as $H(s)e^{st}$ is the response to $e^{st}$ in a constant network, the response to $e^{st}$ in a time-varying network is $H(s, t)e^{st}$. In terms of Eq. (15-14) we have

$$L(p; t)\{H(s, t)e^{st}\} = K(p; t)\{e^{st}\} \tag{15-26a}$$

Although this equation does not seem at first to be very easy to solve, it can be verified (see problem 15-7) that the left-hand side is equivalent to $e^{st}L(p + s; t)H(s, t)$.

Therefore, we have

$$e^{st}L(p + s; t)\{H(s, t)\} = K(p; t)\{e^{st}\} \tag{15-26b}$$

and therefore

$$L(p + s; t)\{H(s, t)\} = K(p; t) \tag{15-26c}$$

This is the differential equation to be solved for $H(s, t)$. Although a general solution is not available, it can be solved for certain cases. It is important to note that only its particular solution, and not the complementary function, is sufficient for $H(s, t)$.

□   **Prob. 15-7**

---

## Example 4

Find $H(s, t)$ for the network of the previous example using the differential equation. Here

$$t^2 \frac{d^3r}{dt^3} + 6t \frac{d^2r}{dt^2} + 6 \frac{dr}{dt} = e(t) \tag{15-27a}$$

therefore,

$$L(p; t) = t^2 p^3 + 6tp^2 + 6p \tag{15-27b}$$

$$K(p; t) = 1 \tag{15-27c}$$

and Eq. (15-26c) becomes

$$\{t^2(p + s)^3 + 6t(p + s)^2 + 6(p + s)\}H(s, t) = 1 \tag{15-27d}$$

That is,

$$t^2 \frac{d^3H}{dt^3} + (3t^2s + 6t) \frac{d^2H}{dt^2} + (3t^2s^2 + 12ts + 6) \frac{dH}{dt} + (t^2s^3 + 6ts^2 + 6s) H = 1 \tag{15-27e}$$

It is readily verified that

$$H(s, t) = \frac{1}{t^2 s^3} \tag{15-27f}$$

as found in the previous example satisfies Eq. (15-27e).    □   **Prob. 15-8**

---

Note that, in general, the order of the differential equation for $H(s, t)$ — Eq. (15-26c) — is the same as that of the differential equation for the network, Eq. (15-14).

It is instructive, at this point, to compare this method with the conventional $H(s)$ for constant networks. As noted earlier, the differential operators $L$ and $K$ have constant coefficients in the case of a constant network. Therefore Eq. (15-26a) becomes

$$L(p)\{H(s)e^{st}\} = K(p)\{e^{st}\} \tag{15-28a}$$

and since $H(s)$ is not a function of time, we have

$$H(s)L(p)\{e^{st}\} = K(p)\{e^{st}\} \tag{15-28b}$$

Now

$$L(p)\{e^{st}\} = \left(d_0 \frac{d^n}{dt^n} + d_1 \frac{d^{n-1}}{dt^{n-1}} + \cdots + d_n\right)\{e^{st}\}$$
$$= d_0 s^n e^{st} + d_1 s^{n-1} e^{st} + \cdots + d_n e^{st} \tag{15-28c}$$

with similar results for $K(p)\{e^{st}\}$. Hence Eq. (15-28b) yields

$$H(s) = \frac{c_0 s^m + c_1 s^{m-1} + \cdots + c_{m-1}s + c_m}{d_0 s^n + d_1 s^{n-1} + \cdots + d_{n-1}s + d_n} \tag{15-28d}$$

as studied in Chapter 7 (see Eq. (7-9)).

The result obtained in Eq. (15-26c) may be generalized to express successive operations on a function of time.[2] For example, let $f_2(t)$ be the result of $M_1(p; t)$ operating on $f_1(t)$:

$$M_1(p; t)\{f_1(t)\} = f_2(t) \tag{15-29a}$$

and then $f_2(t)$ is operated on by $M_2(p; t)$, resulting in $f_3(t)$;

$$M_2(p; t)\{f_2(t)\} = f_3(t) \tag{15-29b}$$

See Fig. 15-3, where the two operators are shown in cascade. We have from Eqs. (15-29a) and (b)

$$M_2(p; t)\{M_1(p: t)\{f_1(t)\}\} = f_3(t) \tag{15-29c}$$

The two successive operations in Eq (15-29c) may be replaced by a single operation with an operator $M_3(p; t)$ such that

$$M_3(p; t)\{f_1(t)\} = f_3(t) \tag{15-29d}$$

It can be shown that, similarly to Eq. (15-26), the operator $M_3(p; t)$ is given by

$$M_3(p; t) = M_2(p + s; t)\{M_1(s, t)\} \tag{15-29e}$$

where $M_2(p + s; t)$ is the operator and $M_1(s, t)$ is considered a function of time involving $s$ as a parameter.

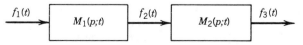

**FIGURE 15-3.** Cascade operators.

***Example 5***[(2)]

The driving-point impedance of a time-varying capacitor $C(t)$ is

$$Z_{dp}(s, t) = \frac{v(t)}{i(t)}\bigg]_{i(t) = e^{st}} \tag{15-30a}$$

Similarly, its driving-point admittance is

$$Y_{dp}(s, t) = \frac{i(t)}{v(t)}\bigg]_{v(t) = e^{st}} \tag{15-30b}$$

The defining relationship between $v(t)$ and $i(t)$ for the capacitor is

$$i(t) = \frac{d}{dt}\{C(t)v(t)\} = p\{C(t)v(t)\} \tag{15-31}$$

Applying Eq. (15-20), we find therefore

$$Z_{dp}(s, t) = \frac{1}{C(t)s} \tag{15-32}$$

and

$$Y_{dp}(s, t) = C(t)s + \dot{C}(t) \tag{15-33}$$

where the dot indicates differentiation with respect to time. We recognize therefore that

$$Y_{dp}(p + s; t)\{Z_{dp}(s, t)\} = 1 \tag{15-34}$$

and it is *in this sense* that $Z_{dp}(s, t)$ is the inverse of $Y_{dp}(s, t)$. For a constant network, this operational relation becomes the familiar, purely algebraic relation,

$$Y_{dp}(s)Z_{dp}(s) = 1 \tag{15-35}$$

☐ ***Probs. 15-9***
***15-10***
***15-11***

## 15-4   STATE VARIABLES

For a linear time-varying network the state and output equations are

$$\dot{\mathbf{x}} = \mathbf{A}(t)\mathbf{x} + \mathbf{B}(t)\mathbf{e}(t) \tag{15-36}$$

and

$$\mathbf{r} = \mathbf{C}(t)\mathbf{x} + \mathbf{D}(t)\mathbf{e}(t) \tag{15-37}$$

Compare with Eqs. (10-69) and (10-70). Consider first the homogeneous, zero input state equation,

$$\dot{\mathbf{x}} = \mathbf{A}(t)\mathbf{x} \tag{15-38}$$

For a first-order network it reduces to the scalar equation,

$$\dot{x} = a(t)x \tag{15-39a}$$

the solution to which is obtained, as usual, by separation of variables

$$\frac{dx}{x} = a(t)\, dt \tag{15-39b}$$

Integration from an initial time $t_0$ to $t$ yields

$$\log \frac{x(t)}{x(t_0)} = \int_{t_0}^{t} a(\tau)\, d\tau \tag{15-39c}$$

and finally,

$$x(t) = \exp\left[\int_{t_0}^{t} a(\tau)\, d\tau\right] x(t_0) \tag{15-39d}$$

where $x(t_0)$ is the given initial state. Therefore, we recognize the state transition matrix (here a scalar) for this case as

$$\boldsymbol{\phi}(t, t_0) = \exp\left[\int_{t_0}^{t} a(\tau)\, d\tau\right] \tag{15-39e}$$

Naturally, we are tempted to extend this solution to higher-order networks as follows

$$\mathbf{x} = \exp\left[\int_{t_0}^{t} \mathbf{A}(\tau)\, d\tau\right] \mathbf{x}(t_0) \tag{15-40}$$

However, if this trial solution is substituted into Eq. (15-38), we see that it is the correct solution *if and only if*

$$\frac{d}{dt}\left\{\exp\left[\int_{t_0}^{t} \mathbf{A}(\tau)\, d\tau\right]\right\} = \left(\frac{d}{dt}\int_{t_0}^{t} \mathbf{A}(\tau)\, d\tau\right)\exp\left[\int_{t_0}^{t} \mathbf{A}(\tau)\, d\tau\right] \tag{15-41}$$

It is not difficult to show that Eq. (15-41) does *not* hold in general. Also it can be shown that Eq. (15-41) is satisfied provided[4]

$$\mathbf{A}(t_1)\mathbf{A}(t_2) = \mathbf{A}(t_2)\mathbf{A}(t_1) \tag{15-42}$$

for all $t_1$ and $t_2$. In this case Eq. (15-40) is indeed the solution to Eq. (15-38), and the state transition matrix is defined as

$$\boldsymbol{\phi}(t, t_0) = \exp\left[\int_{t_0}^{t} \mathbf{A}(\tau)\, d\tau\right] \tag{15-43}$$

so that the zero-input state response is

$$\mathbf{x}(t) = \boldsymbol{\phi}(t, t_0)\mathbf{x}(t_0) \tag{15-44}$$

**Probs. 15-12**
**15-13**
**15-14**
**15-15**

In the case of constant networks the state transition matrix depends, again, only on the *difference* $t - t_0$ so that $\boldsymbol{\phi}(t, t_0) = \boldsymbol{\phi}(t - t_0)$. For time-varying networks it is a function of both the initial time $t_0$ and the observation time $t$. Its meaning, evident from Eq. (15-44), is the same as in constant networks. Namely, it maps the initial state $\mathbf{x}(t_0)$ into the state $\mathbf{x}(t)$; that is, it performs a transition of the state of the network.

If Eq. (15-41) [or Eq. (15-42)] is not satisfied, the zero-input state response is still given by Eq. (15-44), but $\boldsymbol{\phi}(t, t_0)$ is no longer given by Eq. (15-43). The complete solution to Eq. (15-36) is

$$\mathbf{x}(t) = \boldsymbol{\phi}(t, t_0)\mathbf{x}(t_0) + \int_{t_0}^{t} \boldsymbol{\phi}(t, \tau)\mathbf{B}(\tau)\mathbf{e}(\tau) \, d\tau \tag{15-45}$$

which is a result similar to Eq. (11-10). The second term is the zero-state response, expressed as a superposition integral. In constant networks this becomes the convolution integral.

If Eq. (15-41) [or Eq. (15-42)] is not satisfied, $\boldsymbol{\phi}(t, t_0)$ is no longer given by Eq. (15-43). One method to evaluate $\boldsymbol{\phi}(t, t_0)$ is to express $\mathbf{A}(t)$ as a sum of two matrices[4]

$$\mathbf{A}(t) = \mathbf{A}_0(t) + \bar{\mathbf{A}}(t) \tag{15-46}$$

where $\mathbf{A}_0(t)$ satisfies the commutativity condition of Eq. (15-42). Substituting into Eq. (15-38), we obtain

$$\dot{\mathbf{x}} = \mathbf{A}_0(t)\mathbf{x} + \bar{\mathbf{A}}(t)\mathbf{x} \tag{15-47}$$

where the term $\bar{\mathbf{A}}(t)\mathbf{x}$ may now be considered an "input." The state transition matrix for $\mathbf{A}_0(t)$ is

$$\boldsymbol{\phi}_0(t, t_0) = \exp\left[\int_{t_0}^{t} \mathbf{A}_0(\tau) \, d\tau\right] \tag{15-48}$$

since, by assumption, $\mathbf{A}_0(t)$ obeys Eq. (15-42). The complete solution to Eq. (15-47) is, according to Eqs. (15-44) and (15-45)

$$\mathbf{x}(t) = \boldsymbol{\phi}_0(t, t_0)\mathbf{x}(t_0) + \int_{t_0}^{t} \boldsymbol{\phi}_0(t, \tau)\bar{\mathbf{A}}(\tau)\mathbf{x}(\tau) \, d\tau \tag{15-49}$$

In this *integral equation* the unknown $\mathbf{x}(t)$ appears on the left side and under the integral on the right side. The solution for $\mathbf{x}(t)$ may be obtained by successive iterations (corrections) as follows: We begin with a trial solution,

$$\mathbf{x}_0(t) = \boldsymbol{\phi}_0(t, t_0)\mathbf{x}(t_0) \tag{15-50a}$$

and substitute it for $\mathbf{x}$ under the integral sign. Then Eq. (15-49) yields a first approximation $\mathbf{x}_1(t)$:

$$\mathbf{x}_1(t) = \boldsymbol{\phi}_0(t, t_0)\mathbf{x}(t_0) + \int_{t_0}^{t} \boldsymbol{\phi}_0(t, \tau)\bar{\mathbf{A}}(\tau)\mathbf{x}_0(\tau) \, d\tau \tag{15-50b}$$

This approximation, in turn, is substituted for $\mathbf{x}$ under the integral in Eq. (15-49) to obtain the second approximation $\mathbf{x}_2(t)$ as

$$\mathbf{x}_2(t) = \boldsymbol{\phi}_0(t, t_0)\mathbf{x}(t_0) + \int_{t_0}^{t} \boldsymbol{\phi}_0(t, \tau)\bar{\mathbf{A}}(\tau)\mathbf{x}_1(\tau)\, d\tau \tag{15-50c}$$

In general, the $n$th-approximation solution is

$$\mathbf{x}_n(t) = \boldsymbol{\phi}_0(t, t_0)\mathbf{x}(t_0) + \int_{t_0}^{t} \boldsymbol{\phi}_0(t, \tau)\bar{\mathbf{A}}(\tau)\mathbf{x}_{n-1}(\tau)\, d\tau \tag{15-50d}$$

In this method, the matrix $\bar{\mathbf{A}}(t)$ may be considered a *perturbation* on $\mathbf{A}_0(t)$. The successive approximations $\mathbf{x}_1(t)$, $\mathbf{x}_2(t)$, ... are improvements of the unperturbed solution $\mathbf{x}_0(t)$. If $\bar{\mathbf{A}}(t)$ is relatively small, the first or second approximations will provide a good solution.

---

## *Example 6*

To illustrate this perturbation method, consider the scalar differential equation,

$$\dot{x} = (1 + \alpha e^{-t})x \tag{15-51a}$$

with $x(t_0) = K$ (given), and $\alpha \ll 1$. Here we have

$$\mathbf{A}(t) = a(t) = 1 + \alpha e^{-t} \tag{15-51b}$$

and we can consider $\mathbf{A}_0(t) = a_0(t)$ to be

$$a_0(t) = 1 \tag{15-51c}$$

And the perturbation $\bar{\mathbf{A}}(t) = \bar{a}(t)$ is then

$$\bar{a}(t) = \alpha e^{-t} \tag{15-51d}$$

Accordingly,

$$\boldsymbol{\phi}_0(t, t_0) = \exp\left[\int_{t_0}^{t} a_0(\tau)\, d\tau\right] = e^{t-t_0} \tag{15-51e}$$

(Recall: $\boldsymbol{\phi}$ is a scalar in this case). Then the unperturbed solution, as in Eq. (15-50a), is

$$x_0(t) = \boldsymbol{\phi}_0(t, t_0)x(t_0) = Ke^{t-t_0} \tag{15-51f}$$

The first approximation $x_1(t)$ is:

$$x_1(t) = \boldsymbol{\phi}_0(t, t_0)x_0(t) + \int_{t_0}^{t} \boldsymbol{\phi}_0(t, \tau)\bar{a}(\tau)x_0(\tau)\, d\tau$$

$$= Ke^{t-t_0} + \int_{t_0}^{t} e^{t-\tau}\alpha e^{-\tau}Ke^{\tau-t_0}\, d\tau$$

$$= Ke^{t-t_0} + K\alpha e^{t-t_0}\int_{t_0}^{t} e^{-\tau}\, d\tau$$

$$= Ke^{t-t_0}[1 + \alpha(e^{-t_0} - e^{-t})] \tag{15-51g}$$

Proceeding with Eq. (15-50c) we find here

$$x_2(t) = Ke^{t-t_0}[1 + \alpha(e^{-t_0} - e^{-t}) + \frac{\alpha^2}{2}(e^{-t_0} - e^{-t})^2] \qquad (15\text{-}51h)$$

On the other hand, Eq. (15-51a) can be solved directly by separating variables and integrating, as in Eq. (15-39). The result is

$$x(t) = K \exp\left[\int_{t_0}^{t}(1 + \alpha e^{-\tau}) d\tau\right] = Ke^{t-t_0} \exp[e^{\alpha(e^{-t_0} - e^{-t})}] \qquad (15\text{-}52)$$

and in Eqs. (15-51g) and (h) we recognize indeed the power series expansion of the last term in Eq. (15-52).    $\square$   **Prob. 15-16**

In summary, as mentioned earlier in this chapter, analytic methods for the solution of time-varying networks are not universal. Quite often, only approximate solutions are obtained, and numerical methods using computers may be needed. In this chapter we have reviewed several approaches typical for analyzing specific problems. More detailed discussions may be found in the references listed at the end of this chapter, and the interested reader is encouraged to consult them.

## PROBLEMS

15-1  Use Laplace transform methods to solve the differential equation

$$t\frac{d^2y}{dt^2} + (1 - t)\frac{dy}{dt} + 2y = 0$$

given $y(0) = 1$. *Hint*: In solving for $Y(s)$, separate variables and expand in partial fractions the coefficient of $ds$. Then integrate.

15-2  Use Laplace transform methods to solve the differential equation

$$t\frac{d^2y}{dt^2} + (2t + 1)\frac{dy}{dt} + (t + 3)y = 0$$

given $y(0) = 1$.

15-3  Obtain the differential equation for $Y(s)$ in the Bessel equation of index $n$ given by

$$t^2\frac{d^2y}{dt^2} + t\frac{dy}{dt} + (t^2 - n^2)y = 0$$

15-4  It is proposed to solve the differential equation

$$\frac{d^2f}{dt^2} - (t^2 + k)f = 0$$

given $f(0) = f'(0) = 0$ by Laplace transforming it. Here $k$ is a given constant. Take the Laplace transform of this equation and, without actually solving for $f(t)$, discover an interesting property of this function.

15-5 The Mellin transform of $f(t)$ is defined as

$$\mathcal{M}f(t) = \int_0^\infty f(t)t^{s-1}\,dt = F(S)$$

where $S$ is the complex variable of the Mellin transform. Investigate the application of the Mellin transform to differential equations of the form,

$$t^n \frac{d^n y}{dt^n} + t^{n-1}\frac{d^{n-1}y}{dt^{n-1}} + \cdots + t\frac{dy}{dt} + y = \cdots$$

by deriving the following results:

a) $\mathcal{M}\left(t\,\dfrac{dy}{dt}\right) = -SY(S)$

b) $\mathcal{M}\left(t^2\,\dfrac{d^2y}{dt^2}\right) = S(S+1)\,Y(S)$

and generalizing to $\mathcal{M}\left[t^n(d^ny/dt^n)\right]$.

15-6 Prove that the method using an integrating factor, described in an earlier chapter for a first-order constant network, is valid also for a time-varying network. That is, consider the equation

$$\frac{dr}{dt} + a(t)r = f(t)$$

where $f(t) = K(p; t)e(t)$ is known (see Eq. (15-14)).

a) Show that a suitable integrating factor is $e^{\int a(t)\,dt}$, and, after multiplying the given equation by it, we obtain a total differential on the left-hand side, $(d/dt)\,(re^{\int a(t)dt})$.

b) Integrate both sides.

c) Solve for $r(t)$.

d) Solve by this method:

$$\frac{dr}{dt} + \frac{2t}{t^2+1}\,r = 1 \qquad r(0) = -2$$

15-7 Verify the equivalence of Eqs. (15-26a) and (15-26b); that is, show that $L(p; t)\{H(s, t)e^{st}\} = e^{st}L(p + s; t)H(s, t)$. *Hint:* Apply $L(p; t)$ as given in Eq. (15-14) to the product $H(s, t)e^{st}$ and note that

$$\frac{d}{dt}\,(H(s, t)e^{st}) = e^{st}(p + s)H(s, t)$$

$$\frac{d^2}{dt^2}\,(H(s, t)e^{st}) = e^{st}(p + s)^2 H(s, t)$$

etc.

15-8  Find $H(s, t)$ for a network characterized by the differential equation

$$\frac{dr}{dt} + tr = 1$$

*Hint:* The differential equation for $H$ can be solved by an integrating factor.

15-9  Investigate the commutativity of two differential operators $M_1(p; t)$ and $M_2(p; t)$ by considering the combination shown in part (a) and next, that shown in part (b). Is $f_3$ the same in both cases? Draw conclusions about the commutativity of these operators. Now repeat with two constant differential operators, say $M_1(p) = p + 2$, $M_2(p) = p^2 + p + 1$, and draw conclusion about their commutativity.

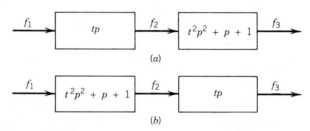

(a)

(b)

**PROBLEM 15-9.**

15-10  Find $Z_{dp}(s, t)$ and $Y_{dp}(s, t)$ of a time-varying inductor, $L(t)$.

15-11  Find the driving-point impedance and admittance of a series $RL$ circuit where $R = 1$ ohm and $L = L(t)$.

15-12  Verify that the commutativity condition on $\mathbf{A}(t)$, Eq. (15-42), is satisfied for any $\mathbf{A}(t)$ that is diagonal. Hence obtain the state transition matrix and the solution to

$$\dot{\mathbf{x}} = \begin{bmatrix} -t & 0 \\ 0 & t \end{bmatrix} \mathbf{x} \qquad \mathbf{x}(t_0) = \begin{bmatrix} 1 \\ -1 \end{bmatrix}$$

15-13  Verify that the commutativity condition on $\mathbf{A}(t)$, Eq. (15-42), is satisfied for any $\mathbf{A}(t)$ that can be written as the product of a *constant* matrix and a single, scalar time function,[4]

$$\mathbf{A}(t) = \mathbf{A}f(t).$$

Hence obtain the state transition matrix

$$\dot{\mathbf{x}} = \begin{bmatrix} -2t & -t \\ 2t & -5t \end{bmatrix} \mathbf{x} \qquad \mathbf{x}(t_0) = \begin{bmatrix} -1 \\ 1 \end{bmatrix}$$

15-14  Derive the following properties of $\boldsymbol{\phi}(t, t_0)$:
  a) $\boldsymbol{\phi}(t_0, t_0) = \mathbf{U}$
  b) $\boldsymbol{\phi}(t_0, t_1)\boldsymbol{\phi}(t_1, t_2) = \boldsymbol{\phi}(t_0, t_2)$
  c) $\boldsymbol{\phi}^{-1}(t_0, t_1) = \boldsymbol{\phi}(t_1, t_0)$. *Hint:* Use (b)

15-15 Verify problem 15-14 for $\boldsymbol{\phi}(t, t_0)$ in problems 15-12 and 15-13.

15-16 Solve by the perturbation method:

$$\dot{\mathbf{x}} = \begin{bmatrix} t + \alpha e^{-t} & t \\ -t & t \end{bmatrix} \mathbf{x} \qquad \mathbf{x}(t_0) = \begin{bmatrix} 1 \\ 0 \end{bmatrix}$$

and compare with the exact solution. *Hint*: $\mathbf{A}_0(t)$ satisfies the condition of problem 15-13.

# REFERENCES

1. L. A. Zadeh, "Frequency analysis of variable networks," *Proceedings of the IRE*, vol. 38, March 1950, pp. 291–299.
2. ———, "Constant resistance networks of the linear varying-parameter type," *Proceedings of the IRE*, vol. 39, June 1951, pp. 688–691.
3. D. Graham et al., "Engineering analysis methods for linear time varying systems," Technical report ASD-TDR-62-362, Office of Technical Services, U.S. Department of Commerce, Washington, D.C., January 1963.
4. B. K. Kinariwala, "Analysis of time-varying networks," *IRE International Conventional Record*, part 4, March 1961, pp. 268–276.

# ADDITIONAL READING

5. H. d'Angelo, *Linear Time-Varying Systems*. Boston: Allyn and Bacon, Inc. 1970.
6. F. R. Gerardi, "Application of Mellin and Hankel transform to networks with time-varying parameters," *IRE Trans. on Circuit Theory*, vol. CT-6, no. 2, June 1959, pp. 197–208.
7. T. Kailath, *Linear Systems*. Englewood Cliffs, N.J.: Prentice-Hall, Inc. chapter 9, 1980.
8. D. G. Luenberger, *Introduction to Dynamic Systems*. New York: John Wiley & Sons, Inc., chapter 4, 1979.
9. P. M. DeRusso, R. J. Roy, C. M. Close, *State Variables for Engineers*. New York: John Wiley & Sons, Inc. 1965.

# MATRICES

## A-1 BASIC CONCEPTS AND TYPES OF MATRICES

A $m \times n$ *matrix* is a rectangular array of scalars, arranged in $m$ rows and $n$ columns. The notation of such a matrix is:

$$\mathbf{A} = [a_{pq}] \qquad 1 \leq p \leq m \qquad 1 \leq q \leq n \tag{A-1}$$

where $a_{pq}$ is the *pq element* of $\mathbf{A}$, occupying the position in row $p$ and column $q$. The *order* or *dimension* of $\mathbf{A}$ is given by $m \times n$.

A *square matrix* has the same number of rows and columns, $m = n$.

A *row matrix*, or *row vector*, has one row, and its dimension is $1 \times n$. A *column matrix*, or *column vector*, has one column, and its dimension is $m \times 1$. A matrix can be thought of as containing $m$ row vectors $\mathbf{A}_p$ or $n$ column vectors $\mathbf{A}_q$.

A *diagonal matrix* is square with the elements off its main diagonal all zero, $a_{pq} = 0$ for $p \neq q$.

A matrix is *triangular* if $a_{pq} = 0$ for $p \geq q$ or if $a_{pq} = 0$ for $p \leq q$. The former is upper triangular, the latter is lower triangular.

The *unit matrix* (also called the *identity matrix*), denoted by $\mathbf{U}$, is a diagonal matrix[†] whose diagonal elements are 1, $u_{pp} = 1$.

The *null* (or *zero*) *matrix* $\mathbf{0}$ has *all* its elements zero.

The *trace* of a square matrix is the sum of its diagonal elements

$$\mathrm{tr}\,\mathbf{A} = \sum a_{pp} \tag{A-2}$$

Two matrices are *equal* if (1) they have the same order, and (2) their corresponding elements are equal one by one. That is, $\mathbf{A} = \mathbf{B}$ means $a_{pq} = b_{pq}$ for $1 \leq p \leq m$, $1 \leq q \leq n$.

---

[†] Other common notations for the unit matrix are $\mathbf{I}$ (for "Identity") or $\mathbf{E}$ (German "Einheit").

The *transpose* of **A**, denoted by $\mathbf{A}^T$, is formed by interchanging rows and columns,

$$\mathbf{A}^T = [a_{qp}] \tag{A-3}$$

---

**Example 1**

$$\begin{bmatrix} 2 & -1 & 0 \\ 3 & 1 & -2 \end{bmatrix}^T = \begin{bmatrix} 2 & 3 \\ -1 & 1 \\ 0 & -2 \end{bmatrix} \qquad \square$$

---

The *complex conjugate* of **A**, denoted by **A***, is formed by taking the conjugate of every element in **A**.

$$\mathbf{A}^* = [a_{pq}^*] \tag{A-4}$$

---

**Example 2**

$$\begin{bmatrix} 3 + j & -2 - j \\ 4 & 6j \end{bmatrix}^* = \begin{bmatrix} 3 - j & -2 + j \\ 4 & -6j \end{bmatrix} \qquad \square$$

---

A *real* matrix satisfies the equation,

$$\mathbf{A} = \mathbf{A}^* \tag{A-5}$$

An *imaginary* matrix satisfies the equation,

$$\mathbf{A} = -\mathbf{A}^* \tag{A-6}$$

A *symmetric* matrix satisfies the equation,

$$\mathbf{A} = \mathbf{A}^T \tag{A-7}$$

A *skew-symmetric* matrix satisfies the equation,

$$\mathbf{A} = -\mathbf{A}^T \tag{A-8}$$

A *hermitian* matrix satisfies the equation,

$$\mathbf{A} = \mathbf{A}^T* \tag{A-9}$$

A *skew-hermitian* matrix satisfies the equation,

$$\mathbf{A} = -\mathbf{A}^T* \tag{A-10}$$

## A-2   ADDITION, SUBTRACTION, AND MULTIPLICATION; PARTITIONING

Matrices can be *added* (or *subtracted*) only if they have the same order. Then, addition (subtraction) is performed element by element. That is,

$$\mathbf{A} + \mathbf{B} - \mathbf{C} = [a_{pq} + b_{pq} - c_{pq}] \tag{A-11}$$

In particular $\mathbf{A} - \mathbf{A} = \mathbf{0}$, the null matrix. The commutative and associative laws of addition are valid, namely

$$\mathbf{A} + \mathbf{B} = \mathbf{B} + \mathbf{A} \tag{A-12}$$

$$\mathbf{A} + (\mathbf{B} + \mathbf{C}) = (\mathbf{A} + \mathbf{B}) + \mathbf{C} \tag{A-13}$$

A *matrix*, when multiplied by a scalar $\alpha$, yields another matrix of the same order whose elements are all multiplied by $\alpha$

$$\alpha\mathbf{A} = \mathbf{A}\alpha = [\alpha a_{pq}] \tag{A-14}$$

The *product* of *two matrices*, $\mathbf{AB}$, is defined only if the number of columns in $\mathbf{A}$ equals the number of rows in $\mathbf{B}$. That is, if the order of $\mathbf{A}$ is $(m \times r)$, then the order of $\mathbf{B}$ must be $(r \times n)$. Such matrices are said to be *compatible* (or *conformable*). The result of the product $\mathbf{AB}$ is another matrix $\mathbf{C}$, of order $(m \times n)$, and every element in $\mathbf{C}$ is given by

$$c_{ik} = \sum_{t=1}^{r} a_{it} b_{tk} \qquad \begin{matrix} 1 \le i \le m \\ 1 \le k \le n \end{matrix} \tag{A-15}$$

In words, "multiply the elements or row $i$ of $\mathbf{A}$ by the corresponding elements of column $k$ of $\mathbf{B}$, then add these products to obtain the $(i, k)$ element of $\mathbf{AB}$."

---

*Example 3*

$$\mathbf{A} = \begin{bmatrix} 3 & -1 & 0 \\ 2 & 1 & 4 \end{bmatrix} \qquad \mathbf{B} = \begin{bmatrix} 2 & 1 \\ 0 & -1 \\ 3 & 2 \end{bmatrix}$$

Then,

$$\mathbf{AB} = \begin{bmatrix} (3 \times 2) + (-1 \times 0) + (0 \times 3) & (3 \times 1) + (-1 \times -1) + (0 \times 2) \\ (2 \times 2) + (1 \times 0) + (4 \times 3) & (2 \times 1) + (1 \times -1) + (4 \times 2) \end{bmatrix}$$

$$= \begin{bmatrix} 6 & 4 \\ 16 & 9 \end{bmatrix}$$

The product $\mathbf{BA}$ is also defined in this case and is found to be

$$\mathbf{BA} = \begin{bmatrix} 8 & -1 & 4 \\ -2 & -1 & -4 \\ 13 & -1 & 8 \end{bmatrix} \qquad \square$$

In general, the commutative law of multiplications does *not* hold, since it is quite possible that the product **AB** exists but the product **BA** may not even be defined! (Even if that product is defined, as in the previous example, the result may not be equal to the result of **AB**). For these reasons, it is important to state clearly which is the first factor or the second factor in the product: We say that **A** is *postmultiplied* by **B** in the product **AB**, or, alternately, that **B** is *premultiplied* by **A**.

The distributive and associative laws of multiplication are valid:

$$\mathbf{A}(\mathbf{B} + \mathbf{C}) = \mathbf{A}\mathbf{B} + \mathbf{A}\mathbf{C} \tag{A-16}$$

$$(\mathbf{A} + \mathbf{B})\mathbf{D} = \mathbf{A}\mathbf{D} + \mathbf{B}\mathbf{D} \tag{A-17}$$

$$(\mathbf{A}\mathbf{B})\mathbf{C} = \mathbf{A}(\mathbf{B}\mathbf{C}) \tag{A-18}$$

Also

$$\mathbf{A}\mathbf{U} = \mathbf{U}\mathbf{A} = \mathbf{A} \tag{A-19}$$

where the order of **U** is chosen to be compatible with **A**.

Positive integer *powers* of a square matrix are defined by

$$\mathbf{A}^2 = \mathbf{A}\mathbf{A} \tag{A-20}$$

$$\mathbf{A}^n = \mathbf{A}^{n-1}\mathbf{A} \tag{A-21}$$

Also

$$\mathbf{A}^0 = \mathbf{U} \tag{A-22}$$

The *transpose of a product* is the product of the individual transposed matrices, in *reversed order*

$$(\mathbf{A}\mathbf{B})^T = \mathbf{B}^T\mathbf{A}^T \tag{A-23}$$

The *scalar product* of two column vectors **X** and **Y**, each of order $(n \times 1)$ is defined as

$$\mathbf{X}^{*T}\mathbf{Y} = \sum_{i=1}^{n} x_i^* y_i \tag{A-24}$$

and is a scalar $(1 \times 1$ matrix). If **X** and **Y** are both real, Eq. (A-24) becomes

$$\mathbf{X}^T\mathbf{Y} = \sum_{i=1}^{n} x_i y_i = x_1 y_1 + x_2 y_2 + \cdots + x_n y_n \tag{A-25}$$

the familiar "dot product" of vectors.

The *length (norm)* of a vector $\mathbf{X} \neq \mathbf{0}$ is denoted by $\|\mathbf{X}\|$ and is given by

$$\|\mathbf{X}\| = +\sqrt{\mathbf{X}^{*T}\mathbf{X}} > 0 \tag{A-26}$$

and $\|\mathbf{0}\| = 0$. A vector of unit length is a *unit vector*.

Two column vectors are *orthogonal* if their scalar product vanishes:

$$\mathbf{X}^T\mathbf{Y} = 0 \tag{A-27}$$

Often, it is convenient to partition a given matrix into *submatrices*. These submatrices are denoted by single letters and appropriate subscripts.

---

*Example 4*

$$\mathbf{A} = \begin{bmatrix} 2 & 0 & 1 \\ -1 & 3 & 2 \end{bmatrix} = \begin{bmatrix} 2 & 0 & \vdots & 1 \\ -1 & 3 & \vdots & 2 \end{bmatrix} = [\mathbf{A}_{11} \quad \vdots \quad \mathbf{A}_{12}] \qquad \square$$

---

Such a partitioning is not unique, of course, and is chosen conveniently. However, if partitioning is to be used on a product of matrices, the submatrices must retain *compatibility*. This is illustrated by the following example.

---

*Example 5*

Assume that matrix **A** of the previous example is to be postmultiplied by

$$\mathbf{B} = \begin{bmatrix} -3 \\ 1 \\ -1 \end{bmatrix}$$

$$\therefore \mathbf{AB} = \begin{bmatrix} 2 & 0 & 1 \\ -1 & 3 & 2 \end{bmatrix} \begin{bmatrix} -3 \\ 1 \\ -1 \end{bmatrix} = \begin{bmatrix} -7 \\ 4 \end{bmatrix}$$

Now, if we want to partition **A** as shown previously (two columns), then **B** must be partitioned to have two rows. That is,

$$\mathbf{B} = \begin{bmatrix} -3 \\ 1 \\ -1 \end{bmatrix} = \begin{bmatrix} \mathbf{B}_{11} \\ \mathbf{B}_{21} \end{bmatrix}$$

and the product **AB** is formed:

$$\mathbf{AB} = [\mathbf{A}_{11} \quad \vdots \quad \mathbf{A}_{12}] \begin{bmatrix} \mathbf{B}_{11} \\ \mathbf{B}_{21} \end{bmatrix} = \mathbf{A}_{11}\mathbf{B}_{11} + \mathbf{A}_{12}\mathbf{B}_{21}$$

where $\mathbf{A}_{11}\mathbf{B}_{11}$ is compatible, as is $\mathbf{A}_{12}\mathbf{B}_{21}$. Performing these products, we obtain

$$\mathbf{A}_{11}\mathbf{B}_{11} + \mathbf{A}_{12}\mathbf{B}_{21} = \begin{bmatrix} 2 & 0 \\ -1 & 3 \end{bmatrix} \begin{bmatrix} -3 \\ 1 \end{bmatrix} + \begin{bmatrix} 1 \\ 2 \end{bmatrix} [-1] = \begin{bmatrix} -7 \\ 4 \end{bmatrix}$$

as before. $\qquad \square$

## A-3 INVERSION; DETERMINANTS

The *inverse matrix* of a square matrix $\mathbf{A}$, denoted by $\mathbf{A}^{-1}$, is such that

$$\mathbf{A}^{-1}\mathbf{A} = \mathbf{A}\mathbf{A}^{-1} = \mathbf{U} \tag{A-28}$$

when this inverse exists. Not all square matrices have an inverse. A matrix is *nonsingular* if its inverse exists, *singular* otherwise.

The *determinant* of a square matrix can be evaluated by familiar methods, and is a single scalar. (A matrix, on the other hand, has no value; it is merely an array of elements.) Several properties of determinants are as follows:

1. If $\mathbf{A}$ is a *diagonal* matrix, its determinant is given by the product of its elements,

$$\det \mathbf{A} = \prod_{p=1}^{n} a_{pp} \tag{A-29}$$

   In particular, $\det \mathbf{U} = 1$.

2. The determinant of a product of square matrices is equal to the product of their determinants

$$\det(\mathbf{AB}) = (\det \mathbf{A})(\det \mathbf{B}) \tag{A-30}$$

3. The determinant of $\mathbf{A}$ is the same as that of $\mathbf{A}^T$

$$\det \mathbf{A} = \det \mathbf{A}^T \tag{A-31}$$

4. The determinant of $\mathbf{A}$ is zero if $\mathbf{A}$ contains a null column vector, or if two column vectors in $\mathbf{A}$ are proportional.

5. If one column vector in $\mathbf{A}$ is multiplied by a scalar $\alpha$, the value of the determinant is multiplied by $\alpha$.

6. The determinant of $\mathbf{A}$ changes sign when any two column vectors in $\mathbf{A}$ interchange places.

7. The determinant of $\mathbf{A}$ remains unchanged if a column vector is multiplied by $\alpha$ and then added to another column vector.

Note: Properties 4–7 hold also true for "row vector" instead of "column vector", due to property 3.

The determinant $m_{kl}$ of the $(n-1) \times (n-1)$ matrix obtained from $\mathbf{A}$ by deleting row $k$ and column $l$ is called the *minor* of the element $a_{kl}$, or, briefly, the $kl$ *minor*.

The *cofactor* $c_{kl}$ is a "signed" minor and is given by

$$c_{kl} = (-1)^{k+l} m_{kl} \tag{A-32}$$

The *adjoint matrix* of $\mathbf{A}$ is a matrix whose $pq$ element is $c_{qp}$

$$\text{Adj } \mathbf{A} = [c_{qp}] \tag{A-33}$$

and is formed by, first, replacing every element $a_{pq}$ by its cofactor and then transposing the result.

---

**Example 6**

$$\mathbf{A} = \begin{bmatrix} 2 & -1 & 3 \\ 1 & 2 & -2 \\ 0 & 1 & -1 \end{bmatrix}$$

$$\det \mathbf{A} = \begin{vmatrix} 2 & -1 & 3 \\ 1 & 2 & -2 \\ 0 & 1 & -1 \end{vmatrix} = 2$$

$$\det \mathbf{A}^T = \begin{vmatrix} 2 & 1 & 0 \\ -1 & 2 & 1 \\ 3 & -2 & -1 \end{vmatrix} = 2$$

The nine minors are:

$$m_{11} = \begin{vmatrix} 2 & -2 \\ 1 & -1 \end{vmatrix} = 0; \quad m_{12} = \begin{vmatrix} 1 & -2 \\ 0 & -1 \end{vmatrix} = -1; \quad m_{13} = \begin{vmatrix} 1 & 2 \\ 0 & 1 \end{vmatrix} = 1$$

$$m_{21} = \begin{vmatrix} -1 & 3 \\ 1 & -1 \end{vmatrix} = -2; \quad m_{22} = \begin{vmatrix} 2 & 3 \\ 0 & -1 \end{vmatrix} = -2; \quad m_{23} = \begin{vmatrix} 2 & -1 \\ 0 & 1 \end{vmatrix} = 2$$

$$m_{31} = \begin{vmatrix} -1 & 3 \\ 2 & -2 \end{vmatrix} = -4; \quad m_{32} = \begin{vmatrix} 2 & 3 \\ 1 & -2 \end{vmatrix} = -7; \quad m_{33} = \begin{vmatrix} 2 & -1 \\ 1 & 2 \end{vmatrix} = 5$$

and the nine cofactors are:

$$\begin{array}{lll} c_{11} = 0 & c_{12} = 1 & c_{13} = 1 \\ c_{21} = 2 & c_{22} = -2 & c_{23} = -2 \\ c_{31} = -4 & c_{32} = 7 & c_{33} = 5 \end{array}$$

$$\mathrm{Adj}\,\mathbf{A} = \begin{bmatrix} 0 & 1 & 1 \\ 2 & -2 & -2 \\ -4 & 7 & 5 \end{bmatrix}^T = \begin{bmatrix} 0 & 2 & -4 \\ 1 & -2 & 7 \\ 1 & -2 & 5 \end{bmatrix} \qquad \square$$

---

It can be shown that

$$\mathbf{A}\,(\mathrm{Adj}\,\mathbf{A}) = (\det \mathbf{A})\mathbf{U} \qquad \text{(A-34)}$$

i.e., the product of a matrix and its adjoint is a diagonal matrix whose elements are all equal to the value of the determinant of that matrix.

*Example 7*

From the previous example, we find that

$$\mathbf{A}(\mathrm{Adj}\,\mathbf{A}) = \begin{bmatrix} 2 & -1 & 3 \\ 1 & 2 & -2 \\ 0 & 1 & -1 \end{bmatrix} \begin{bmatrix} 0 & 2 & -4 \\ 1 & -2 & 7 \\ 1 & -2 & 5 \end{bmatrix} = \begin{bmatrix} 2 & 0 & 0 \\ 0 & 2 & 0 \\ 0 & 0 & 2 \end{bmatrix}$$

$$= \det \mathbf{A} \begin{bmatrix} 1 & 0 & 0 \\ 0 & 1 & 0 \\ 0 & 0 & 1 \end{bmatrix} \qquad \square$$

If we now divide Eq. (A-34) by det $\mathbf{A}$ (provided det $\mathbf{A} \neq 0$, of course!), we get

$$\mathbf{A}\left(\frac{\mathrm{Adj}\,\mathbf{A}}{\det \mathbf{A}}\right) = \mathbf{U} \qquad (\text{A-35})$$

which, compared with Eq. (A-28), yields the inverse of $\mathbf{A}$:

$$\mathbf{A}^{-1} = \frac{1}{\det \mathbf{A}}\,\mathrm{Adj}\,\mathbf{A} \qquad (\text{A-36})$$

*Example 8*

In the previous example, we get

$$\mathbf{A}^{-1} = \begin{bmatrix} 2 & -1 & 3 \\ 1 & 2 & -2 \\ 0 & 1 & -1 \end{bmatrix}^{-1} = \frac{1}{2} \begin{bmatrix} 0 & 2 & -4 \\ 1 & -2 & 7 \\ 1 & -2 & 5 \end{bmatrix} = \begin{bmatrix} 0 & 1 & -2 \\ 0.5 & -1 & 3.5 \\ 0.5 & -1 & 2.5 \end{bmatrix} \qquad \square$$

In summary, the inverse of a square matrix exists provided its determinant is nonzero (the matrix is nonsingular). This inverse is given by Eq. (A-36).

The inverse of a product of matrices is the product of the individual inverse matrices, taken in reverse order,

$$(\mathbf{AB})^{-1} = \mathbf{B}^{-1}\mathbf{A}^{-1} \qquad (\text{A-37})$$

provided, of course, each of the inverses exist.

## A-4 RANK AND LINEAR DEPENDENCE

A matrix of order $(m \times n)$ is of *rank* $r$ if it contains at least one $(r \times r)$ nonzero minor, while all possible minors of higher order, $(r + 1) \times (r + 1)$, $(r + 2) \times (r + 2)$ etc., are zero. (Note: The rank $r$ cannot exceed $m$ or $n$, whichever is smaller. However, it may be less.)

A finite set of $n$ vectors $\mathbf{A}_k$ is *linearly dependent* if there is a set of scalars $x_k$, not all zero, such that

$$\sum \mathbf{A}_k x_k = \mathbf{A}_1 x_1 + \mathbf{A}_2 x_2 + \cdots + \mathbf{A}_n x_n = 0 \tag{A-38}$$

In other words, one of these vectors, say $\mathbf{A}_i$, can be expressed as a linear combination of some, or all, of the other vectors. For then we have (with $x_i \neq 0$) from Eq. (A-38)

$$\mathbf{A}_i = -\frac{1}{x_i}(x_1 \mathbf{A}_1 + x_2 \mathbf{A}_2 + \cdots) \tag{A-39}$$

If no such scalars $x_k$ exist, that is, if Eq. (A-38) is satisfied only when

$$x_1 = x_2 = \cdots = x_n = 0 \tag{A-40}$$

then these vectors are *linearly independent*.

In order to investigate the existence of these scalars $x_k$, Eq. (A-38) is written out fully, in matrix form as follows

$$\begin{bmatrix} a_{11} & a_{12} & a_{13} & \cdots & a_{1n} \\ a_{21} & a_{22} & a_{23} & \cdots & a_{2n} \\ \vdots & \vdots & \vdots & & \vdots \\ a_{m1} & a_{m2} & a_{m3} & \cdots & a_{mn} \end{bmatrix} \begin{bmatrix} x_1 \\ x_2 \\ \vdots \\ x_n \end{bmatrix} = \begin{bmatrix} 0 \\ 0 \\ \vdots \\ 0 \end{bmatrix} \tag{A-41}$$

where each $\mathbf{A}_k$ is written as a column vector with $m$ components.

$$\mathbf{A}_k = [a_{1k} a_{2k} \cdots a_{mk}]^T \tag{A-42}$$

Therefore, we have in Eq. (A-41) a set of $m$ linear homogeneous equations in the $n$-unknowns $x_1, x_2, \ldots, x_n$. Obviously $x_1 = x_2 = \cdots = 0$ is a trivial solution. The existence of nontrivial solutions will depend on the given matrix $\mathbf{A}$, formed by the given vectors

$$\mathbf{A} = \begin{bmatrix} a_{11} & a_{12} & a_{13} & \cdots & a_{1n} \\ a_{21} & a_{22} & a_{23} & \cdots & a_{2n} \\ \cdots & \cdots & \cdots & \cdots & \cdots \\ a_{m1} & a_{m2} & a_{m3} & \cdots & a_{mn} \end{bmatrix} \tag{A-43}$$

Nontrivial solutions exist if

$$r < n \tag{A-44}$$

i.e., if the rank is less than the number of unknowns. In this case, only $r$ vectors of the given $n$ are linearly independent, and the other $(n - r)$ vectors are linear combinations of these $r$ vectors.

In summary: The given $n$ vectors are *linearly independent* if, and only if,

$$r = n \tag{A-45}$$

If $n > m$, these $n$ vectors are linearly dependent.

*Gauss's elimination method* serves as a basis for efficient computer evaluation of the rank of a given matrix. In addition, it is very convenient for solving non-homogeneous linear algebraic equations.[‡] For these reasons, we outline it here by means of two examples.

---

### Example 9

Given three equations with four unknowns:

$$2x_1 - 6x_2 + x_3 - 2x_4 = 8$$

$$x_1 + 2x_2 - 2x_3 + 4x_4 = 9$$

$$-3x_1 + x_2 + 6x_3 + 5x_4 = 0$$

We shall use the first equation (to be called "the slave equation"[†]) to eliminate $x_1$ from all the equations that follow: Multiply the slave equation by $(-1/2)$ and add to the second equation, then multiply it by $3/2$ and add to the third equation. The results are:

$$2x_1 - 6x_2 + x_3 - 2x_4 = 8$$

$$5x_2 - 2.5x_3 + 5x_4 = 5$$

$$-8x_2 + 7.5x_3 + 2x_4 = 12$$

In this new set, use the second equation as a "slave" to eliminate $x_2$ from all the equations that follow: Multiply the slave equation by $8/5$ and add to the third one. The results are:

$$2x_1 - 6x_2 + x_3 - 2x_4 = 8$$

$$5x_2 - 2.5x_3 + 5x_4 = 5$$

$$3.5x_3 + 10x_4 = 20$$

The process is finished, and, in this case, we can choose one unknown, say $x_4$, arbitrarily. Then $x_3$ is found from the last equation. With these values of $x_4$ and $x_3$, we determine $x_2$ from the second equation; finally, $x_1$ is determined from the first equation. For example, let $x_4 = 2$; then $x_3 = 0$, and $x_2 = -1$, and $x_1 = 3$.

A variation on this method, known as the *Gauss-Jordan elimination*, utilizes the $k^{\text{th}}$ slave equation to eliminate $x_k$ from all the equations that follow *and that precede* the $k^{\text{th}}$ equation. Thus, in our example, the final form of the Gauss–Jordan elimination will yield

$$2x_1 \qquad\qquad = 6$$

$$5x_2 \qquad\qquad = -5$$

$$3.5x_3 + 10x_4 = 20$$

---

[‡] See Appendix E.

[†] Although not standard, this term describes well the role of this equation!

When there are $n$ independent equations with $n$ unknowns, the Gauss–Jordan method yields the answers immediately, without the need of substituting backwards, from the last equation to the first one.   □

The student will undoubtedly recognize that the operations outlined previously were carried out on the matrix,

$$\left[\begin{array}{cccc:c} 2 & -6 & 1 & -2 & 8 \\ 1 & 2 & -2 & 4 & 9 \\ -3 & 1 & 6 & 5 & 0 \end{array}\right]$$

which is the matrix of the coefficients *augmented* by the column of the right-hand side. These operations are called *elementary matrix transformations*:

a)  Interchange two rows (or two columns).

b)  Multiply a row (or a column) by a non-zero constant.

c)  Add a multiplied row (or column) to another row (or column).

Two matrices are *equivalent* (*not* equal!) if one is obtained from the other by a sequence of elementary matrix transformations. It is important to stress that *the rank of a matrix does not change under elementary matrix transformations.* Equivalent matrices have the same rank. In the following example, we shall perform Gauss's elimination method using elementary transformations on the matrix.

## Example 10

$$x_1 + 2x_2 + 6x_3 = 2$$
$$3x_1 - 2x_2 - x_3 = 1$$
$$4x_1 - 3x_2 + 2x_3 = -3$$

Equivalent matrices will be separated by $\sim$.

$$\left[\begin{array}{ccc:c} 1 & 2 & 6 & 2 \\ 3 & -2 & -1 & 1 \\ 4 & -3 & 2 & -3 \end{array}\right] \sim \left[\begin{array}{ccc:c} 1 & 2 & 6 & 2 \\ 0 & -8 & -19 & -5 \\ 0 & -11 & -22 & -11 \end{array}\right]$$

$$\sim \left[\begin{array}{ccc:c} 1 & 2 & 6 & 2 \\ 0 & 1 & \dfrac{19}{8} & \dfrac{5}{8} \\ 0 & 0 & \dfrac{33}{8} & \dfrac{-33}{8} \end{array}\right]$$

Therefore,

$$\frac{33}{8} x_3 = \frac{-33}{8} \qquad \therefore x_3 = -1$$

and

$$x_2 + \frac{19}{8}(-1) = \frac{5}{8} \qquad \therefore x_2 = 3$$

and

$$x_1 + (2)(3) + (6)(-1) = 2 \qquad \therefore x_1 = 2 \qquad \qquad \square$$

## ADDITIONAL READING

1. S. J. Frame, "Matrix Functions and Applications," (parts I through V), *IEEE Spectrum*, pp. 209–220, March 1964; pp. 102–108, April 1964; pp. 100–109, May 1964; pp. 123–131, June 1964; pp. 103–109, July 1964.
2. A. C. Aitken, *Determinants and Matrices*. New York: Interscience, 1958.
3. H. Anton, *Elementary Linear Algebra*. New York: John Wiley & Sons, Inc., 1977.
4. F. Ayres, *Theory and Problems of Matrices*. New York: Schaum Publishing Co. 1962.
5. A. Graham, *Matrix Theory and Applications for Engineers and Mathematicians*. New York: Halstead Press, 1979.
6. E. Guillemin, *The Mathematics of Circuit Analysis*. New York: John Wiley & Sons, 1958, Chapters 2, 3.
7. S. Karni, *Network Theory: Analysis and Synthesis*. Boston: Allyn and Bacon, Inc., 1966, pp. 443–457.
8. L. A. Pipes, *Matrix Methods for Engineering*. Englewood Cliffs, N.J.: Prentice-Hall, Inc., 1963.

# THE OPERATIONAL AMPLIFIER

The operational amplifier (*op amp* for short) is a very useful and versatile element. It is essentially a voltage-controlled voltage source consisting of many transistors, resistors and capacitors. Numerous commerical op amps are available; in Fig. B-1 we show diagrams of the Fairchild Corporation's $\mu$A741. As a circuit model, the op amp is shown in Fig. B-2 together with its familiar triangular symbol. This input terminals are called the inverting terminal $(v_-)$ and the non-inverting terminal $(v_+)$. The op amp's input–output relation is given by

$$v_0 = A(v_+ - v_-)$$

in its linear region, as shown in Fig. B-3. The input voltage to the op amp is $(v_+ - v_-)$ and is of the order of several millivolts (mV), while $A$, the open-loop gain, is in the range $10^4 - 10^6$. Typically, the input resistance is $R_i = 100$ K$\Omega$, while the output resistance is small, $R_0 = 100$ $\Omega$.

The ideal model, quite useful in most network applications, is shown in Fig. B-4 and is defined by

$$v_+ - v_- = 0$$

$$A = \infty$$

$$R_i = \infty$$

$$R_0 = 0$$

Consequently,

$$v_+ = v_-$$

$$i_+ = i_- = 0$$

and $v_0$ is finite, nonzero.

μA741
Operational Amplifier

Description

The μA741 is a high performance Monolithic Operational Amplifier constructed using the Fairchild Planar epitaxial process. It is intended for a wide range of analog applicaitons. High common mode voltage range and absence of latch-up tendencies make the μA741 ideal for use as a voltage follower. The high gain and wide range of operating voltage provides superior performance in integrator, summing amplifier, and general feedback applications.

- NO FREQUENCY COMPENSATION REQUIRED
- SHORT-CIRCUIT PROTECTION
- OFFSET VOLTAGE NULL CAPABILITY
- LARGE COMMON MODE AND DIFFERENTIAL VOLTAGE RANGES
- LOW POWER CONSUMPTION
- NO LATCH-UP

Connection Diagram
8-Pin Metal Package

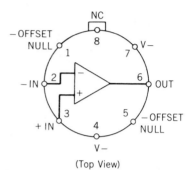

(Top View)

Pin 4 connected to case

Connection Diagram
10-Pin Flatpak

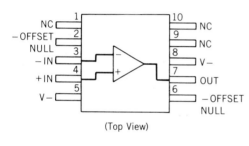

(Top View)

Connection Diagram
8-Pin DIP

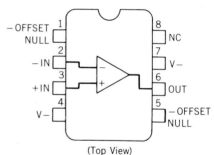

(Top View)

**FIGURE B-1.** The μA741 Operational Amplifier. Reprinted with permission of Fairchild Camera and Instrument Corporation. © 1984.

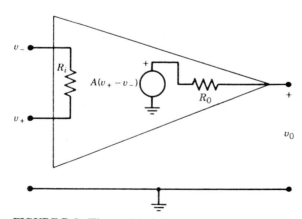

**FIGURE B-2.** The model of the op amp.

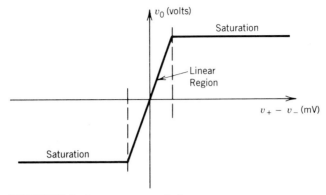

**FIGURE B-3.** Input–output relations.

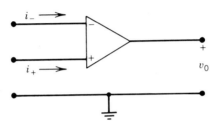

**FIGURE B-4.** Ideal op amp.

The various pins (connections) in Fig. B-1 are used for power supply and "offset null." They are of no immediate interest in analyzing the model.

The equations of an op amp circuit are obtained by applying Kirchhoff's current laws (KCL) to the inverting or the noninverting terminals, and Kirchhoff's voltage law (KVL) to the input and/or output loops, as appropriate. For example, in the voltage adder circuit of Fig. B-5, we apply KCL at the inverting terminal

$$\frac{v_1}{R_1} + \frac{v_2}{R_2} + \frac{v_3}{R_3} + \frac{v_0}{R_f} = 0$$

noting that the voltage across $R_1$ is $(v_1 - v_- - v_+) = v_1$ because $v_+ = 0$ and $v_- = v_+$. Similar expressions hold for the other voltages. The resulting equation is solved for $v_0$, as given in Fig. B-5.

Several common circuits using the op amp are shown in Fig. B-5, together with their equations.

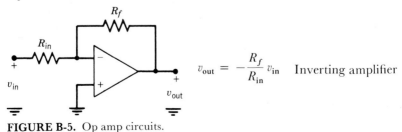

$$v_{\text{out}} = -\frac{R_f}{R_{\text{in}}} v_{\text{in}} \quad \text{Inverting amplifier}$$

**FIGURE B-5.** Op amp circuits.

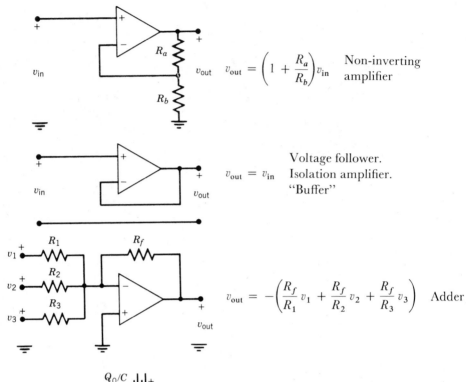

$$v_{out} = \left(1 + \frac{R_a}{R_b}\right)v_{in}$$ Non-inverting amplifier

$$v_{out} = v_{in}$$ Voltage follower. Isolation amplifier. "Buffer"

$$v_{out} = -\left(\frac{R_f}{R_1}v_1 + \frac{R_f}{R_2}v_2 + \frac{R_f}{R_3}v_3\right)$$ Adder

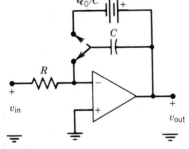

$$v_{out} = -\frac{1}{RC}\int v_{in}\,dt$$ Integrator (Battery accounts for initial conditions)

**FIGURE B-5** (*continued*)

# C

# MAGNITUDE AND FREQUENCY SCALING

In practice, element values as well as frequencies range widely. Typically, such values may be, for example, $R = 600\ \Omega$, $L = 0.8$ mH, $C = 3.5\ \mu\mu$F, $\omega = 2\pi \times 10^6$ rad/sec. Computations (and examples throughout this book) would be cumbersome indeed if scaling (normalization) were not used.

## C-1 MAGNITUDE SCALING

If every $R$, $L$, and $1/C$ is divided by a positive constant $A$ ($A$ may be greater or less than unity), the magnitude-scaled impedance of each element becomes

$$Z_{Rm}(s) = \frac{R}{A} \qquad Z_{Lm}(s) = \frac{Ls}{A} \qquad Z_{Cm}(s) = \frac{1}{ACs} \qquad \text{(C-1)}$$

where the subscript "$m$" reminds us of "magnitude scaling." As a result of this scaling, every network function that is a ratio of two determinants of *different* orders (say, a $p \times p$ determinant divided by a $(p-1) \times (p-1)$ determinant) is affected by magnitude scaling. For example, the driving-point impedance of a one-port network becomes

$$Z_m(s) = \frac{1}{A} Z(s) \qquad \text{(C-2)}$$

However, network functions that are dimensionless (such as a current transfer or a voltage transfer function) remain unchanged by magnitude scaling. For example,

$$\alpha_m(s) = \alpha(s) \qquad \text{(C-3)}$$

is the expression for a current transfer function after and before magnitude scaling.

## C-2   FREQUENCY SCALING

Let us divide $s$, the complex frequency variable, by a positive constant $B$ (greater or less than unity)

$$s_f = \frac{s}{B} \tag{C-4}$$

where the subscript "$f$" indicates "frequency-scaled." Quite obviously, the actual impedance of each element in the network is not affected by this mathematical step. Therefore, we have for a resistor

$$Z_R(s) = R \qquad \text{(resistance is frequency-independent)} \tag{C-5}$$

for an inductor

$$Z_L(s) = Ls = LB\frac{s}{B} = L_f s_f \tag{C-6}$$

and for a capacitor

$$Z_c(s) = \frac{1}{Cs} = \frac{1}{CB}\frac{B}{s} = \frac{1}{CB}\frac{1}{s_f} = \frac{1}{C_f s_f} \tag{C-7}$$

where $L_f$ and $C_f$ are frequency-scaled values of inductance and capacitance.

## C-3   MAGNITUDE AND FREQUENCY SCALING

If we scale both magnitude and frequency, the total result is the combined effect of these scalings as given in Eqs. (C-1), (C-2), (C-4) through (C-7). These are summarized in Table C-1.

**TABLE C-1  Magnitude and Frequency Scaling**

| Name | Scaled<br>$m$ = magnitude<br>$f$ = frequency | Unscaled<br>(Actual) |
|---|---|---|
| Resistance | $R_{m,f}$ | $R = AR_{m,f}$ |
| Inductance | $L_{m,f}$ | $L = \dfrac{A}{B} L_{m,f}$ |
| Capacitance | $C_{m,f}$ | $C = \dfrac{1}{AB} C_{m,f}$ |
| Frequency | $s_f$ | $s = Bs_f$ |
| Driving-point impedance | $Z_m(s_f)$ | $Z = AZ_m\left(\dfrac{s}{B}\right)$ |
| Current transfer | $\alpha_m(s_f)$ | $\alpha = \alpha_m\left(\dfrac{s}{B}\right)$ |

In conclusion, it might be pointed out that in addition to the advantage of working with convenient numerical values, a scaled network can serve as a prototype (for example in design): Its basic properties are unaffected by scaling. Thus, a large number of practical (unscaled) networks can be obtained from it by a simple unscaling operation.

Unless stated otherwise, all networks are scaled in this book.

## ADDITIONAL READING

1. G. C. Temes and J. W. LaPatra, *Introduction to Circuit Synthesis and Design*. New York: McGraw-Hill Book Co., 1977. pp. 18–25.
2. M. E. Van Valkenburg, *Introduction to Modern Network Synthesis*. New York: John Wiley & Sons, Inc., 1960.
3. G. S. Moschytz and P. Horn, *Active Filter Design Handbook*. New York: John Wiley & Sons, Inc. 1981. Chapter 7.

# APPENDIX D

---

# SIGNAL FLOW GRAPHS

Undoubtedly, we have realized by now that the solution of a linear network amounts to handling a set of simultaneous algebraic (Laplace transformed) equations. Using matrix algebra, determinants, or successive elimination of variables, we solve for the required unknowns (output variables), or else we find relationships between an output and an input (network functions).

The technique of signal flow graphs is an additional method for handling a set of simultaneous linear equations. It has several attractive features:

1. It gives an insight into the relationships among variables.

2. Whereas these relationships may get "lost" during the process of an analytic solution (say, by matrix inversion, the different co-factors that are computed convey little, if any information), the insight provided by a signal flow graph is not obscured during the steps of the solution. In other words, the signal flow graph retains throughout a clear picture of "cause-and-effect."

3. The formulation of the problem and the solution can be done in easy graphical steps.

4. Signal flow graphs are helpful in the preparation of a solution on a computer.

## D-1  DEFINITIONS AND RELATIONS[1,2]

A *signal flow graph* describes a set of simultaneous equations in the form of oriented branches connected at nodes.[†] Branch $jk$ leaves node $j$ and enters node $k$, with its orientation indicated by an arrow (see Fig. D-1). The signal flow graph can be

---

[†] Although certain terms resemble those used in network topology (Chapter 6), their meanings are quite different here. Such differences will be clear from the context of the discussion.

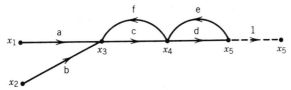

**FIGURE D-1.** Signal flow graph.

considered as a signal transmission system, where each node is a station that receives signals via incoming branches, combines them, then sends out the result along each outgoing branch. Each branch $jk$ has a *branch transmittance* $t_{jk}$ (also called *gain*). A signal traverses a branch only in the direction of the arrowhead and is multiplied by the transmittance of that branch. Each node adds algebraically all the signals entering it, and then transmits the resulting signal along each outgoing branch. Thus, in Fig. D-1 we have[†]

$$x_3 = ax_1 + bx_2 + fx_4 \tag{D-1a}$$

$$x_4 = cx_3 + ex_5 \tag{D-1b}$$

$$x_5 = dx_4 \tag{D-1c}$$

As a signal traverses some successive branches along their indicated orientations, it traces a *path*. In Fig. D-1, $x_1x_3x_4x_5$ is a path, and so is $x_2x_3x_4$. Generally, there may be one, several, or no path at all starting from node $j$ and terminating at node $k$. For example, there is no path from $x_2$ to $x_1$ in Fig. D-1. A *forward path* is a path along which the same node is not encountered more than once. For example, $x_1x_3x_4x_5$ is a forward path. A *loop* is a path which returns to its starting node, along which the same node is not encountered more than once; for example, $x_3x_4x_3$ (via $c$ and $f$) is a loop in Fig. D-1. However, $x_3x_4x_5x_4x_3$ (via $c, d, e, f$) is not a loop.

The *path gain* (or *path transmittance*) is the product of the gains of all the branches in that path. For example, the *forward path gain* between $x_1$ and $x_5$ in Fig. D-1 is *acd* while the *loop gain* of the loop $x_3x_4x_3$ is *cf*.

A node is called a *source* if all its branches leave it. In other words, the variable associated with a source is considered to be the excitation, i.e., the "cause." For example, $x_1$ and $x_2$ in Fig. D-1 are sources. A *sink*, on the other hand, is a node having only incoming but no outgoing branches. Such a node is associated with a response, that is, the "effect." Although no node in Fig. D-1 seems to be a sink, any node $x_j$ (other than a source) can be made into a sink by merely adding a new branch of unity gain directed from $x_j$ to a new node labelled also $x_j$. This amounts to the trivial equation $x_j = x_j$, and is shown in dotted lines in Fig. D-1: The "new" (dotted) node $x_5$ is a sink.

---

[†] In simple and specific examples, we shall use for convenience the letters $a, b, c, d, \ldots$, instead of $t_{jk}$. Here we write $a = t_{13}, b = t_{23}, c = t_{34}, d = t_{45}, e = t_{54}$, and $f = t_{43}$.

## Example 1

Draw the signal flow graph for the equations

$$2x_1 + 3x_2 - x_3 = q_1$$
$$-3x_1 + 3x_2 + 2x_3 = q_2$$
$$4x_1 - 2x_2 - 4x_3 = q_3$$

Do it by expressing each variable ($x_1$, $x_2$, and $x_3$) as a linear combination of all the other variables, including itself and the known quantities $q_1$, $q_2$, and $q_3$.

We rewrite the given equations as required, by transposing the necessary terms:

$$x_1 = -x_1 - 3x_2 + x_3 + q_1$$
$$x_2 = 3x_1 - 2x_2 - 2x_3 + q_2$$
$$x_3 = 4x_1 - 2x_2 - 3x_3 - q_3$$

and the corresponding signal flow graph is shown in Fig. D-2. Note that a *self loop* is formed by a single branch leaving and entering the same node. One self loop is due to the term $-x_1$ in the first equation, another is due to the term $-2x_2$ in the second equation, and the third one is due to $-3x_3$ in the third equation.

The nodes $q_1$, $q_2$ and $q_3$ are clearly sources; in fact, in a network, these terms may represent current or voltage sources. Neither $x_1$, $x_2$ nor $x_3$ are sinks without further modification. If, for example, we want to consider $x_1$ as the desired output ("effect"), we add the trivial equation $x_1 = x_1$, with the resulting branch (in dotted line).

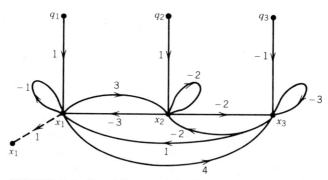

**FIGURE D-2.** Signal flow graph for Example 1.

## D-2   THE ALGEBRA OF SIGNAL FLOW GRAPHS

Having drawn the signal flow graph, we wish to reduce it, just as we would reduce algebraically the corresponding equations. For this purpose we use the following elementary transformations:

*Cascade connection.* As shown in Fig. D-3a, we have

$$x_2 = l_{12}x_1 \qquad x_3 = l_{23}x_2 \qquad \qquad \text{(D-2)}$$

Eliminating $x_2$, we get

$$x_3 = l_{12}l_{23}x_1 \qquad \qquad \text{(D-3)}$$

and therefore the rule is: *the gain of a cascade connection is the product of the gains of branches.* Note that in the resulting graph, as in the corresponding equation, node $x_2$ is eliminated.

*Parallel connection.* As shown in Fig. D-3b we have

$$x_2 = l_{12}x_1 + l'_{12}x_1 = (l_{12} + l'_{12})x_1 \qquad \qquad \text{(D-4)}$$

and the rule is: *the gains of branches in parallel are added.*

*Star-mesh connection.* This connection is shown in Fig. D-3c. Node $x_0$ is eliminated between $x_1$ and $x_2$ as follows:

$$l_{12}x_1 = x_0 \qquad l_{04}x_0 = x_4 \qquad \therefore x_4 = l_{12}l_{04}x_1 \qquad \qquad \text{(D-5)}$$

then between $x_2$ and $x_3$ as

$$l_{30}x_3 = x_0 \qquad l_{02}x_0 = x_2 \qquad \therefore x_2 = l_{30}l_{02}x_3 \qquad \qquad \text{(D-6)}$$

and between $x_3$ and $x_4$ as

$$l_{30}x_3 = x_0 \qquad l_{04}x_0 = x_4 \qquad \therefore x_4 = l_{30}l_{04}x_3 \qquad \qquad \text{(D-7)}$$

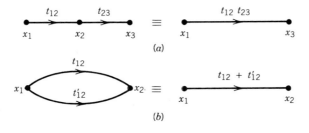

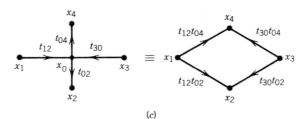

**FIGURE D-3.** Elementary transformations.

Here the rule is: *to eliminate a node (without a self loop), replace it by one branch for every possible forward path that goes through that node. The gain of such branch is the gain of that path.* Note that the cascade connection, Fig. D-3a, is a special case of the star-mesh connection.

---

### Example 2

Use the elementary transformations to eliminate $x_3$ and reduce the signal flow graph shown in Fig. D-4a. The star-mesh transformation eliminates node $x_3$ immediately, as shown in Fig. D-4b, where the loop $x_2 x_3 x_2$ in Fig. D-4a becomes a self-loop $x_2 x_2$ with a gain $be$. Next, the parallel connection of branches between $x_2$ and $x_4$ yields the result shown in Fig. D-4c.

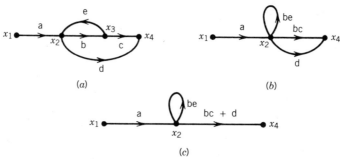

(a)                                        (b)

(c)

**FIGURE D-4.**  Reduction of a signal flow graph.                            ☐

---

*Self-loop.* As illustrated in the previous two examples (Figs. D-2 and D-4), a self-loop occurs when a variable, say $x_p$, is written as a linear combination of the other variables including itself. Consider, in general, the set of linear algebraic equations

$$\mathbf{AX} = \mathbf{Q} \tag{D-8}$$

where

$$\mathbf{A} = \begin{bmatrix} a_{11} & a_{12} & \cdots & a_{1n} \\ & \ddots & & \\ a_{n1} & a_{n2} & \cdots & a_{nn} \end{bmatrix} \qquad \mathbf{X} = \begin{bmatrix} x_1 \\ \vdots \\ x_n \end{bmatrix} \qquad \mathbf{Q} = \begin{bmatrix} q_1 \\ \vdots \\ q_n \end{bmatrix} \tag{D-9}$$

Example 1 of the previous section is such a set. In general, Eq. (D-8) is recognized as the set of equations for network analysis, where $\mathbf{X}$ is the response ("effect") matrix and $\mathbf{Q}$ is the excitation ("cause") matrix.

As illustrated earlier, Eq. (D-8) must be modified slightly before drawing the corresponding signal flow graph. That is, every element of $\mathbf{X}$ must be expressed as a

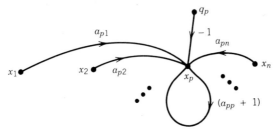

**FIGURE D-5.** Formation of a self-loop.

linear combination of all the others including itself. Consider specifically the $p$th equation $(1 \le p \le n)$ in Eq. (D-8); it is

$$a_{p_1}x_1 + a_{p_2}x_2 + \cdots + a_{pp}x_p + \cdots + a_{pn}x_n = q_p \qquad (D\text{-}10)$$

The desired modification is done by adding $x_p$ to both sides of Eq. (D-10) and rearranging, as follows:

$$x_p = a_{p_1}x_1 + a_{p_2}x_2 + \cdots + (a_{pp} + 1)x_p + \cdots + x_{pn}x_n - q_p \qquad (D\text{-}11)$$

This is suitable for drawing the signal flow graph, as shown in Fig. D-5.

The method is clear now: Add $x_1$ to the first equation in Eq. (D-3), $x_2$ to the second one, etc., and rearrange. In other words,

$$\mathbf{AX} + \mathbf{X} = \mathbf{Q} + \mathbf{X} \qquad (D\text{-}12)$$

and

$$\mathbf{X} = (\mathbf{A} + \mathbf{U})\mathbf{X} - \mathbf{Q} \qquad (D\text{-}13)$$

which is the form suitable for drawing the signal flow graph.

The elimination of a self-loop (and, thus, a further reduction of the signal flow graph) can be done as follows: Rewrite Eq. (D-11), using the general notation $t_{jk}$ for transmittances,

$$x_p = t_{1p}x_1 + t_{2p}x_2 + \cdots + t_{p-1,p}x_{p-1} + t_{pp}x_p + \cdots + t_{np}x_n + t_{qp}q_p \qquad (D\text{-}14)$$

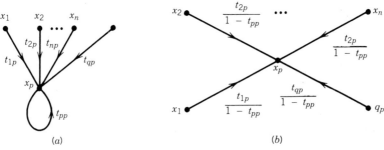

(a)                (b)

**FIGURE D-6.** Elimination of a self-loop.

See Fig. D-6a. Transpose the term $l_{pp}x_p$ (representing the self-loop) to the left side and solve for $x_p$:

$$x_p = \frac{l_{1p}}{1 - l_{pp}} x_1 + \frac{l_{2p}}{1 - l_{pp}} x_2 + \cdots + \frac{l_{p-1,p}}{1 - l_{pp}} x_{p-1}$$

$$+ \cdots + \frac{l_{np}}{1 - l_{pp}} x_n + \frac{l_{qp}}{1 - l_{pp}} q_p$$

$$(\text{D-15})$$

Equation (D-15) is shown graphically in Fig. D-6b and the self-loop has been eliminated. Therefore the rule is: *to eliminate a self loop at a node* $x_p$, *divide the gain of every branch entering* $x_p$ *by* $1 - l_{pp}$.

---

## Example 3

The elimination of the self-loop at node $x_2$ in Fig. D-4 and the final reduction are shown in Fig. D-7, and we have the input-output relation,

$$\frac{x_4}{x_1} = \frac{a(bc + d)}{1 - be}.$$

**FIGURE D-7.** Complete reduction of Fig. D-4. □

---

A final word of caution: division by $(1 - l_{pp})$ is valid provided $l_{pp} \neq 1$. Two possibilities exist if $l_{pp} = 1$:

a) The corresponding linear equations are independent, and thus solvable. In this case, we can always use the star-mesh transformation to eliminate other nodes, thereby changing the value of $l_{pp}$.

b) The equations are linearly dependent. No unique solution exists and there is no need to pursue it any further.

In summary, successive application of the elementary transformations (Fig. D-3) and elimination of self-loops (Fig. D-6) will reduce the signal flow graph.

---

## Example 4

Consider $x_2 + 2x_3 = 0$, $x_1 - 2x_2 - 3x_3 = 0$, and $2x_1 - x_2 + x_3 = -1$. Rewrite in the form of Eq. (D-13), suitable for drawing the signal flow graph:

$$x_1 = x_1 + x_2 + 2x_3$$

$$x_2 = x_1 - x_2 - 3x_3$$

$$x_3 = -x_1 + \frac{x_2}{2} + \frac{x_3}{2} - \frac{1}{2}$$

The signal flow graph is shown in Fig. D-8a, and it is required to find $x_3$ ("effect"). Note that we designate by $q$ the input ("cause"), where $q = \frac{1}{2}$. The steps are:

1. Eliminate the self-loops of $x_2$ and $x_3$, Fig. D-8b. The self-loop of $x_1$ cannot be eliminated now, since $t_{11} = 1$.

2. Eliminate node $x_2$ using the star-mesh transformation, Fig. D-8c. Note the new self-loops formed at $x_1$ and $x_3$.

3. Combine several parallel branches of Fig. D-8c and eliminate the self-loop of $x_3$, Fig. D-8d.

4. Eliminate the self-loop of $x_1$, as well as $x_1$. See Fig. D-8e and f.

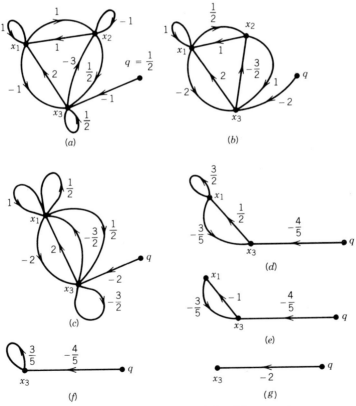

**FIGURE D-8.** Example 4: Reduction of signal flow graphs.

5. Finally, eliminate the self-loop of $x_3$, to obtain Fig. D-8g. From this result we write immediately

$$x_3 = -2q$$

or $x_3 = -1$ since $q = \frac{1}{2}$.   □

## D-3   OVERALL GAIN—MASON'S FORMULA[2]

The reduction of a signal flow graph, as outlined previously, is a straightforward procedure. However, it is lengthy and tedious; sometimes (if not often!) mistakes can occur. When the overall gain $G$ between a source and a sink is required, there is a formula which gives the result almost by inspection. This is Mason's formula,

$$G = \frac{1}{\Delta} \sum_k G_k \Delta_k \qquad (D\text{-}16)$$

where

$G_k$ = gain of the $k$th forward path,   (D-17a)

$\Delta$ = the determinant of the graph

$\quad = 1 - \sum_m P_{m1} + \sum_m P_{m2} - \sum_m P_{m3} + \cdots,$   (D-17b)

$P_{m1}$ = loop gain of the $m$th feedback loop,   (D-17c)

$P_{m2}$ = product of the loop gains of the $m$th nontouching two   (D-17d)
feedback loops,

$P_{m3}$ = product of the loop gains of the $m$th nontouching three   (D-17e)
feedback loops, etc.,

$\Delta_k$ = the determinant of that part of the graph not touching   (D-17f)
the $k$th forward path.

Notes:

1. *Nontouching* loops or parts have no common nodes or branches.

2. The summation over $k$ in Eq. D-16 is for *all* the possible forward paths.

3. $\sum_m P_{m1}$ is the sum of the individual loop gains.

4. $\sum_m P_{m2}$ is the sum of the gain products of all possible combinations of two nontouching loops.

5. $\sum_m P_{mr}$ is the sum of the gain products of all possible combinations of $r$ nontouching loops.

*Example 5*

Consider the signal flow graph in Fig. D-9. It is required to find the relation between $x_1$ (source) and $x_3$ (sink). We write by inspection:

1. There are two forward paths from $x_1$ to $x_3$. Therefore

$$G_1 = ab$$
$$G_2 = ceb$$

2. $P_{m1}$ is obtained by taking all the feedback loops, one at a time. Thus

$$\sum_m P_{m1} = h + fg + de$$

3. $P_{m2}$ takes the loops that do not touch, two at a time. Here,

$$\sum_m P_{m2} = (h)(de) = hde$$

is the only combination permissible. Also, there are no sets of three or more nontouching loops; therefore $\sum_m P_{m3} = 0$, etc.

4. Since the first forward path $(G_1 = ab)$ touches the loop $de$, this term will be missing in $\Delta_1$.

$$\therefore \Delta_1 = 1 - h - fg$$

and similarly, since the second path $(G_2 = ceb)$ touches both $fg$ and $de$, these terms will be missing in $\Delta_2$.

$$\therefore \Delta_2 = 1 - h$$

5. Finally, Mason's formula reads

$$\frac{x_3}{x_1} = G = \frac{ab(1 - h - fg) + ceb(1 - h)}{1 - h - fg - de + hde}$$

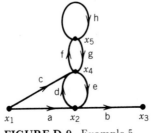

**FIGURE D-9.** Example 5.

### Example 6

Consider the signal flow graph in Fig. D-10. It is required to find the ratio $x_4/x_1$. Notice the additional branch of unity gain between $x_4$ and $x_4$, making $x_4$ a sink. This is *required* because Mason's formula applies only to the overall gain between a source and a sink!

We write by inspection:

$$G_1 = 1 \cdot a \cdot 1 = a$$

$$G_2 = 1 \cdot b \cdot c \cdot 1 = bc$$

$$\sum_m P_{m1} = ad + be + cf + bcd + afe$$

$$\sum_m P_{m2} = 0 \text{ since there are no sets of two nontouching loops. Similarly,}$$

$$\sum_m P_{m3} = 0, \text{ etc.}$$

$$\Delta_1 = 1$$

$$\Delta_2 = 1$$

$$\therefore \frac{x_4}{x_1} = G = \frac{a + bc}{1 - ad - be - cf - bcd - afe}$$

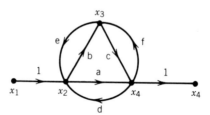

**FIGURE D-10.** Example 6.

## D-4  APPLICATIONS TO NETWORKS

Depending on our choice of variables, different signal flow graphs can be drawn. For example, the set of (the Laplace transformed) loop equations or node equations is of the form given by Eq. (D-8). Their rearrangement, Eq. (D-13), and the drawing of the signal flow graph can be done in a straightforward manner.

Another choice of network variables proves to be more useful for signal flow graphs. The choice of these variables depends on the topology of the network. We recall (see Chapter 6) that the branch voltages of a tree are independent, and so are the link currents. The reasons for this independence, briefly, are: by definition, branches of a tree cannot form a loop, and therefore Kirchhoff's voltage law $\sum v = 0$

does not restrict them. Similarly, link currents are not restricted by Kirchhoff's current law, $\sum i = 0$, since links alone cannot form a cut set. Also, once the branch voltages are known, the voltages of the links can be evaluated; likewise all branch currents can be found from the knowledge of link currents.

Our choice of independent variables, then, will be *the branch voltages* and *the link currents*, a total of $e$ variables. The tree will include all the voltage sources, and its co-tree—all the current sources. Then we write the necessary equations for these branch voltages and link currents.

A branch voltage is either: (1) an independent source, (2) a drop across an impedance, or (3) a dependent source. In the first case it is given as the excitation (i.e., the "cause"); in the second case, it can be expressed in terms of that impedance and the branch current which, in turn, is expressed in terms of some link currents; in the third case, if it depends on a current (link current or branch current) it can be expressed in terms of some link currents. Whereas if it depends on a voltage, it reduces eventually to the first or the second case.

Similar remarks hold for a link current. It is either: (1) an independent source— a given excitation, a "cause," (2) a current through an admittance, expressible in terms of that admittance and the link voltage which, in turn, is a linear combination of some branch voltages, or (3) a dependent source, depending on a current or a voltage.

---

## Example 7

Consider the network shown in Fig. D-11$a$. It is a Laplace transform diagram, with elements given by their impedances—$Z_3(s)$, $Z_5(s)$—or their admittances $Y_2(s)$, $Y_4(s)$. Notice the dependent source. In Fig. D-11$b$ the chosen tree is shown in heavy lines $(1, 3, 5)$ and the co-tree in thin lines $(2, 4, 6)$.

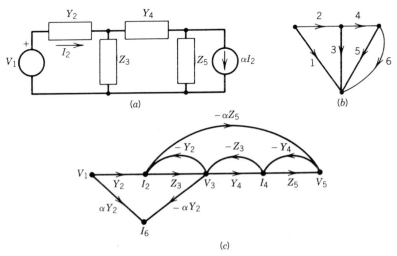

FIGURE D-11. Example 7.

The equations for the branch voltages are:[†]

$$V_1 \text{ is the given source}$$

$$V_3 = Z_3 I_3 = Z_3 (I_2 - I_4)$$

where $I_3$, a branch current, is expressed in terms of the link currents $I_2$ and $I_4$. Note that $I_3 - I_2 + I_4 = 0$ is Kirchhoff's current law for a fundamental cut set.

$$V_5 = Z_5 I_5 = Z_5 (I_4 - I_6) = Z_5 (I_4 - \alpha I_2)$$

Here, again, we have used $I_5 + I_6 - I_4 = 0$ for a fundamental cut set and, in addition, $I_6 = \alpha I_2$.

The equations for the link currents are

$$I_2 = Y_2 V_2 = Y_2 (V_1 - V_3)$$

where the link voltage $V_2$ is expressed in terms of the branch voltages $V_1$ and $V_3$, $V_2 + V_3 - V_1 = 0$. Similarly,

$$I_4 = Y_4 V_4 = Y_4 (V_3 - V_5)$$

and

$$I_6 = \alpha I_2 = \alpha Y_2 (V_1 - V_3)$$

The signal flow graph is shown in Fig. D-11c, and it shows clearly the conceptual relation of cause-effect: the immediate effect due to the source $V_1$ is $I_2$ (and $I_6 = \alpha I_2$), then $V_3$, etc. The signal flow graph also shows the effects of feedback, such as $V_3$ affecting $I_3$ through $-Y_2$, etc.

Finally, if we are asked to find, say, the transfer admittance $I_4(s)/V_1(s)$, we would require $I_4$ to be a sink. We would add an extra branch of unity transmittance; then, using Mason's formula, we would compute the overall gain of the graph between $V_1$ and $I_4$. This gain will be the desired transfer admittance.

In addition to the solution of the specific problem, this example illustrates the flexibility of the method. Rather than apply it formally, we may pick those variables which are obvious and physically important, then build the signal flow graph step by step. Needless to say, there is no unique signal flow graph for a given problem. The graph will depend on our choice of variables. The final answer—the overall gain—is unique, of course.   □

---

## REFERENCES

1. S. J. Mason, "Feedback theory: some properties of signal-flow graphs," *Proceedings of the IRE*, vol. 41, September 1953, pp. 1144–1156.

2. S.J. Mason, "Feedback theory—further properties of signal-flow graphs," *Proceedings of the IRE*, vol. 44, July 1956, pp. 920–926.

---

[†] All are functions of $s$. Thus $V_1 = V_1(s)$, $Z_3 = Z_3(s)$, etc.

# APPENDIX E

## COMPUTER PROGRAMS

### 1   LINEAR EQUATIONS

```
      SUBROUTINE LASSOL(N,A,B,M,X,IFLAG)
C
C LASSOL SOLVES A SYSTEM OF N LINEAR EQUATIONS IN N UNKNOWNS,
C AX=B, USING GAUSSIAN ELIMINATION WITH PARTIAL PIVOTING AND
C ROW EQUILIBRATION.  INPUT QUANTITIES ARE
C
C    N - NUMBER OF EQUATIONS AND UNKNOWNS.
C    A - N BY N MATRIX OF COEFFICIENTS STORED IN AN M BY M
C        ARRAY.
C    B - VECTOR OF RIGHT HAND SIDES.
C    M - DIMENSION OF A IN THE CALLING PROGRAM.
C
C OUTPUT QUANTITIES ARE
C
C    X - SOLUTION VECTOR.
C    IFLAG = 1 FOR NORMAL RETURN.
C          = 2 IF MATRIX APPEARS SINGULAR TO THE CODE.  CONTROL
C              IS RETURNED TO THE CALLING PROGRAM.
C
C THE ARRAYS A, B, AND X MUST BE DIMENSIONED IN THE
C CALLING PROGRAM.
C
      COMPLEX A(M,M),B(N),X(N),TEMPI,QUOT,SUM
C
C LASSOL IS WRITTEN TO SOLVE SYSTEMS OF SIZE UP TO 10 BY 10.
C DIMENSION OF WORKING STORAGE AB MUST BE GREATER THAN OR
C EQUAL TO (N,N+1).  TO CHANGE THE SIZE OF PROBLEMS THAT CAN
C BE HANDLED, ONLY THE NEXT LINE NEED BE MODIFIED.
```

```
C
      COMPLEX AB(10,11)
      REAL*8 SUMR
      NP1=N+1
      NM1=N-1
C
C FORM THE N BY (N+1) MATRIX AB, THE FIRST N COLUMNS OF WHICH
C ARE A AND THE REMAINING COLUMN B.  CALCULATE SCALE FACTORS
C AND SCALE AB.
C
      DO 3 I=1,N
      ROWMAX=0.0
      DO 1 J=1,N
    1 ROWMAX=AMAX1(ROWMAX,CABS(A(I,J)))
      IF(ROWMAX.EQ.0.0)GO TO 14
      SCALE=1.0/ROWMAX
      DO 2 J=1,N
    2 AB(I,J)=A(I,J)*SCALE
    3 AB(I,NP1)=B(I)*SCALE
C
C BEGIN BASIC ELIMINATION LOOP.  ROWS OF AB ARE PHYSICALLY
C INTERCHANGED IN ORDER TO BRING ELEMENT OF LARGEST MAG-
C NITUDE INTO PIVOTAL POSITION.
C
      DO 9 K=1,NM1
      BIG=0.0
      DO 4 I=K,N
      TEMPB=CABS(AB(I,K))
      IF(BIG.GE.TEMPB)GO TO 4
      BIG=TEMPB
      IDXPIV=I
    4 CONTINUE
      IF(BIG.EQ.0.0)GO TO 14
      IF(IDXPIV.EQ.K)GO TO 6
C
C PIVOT IS IN ROW IDXPIV.  INTERCHANGE ROW IDXPIV WITH ROW K.
C
      DO 5 I=K,NP1
      TEMPI=AB(K,I)
      AB(K,I)=AB(IDXPIV,I)
    5 AB(IDXPIV,I)=TEMPI
    6 KP1=K+1
C
C ELIMINATE X(K) FROM EQUATIONS K+1,K+2,...,K+N.
C
      DO 8 I=KP1,N
      QUOT=AB(I,K)/AB(K,K)
      DO 7 J=KP1,NP1
    7 AB(I,J)=AB(I,J)-QUOT*AB(K,J)
```

```
 8   CONTINUE
 9   CONTINUE
C
C BEGIN CALCULATION OF SOLUTION X USING BACK SUBSTITUTION.
C
     IF(CABS(AB(N,N)).EQ.0.0) GO TO 14
     X(N)=AB(N,NP1)/AB(N,N)
     DO 11 IB=2,N
     I=NP1-IB
     IP1=I+1
     SUM=CMPLX(0.,0.)
     DO 10 J=IP1,N
10   SUM=SUM+AB(I,J)*X(J)
11   X(I)=(AB(I,NP1)-SUM)/AB(I,I)
     IFLAG=1
     RETURN
14   IFLAG=2
     RETURN
     END
```

## 2   PARTIAL FRACTION EXPANSION—THE KARNI METHOD[1,2]

```
c      subroutine pfexp (n,p,m,coef,c)


c      partial fraction expansion

c      This subroutine was designed to perform partial fraction
c      expansion of proper fractions.  It will accomodate
c      any number of poles and multiplicities and is bounded
c      only by the dimension statement.  The calling sequence
c      is:
c           call pfexp (n,p,m,coeff,c)
c      where
c           n = the number of distinct poles
c           p = a one dimensionsl array containing
c              the poles
c           m = a one dimensional array containing
c              the multiplicities
c              (note that there is a one to one
c              correspondence between the arrays
c              p and m.  Thus m(1) is the multiplicity
c              of the pole p(1).)
c         coef = a one dimensional array containing
c              the coeficients of the numerator
c              polynomial.  For n distinct poles
c              there must be n coefficiens.  They
```

```
c                 must be specified from order zero
c                 to order (n-1). For example, if the
c                 numerator for a polynomial of n = 5 is
c                    (s**4)+(2*(s**3))-6*s+12
c                 then coef(1)=12,coef(2)=-6,coef(3)=0,
c                 coef(4)=2,coef(5)=1.
c             c = a (6 x n) array used for output from
c                 the subroutine. The row designates the
c                 order of the constant and the column
c                 designates the particular pole. For
c                 example, if c(3,6)=4.33 and the sixth
c                 pole is -2, then the expansion includes
c                    4.33/((s+2)**3)

      dimension a(6,6),c(6,6),p(6),m(6),coef(6)
c     zero fill the work array
      do 10 i=1,6
      do 10 j=1,6
      c(i,j)=0.0
   10 a(i,j)=0.0
c     compute first order coefficients
      do 20 jj=1,n
      hold=1.0
      cnumer=0.0
      s=p(jj)
      do 15 i=1,n
      if(jj.ne.i)hold=hold*(s-p(i))
      if(s.ne.0.0)cnumer=cnumer+coef(i)*(s**(i-1))
   15 continue
      c(1,jj)=cnumer/hold
   20 continue
c     set order pointer: korder
      do 900 korder=2,6
c     set pole pointer: npole
      do 800 npole=1,n
c     if multiplicity of npole < korder try next pole
      if(m(npole).lt.korder) go to 800
c     if multiplicity of npole not < korder then
c        (1) shift npole column of the coeffcient array of
c            output values down one element
c        (2) calculate contribution of npole with all other terms
c            and add contributions to corresponding locations in
c            the coefficient array of output values
      do 50 iwork=1,5
      iw=7-iwork
      jw=iw-1
      a(iw,npole)=c(jw,npole)
   50 c(iw,npole)=c(jw,npole)
      a(1,npole)=0.0
      c(1,npole)=0.0
```

```
      do 700 jpole=1,n
      if (jpole.eq.npole)go to 700
      kwork=korder
      a(kwork,jpole)=0.0
200   continue
      if (c(kwork,jpole).ne.0.0)go to 300
      kwork=kwork-1
      if(kwork.eq.0) go to 700
      go to 200
300   khold=kwork
400   con=c(kwork,jpole)
500   a1=con/(p(npole)-p(jpole))
      b=con/(p(jpole)-p(npole))
      a(kwork,jpole)=a(kwork,jpole)+b
      if(kwork.eq.1)go to 600
      kwork=kwork-1
      con=a1
      go to 500
600   a(1,npole)=a(1,npole)+a1
      khold=khold-1
      if(khold.eq.0)go to 700
      kwork=khold
      go to 400
700   continue

c   replace coeffiecient array for output values with work array
      do 750 i=1,6
      do 750 j=1,n
      c(i,j)=a(i,j)
750   a(i,j)=0.0
800   continue
900   continue
c   print the coefficient array of output values in matrix form


      write(6,955)
955   format('0','coefficient array for partial fraction expansion')
      do 950 i=1,6
950   write(6,960)(c(i,j),j=1,n)
960   format(6f10.3)
      return
      end
```

# 3   STABILITY—ROUTH ARRAY

```
c   this program generates the routh array for a given polynomial.
c   the program asks for the order of the polynomial and then requires
c   that the coefficients be entered when prompted in descending order,
c   i.e. first enter the coefficient of the highest power, then the
c   coefficient of the next highest power, etc. until the coefficient of
```

```
c    the constant term is entered.  note:  a coefficient for each power
c    must be entered.

     program routharray

     integer order , horder,q,x,i,j,k,z,r,current,l,b,m,h
     real  routh(21,11),sum


     print*,'enter order'
     read(*,*)order
     data routh/231*0.0/
     horder=order+1
     l=nint(float(horder/2.0))

c    l is the length of rows 1 and 2 of the routh array.

     q=l-1

c    q is the length of the row to be generated.

     x=1
     i=1
     j=-1
     m=(-1)**order
     if (m.lt.0) then
        r=1
     else
        r=2
     endif

c    horder, as a row name, refers to the first row of the routh array.
c    order, as a row name, refers to the second row of the routh array.

c    read in the given polynomial.

     do 15 k=1,l
        b=2
        if ((r.eq.2).and.(k.eq.l)) b=1
        do 18 z=1,b
        print*,'enter coefficient'
        read(*,*)routh(horder,i)
        x=j*x
        horder=horder+x
18      continue
        i=i+1
15   continue
     horder=order+1
     print*,(routh(horder,j),j=1,q+1)
     print*
```

```
      sum=0
      do 20 j=1,l
        sum=sum+routh(order,j)
20    continue
      h=order

c     check to see if every other power of the given polynomial is missing.
c     if it is, replace the zero row with the derivative of the row above it.

      if(sum.eq.0.0) then
        do 22 j=1,l
          routh(order,j)=routh(horder,j)*h
          h=h-2
22      continue
      endif
      print*,(routh(order,j),j=1,q+1)
      if(routh(order,1).eq.0) routh(order,1)=0.000001
      current=horder-2

c     current, as a row name, refers to the row of the routh array currently
c     being generated.

c     generate the elements of the routh array.

      do 80 w=1,order-1
        do 30 j=1,q
          routh(current,j)=(routh(current+1,1)*routh(current+2,j+1) -
     1      routh(current+2,1)*routh(current+1,j+1))/routh(current+1,1)
30      continue
        sum=0
        do 35 j=1,q
          sum=sum+routh(current,j)
35      continue

c     check to see if an auxiliary polynomial is generated.

        if (sum.eq.0) then
          print*,'the degree of the auxiliary polynomial is ',current
        print*,'the coefficients in descending order with every other'
          print*,'power missing are '
          print*,(routh(current+1,j),j=1,l)
          go to 100
        elseif (routh(current,1).eq.0) then
          routh(current,1)=0.000001
        endif
        print*
        print*,(routh(current,j),j=1,q)
        current=current-1
        r=r+1
        if(mod(r,3).eq.0)then
          q=q-1
          r=r+1
        endif
```

```
80      continue
100  end
```

## 4A  CONVOLUTION WITH FUNCTIONS

```
c     this program performs the numerical convolution of two functions,
c     h and x, defined in the function statements at the end of the program.
c     the program assumes the functions to equal 0 for t<0, therefore the
c     convolution begins for t>0.  the user should define the desired functions
c     properly, using fortran 77 statements.

c     the program also requires that the user input two variables when
c     prompted:
c        1) the sampling interval, t; and
c        2) the number of intervals over which the convolution
c             should take place, k.

      program convolution

      integer l,k,n
      real    y,t,sum,h,x
      common t

      print*,'enter an integer value for k.'
      read*,k
      print*,'enter an value for t.'
      read*,t
      print*
      print*,' t= ',t
      print*

c     perform the convolution

      do 40 l = 1 , k
         sum = 0
         do 30 n = 1 , l-1
             sum=sum+2*h(n)*x(l-n)
30         continue
         y=t*(sum+h(0)*x(l) +h(l)*x(0))/2.0
         print*,'y( ',l,'t)=    ',y
40    continue
      end

      function h(arg)
      common t
      integer arg
         h=exp(-2*arg*t)
      return
      end
```

```
function x(arg)
common t
integer arg
   x=10
return
end
```

## 4B   CONVOLUTION WITH DATA POINTS

```
c   this program performs the numerical convolution of two functions,
c   x and h.  the program assumes the functions to equal 0 for
c   t<0, therefore the convolution begins for t>0.
c   the functions must be given at equally spaced intervals, where t is
c   the spacing.  these data must be stored in a file called convol.dat.
c   the format is one data entry per line.  if the convolution is to take
c   place over k intervals, x(0),x(1t),x(2t),...,x(kt),h(0),h(1t),h(2t),...
c   h(kt) must be given in that order in the file.

c   the program also requires that the user input 2 variables when
c   prompted:
c        1)  the sampling interval, t; and
c        2)  the number of intervals over which the convolution
c               should take place, k.


    program convolution

    integer l,k,n,j
    real    y,t,sum,h(0:20),x(0:20)

    print*,'enter an integer value for k.'
    read*,k
    print*,'enter an value for t.'
    read*,t
    print*
    print*,'  t= ',t
    print*

    open(unit=15,file='convol.dat',status='old')
    rewind(unit=15)


c   read in the data

    do 10 j=0,k
        read(15,*)x(j)
10  continue
    do 20 j=0,k
        read(15,*)h(j)
20  continue
```

```
c     perform the convolution

      do 40 l = 1 , k
         sum = 0
         do 30 n = 1 , l-1
            sum=sum+2*h(n)*x(l-n)
30       continue
         y=t*(sum+h(0)*x(l) +h(l)*x(0))/2.0
         print*,'y( ',l,'t)=   ',y
40    continue
      end
```

## REFERENCES

1. S. Karni, "Easy partial fraction expansion with multiple poles," *Proceedings of the IEEE*, vol. 57, no. 2, February 1969, pp. 231–232.
2. S. Karni and D. Etter, "An algebraic-recursive algorithm for partial fraction expansion with multiple poles," *IEEE Trans. on Education*, vol. 22, no. 1, February 1979, pp. 25–27.

# HINTS AND SOLUTIONS TO SELECTED PROBLEMS†

## CHAPTER 1

1-2. $i(t) = \dot{C}v + C\dot{v}$ $\qquad v(t) = \dfrac{1}{C(t)} \displaystyle\int_0^t i(x)\,dx + v(0_-)$

1-4. $v_3 = L_3 \dfrac{di_3}{dt} - M_{13} \dfrac{di_1}{dt} + M_{23} \dfrac{di_2}{dt}$ $\qquad$ etc.

1-9. Nonlinear! $\qquad y_1 + y_2 \neq m(x_1 + x_2) + b$

1-14. $-v_{in} - \dfrac{1}{R_2 C} \displaystyle\int_0^t v_0\,dx - \dfrac{R_1}{R_2} v_0 = 0$

## CHAPTER 2

2-1. $i(t) = 6u(t) - 12u(t - 2) + 6u(t - 4)$

2-6. $\phi = L$ (webers)

2-10. a) $|e^{j\beta t}| \equiv 1$ $\quad \therefore M = 1$ $\quad \therefore c = 0$ $\quad \therefore \sigma > 0$
$\qquad \therefore \mathscr{L}e^{j\beta t}u(t) = 1/(s - j\beta),$ $\quad \sigma > 0.$

2-11. Set first $st = x^2$ $\quad \therefore t = s^{-1}x^2,$ $\qquad dt = 2s^{-1}x\,dx$

2-12. No, because there is no $c$ such that
$$|e^{t^2}| < Me^{ct}$$

2-13. $\mathscr{L}\delta(t) = s \cdot \dfrac{1}{s} - u(0_-) = 1$

---

† Don't peek!

2-14. $\mathscr{L}\delta^{(n)} = s^n$

2-15. $i(t) = -\dfrac{11}{3} e^{-2/3t} u(t)$

2-18. KCL: $C\dfrac{d}{dt}(v_0 - 0) + \dfrac{v_{in} - 0}{R} = 0$

2-19. $\dfrac{dv_0}{dt} + \dfrac{1}{3RC} v_0 = \dfrac{1}{3RC} v_{in}$

2-20. $s^2 + 7s + 12 = 0 \qquad s_{1,2} = -3, -4$

2-24. $i(t) = \left[\displaystyle\int_0^t e^{-x}\, dx\right]_{\text{shifted by 2 sec.}} = u(t-2) - e^{-(t-2)} u(t-2)$

# CHAPTER 3

3-2. $n = 2$

3-3. $5s^3 - 200s^2 + 180s + 1200 = 0$

3-6. $I(s)$ is in [amp][sec] = [coulomb]
$V(s)$ is in [volt][sec] = [weber]

3-10. $\begin{bmatrix} \dfrac{1}{2s} + 1 & -1 \\ -1 & 1 + 3s \end{bmatrix} \begin{bmatrix} V_1 \\ V_2 \end{bmatrix} = \begin{bmatrix} \dfrac{3}{s+1} - \dfrac{1}{s+3} \\ \dfrac{1}{s+3} + \dfrac{2}{s} \end{bmatrix} + \begin{bmatrix} -\dfrac{1}{2s} \\ 9 \end{bmatrix}$

3-11. $\dfrac{V_0}{V_{in}} = \dfrac{-10(5s+1)}{s^2 + 0.3s}$

3-14. $\left. \begin{aligned} \dfrac{1}{r_0 + \dfrac{1}{C_0 s}} \cdot V_c + \dfrac{V_c}{r_2} + kI_{in} = 0 \\[2em] I_{out} = -\dfrac{1}{r_0 + \dfrac{1}{C_0 s}} V_c \end{aligned} \right\} \begin{aligned} &\therefore \text{ calculate} \\[1em] &\dfrac{I_0}{I_{in}} \text{ with } s = j\omega \end{aligned}$

3-16. 1) $z_1 = -1.64 \qquad z_2 = 0.82 + j1.165 \qquad z_3 = z_2^*$
2) $z_1 = 0.25 \qquad z_2 = z_3 = z_4 = \infty$
$\quad p_1 = -1 \qquad p_2 = -1.11 \qquad p_3 = -0.445 + j2.97 \qquad p_4 = p_3^*$
13) $p_1 = -1 + j3 \qquad p_2 = p_1^* \qquad p_3 = j10 \qquad p_4 = p_3^*$

3-17. b) $v(t) = (0.1e^{-t} - 0.167e^{-3t} + 0.067e^{-6t}) u(t)$

3-22. c) $-2.4$ \quad d) 1

3-24. c) 2 \quad d) 0.25

## CHAPTER 4

4-4.  $Z_R(j\omega) = R$      $Z_L(j\omega) = j\omega L$      $Z_c(j\omega) = \dfrac{1}{j\omega C}$

4-7.  a) Current in $R$ is $I_2 + I_1(1 - \alpha)$
$$z_{11} = R(1 - \alpha) = z_{21}\qquad z_{12} = z_{22} = R$$

4-8.  a) current in $R$ is $I_1(1 - \alpha)$

4-12.  $L_1 L_2 - M^2 > 0$ or else $\mathbf{Z}_{oc}$ is singular

4-13.

$\mathbf{Z}_{oc} = \begin{bmatrix} Z_A & Z_A \\ Z_A & Z_A \end{bmatrix}$ singular

4-15.  $\mathbf{Z}_{oc} = \begin{bmatrix} 0 & -R \\ R & 0 \end{bmatrix}$    $z_{12} \neq z_{21}$

4-16.  $Z = \dfrac{1}{a^2} Z'$

4-19.  $V_L = \dfrac{z_{21}}{z_{22} + 1} I_L$

4-23.  $z_{12} = z_{21}$     reciprocal

4-24.  $V_1 - V_2 = L_1 s I_1 \pm M s I_3$
$V_3 - V_4 = \pm M s I_1 + L_2 s I_3$
$\therefore I_1 = \cdots \qquad \therefore I_3 = \cdots$
and $I_2 = -I_1 \qquad I_4 = -I_3$

4-26.  $y_{21} = -\dfrac{Y_3}{Y_2}(Y_1 + Y_2)$    $V_x = 0$

## CHAPTER 5

5-4.  Need a NIC (negative impedance converter) for the two $Z_A$'s because
$$\mathbf{Z}_{oc} = \frac{1}{2}\begin{bmatrix} Z_B + Z_A & Z_B - Z_A \\ Z_B - Z_A & Z_B + Z_A \end{bmatrix}$$

5-9.  $q$ even    $\begin{bmatrix} A & B \\ C & D \end{bmatrix} = \begin{bmatrix} 1 & 0 \\ 0 & 1 \end{bmatrix}$

$q$ odd    $\begin{bmatrix} A & B \\ C & D \end{bmatrix} = \begin{bmatrix} 0 & R \\ \dfrac{1}{R} & 0 \end{bmatrix}$    $\therefore$ as a single gyrator

5-15. $p < n$     $I_{p+1} = I_{p+2} = \cdots I_n = 0$     $\therefore \mathbf{Y}_i = \begin{bmatrix} p \times p & \vdots & 0 \\ \hdashline 0 & \vdots & 0 \end{bmatrix} \}n-p$
$$\underbrace{\phantom{xxxxxxxxxx}}_{n-p}$$

## CHAPTER 6

6-1.  Count *total* incidence at all nodes $= S$

$$\therefore e = \tfrac{1}{2}S$$

$$e = \tfrac{1}{2}[5 + 6 + 4 + 6 + 5 + 6 + 6 + 6 + 6] = 25$$

6-2.  Let $n_p$ = number of nodes of degree $p$
$\therefore 1p_1 + 2p_2 + 3p_3 + \cdots = 2e$
$\therefore p_1 + p_3 + p_5 + \cdots = 2e - 2p_2 - 2p_3 - 4p_4 - \cdots = \text{even}$

6-9.  c) $Q_f$ is *guaranteed* to have rank $= n - 1$, resulting $(n - 1)$ equations independent and solvable.

6-15.  Check rank of student's **B**. It is $< 4$
OR: student's choice *forces* $i_d = -i_e$, $i_b = -i_g$
$\therefore$ not valid

6-17.  $\mathbf{B} = [\mathbf{B}_{11} \quad \mathbf{B}_{12}]$    $\mathbf{A} = [\mathbf{A}_{11} \quad \mathbf{A}_{12}]$
$\therefore \mathbf{AB}^T = 0 \rightarrow \mathbf{B}_{12}^T = -\mathbf{A}_{12}^{-1}\mathbf{A}_{11}\mathbf{B}_{11}^T$
$\mathbf{I}_e = \mathbf{B}^T\mathbf{I}_m$        $\therefore \mathbf{I}_{link} = \mathbf{B}_{11}^T\mathbf{I}_m$
            $\therefore \mathbf{I}_{br} = -\mathbf{A}_{12}^{-1}\mathbf{A}_{11}\mathbf{B}_{11}^T\mathbf{I}_m$

6-23.  $p = v_1i_1 + v_2i_2 \equiv 0$     $\therefore$ lossless (passive)

## CHAPTER 7

7-4.  $Z_{dp} = Z_1(1 - k)$

7-6.  $Z_{dp} = r_1 - \dfrac{k_1k_2}{G + g_2}$

7-7.  $Z_{dp} = 1$

7-9.  $Z_{dp} = z_{11} - \dfrac{z_{12}z_{21}}{z_{22} + Z_L}$

7-12.  $Z_{dp} = \dfrac{1 - r^2}{Z_0 + r} + r$

7-14.  b) excite with $I_1$    $\alpha = -\dfrac{h_f}{1 + h_oZ_L}$

7-15.  $3/(2s^2 + 5s + 4)$

7-16.  d) $(s + 1)$ is a common factor, and it cancels. You discover it early OR when you calculate the residue for the pole at $s = -1$ (residue $= 0$).

7-17.  $K = 0.3$

7-20.  a)

$$|H|$$

7-21.  b) $H(s) = \dfrac{\text{even polynomial}}{\text{odd polynomial}} \quad \text{or} \quad \dfrac{\text{odd polynomial}}{\text{even polynomial}}$

a) $R(\omega) = \dfrac{P_1 Q_1 + P_2 Q_2}{Q_1^2 + Q_2^2} = \tfrac{1}{2}[H(j\omega) + H(-j\omega)]$

$\phi(\omega) = \tan^{-1}\dfrac{P_2}{P_1} - \tan^{-1}\dfrac{Q_2}{Q_1}$

or

$\tan\phi(\omega) = \dfrac{P_2/P_1 - Q_2/Q_1}{1 + \dfrac{P_2 Q_2}{P_1 Q_1}} = \dfrac{P_2 Q_1 - P_1 Q_2}{P_1 Q_1 + P_2 Q_2}$

7-22.  b) no ch. eq., no ch. values

7-23.  b) $h(t) = \delta(t) \qquad i_{dp} = \delta(t) \qquad \therefore v_{dp} = \delta(t)$

7-27.  $f\left(\dfrac{d}{dt}, \displaystyle\int, \cdots\right) i_1 = v_1 + v_2$

If $f$ is a linear operator then

$f(\cdots)i_A = v_1 \leftarrow$ partial response to $v_1$ alone

$f(\cdots)i_B = v_2 \leftarrow$ partial response to $v_2$ alone

$\therefore f(\cdots)\underbrace{(i_A + i_B)}_{= i_1} = v_1 + v_2$

7-28.  b) $R(s) = \dfrac{1}{s} H(s) \qquad \therefore r(t) = Ke^{-t} + Bu(t) + C\sin(2t + D)$

c) $Ke^{-t}$ transient; all others steady-state.

d) $R(s) = \dfrac{A}{s^2 + 4} H(s)$

$\therefore r(t) = \text{as before} + \underbrace{At\cos(2t + \theta)}_{\substack{\text{resonance, natural}\\ \text{frequency at } s = \pm j2 \text{ excited}}}$

7-30.  $\omega_0 = 1 \qquad \sqrt{3}$

7-33.  $P(s) = \dfrac{1}{s} H(s) \qquad$ with zero-state

$\therefore H(s) = sP(s) - \cancel{p(0_-)} = 0$

$\therefore h(t) = \dfrac{d}{dt} p(t)$

## CHAPTER 8

8-3. $h(t) = \dfrac{d}{dt}(3e^{-2t}u(t)) = 3\delta(t) - 6e^{-2t}u(t)$

$r(t) = (90e^{-3t} - 60e^{-2t})u(t)$

8-5. $v(t) = u(t-1) - u(t-2)$
let $v_1(t) = u(t-1)$ and its response $r_1(t)$
∴ the response to $v_2(t) = u(t-2)$ will be $r_2 = r_1(t-1)$

Finally, $r(t) = r_1(t) - r_1(t-1) = r_1 - r_2$

8-6.
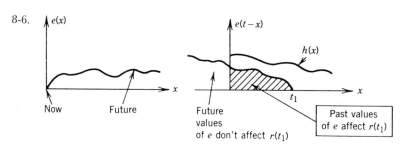

8-11. $r(t) = h * e = \delta * e = e$

8-14. See appendix E

8-16. b) $e^{-at}u(t) * e^{-at}u(t) = te^{-at}u(t)$

8-19. $Y(s) = G(s) + K(s)Y(s)$

∴ $Y(s) = \dfrac{G(s)}{1 - K(s)}$    $y(t) = \mathcal{L}^{-1}Y(s)$

## CHAPTER 9

9-1.

9-3.
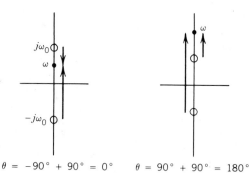

$\theta = -90° + 90° = 0°$     $\theta = 90° + 90° = 180°$
for $0 < \omega < \omega_0$     for $\omega > \omega_0$

9-4.

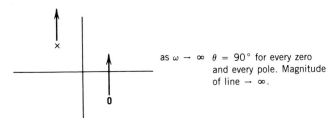

as $\omega \to \infty$  $\theta = 90°$ for every zero and every pole. Magnitude of line $\to \infty$.

9-6.

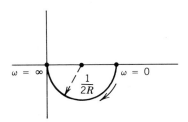

9-7.  $F(s) = \dfrac{A}{s - p_1} + \dfrac{B}{s - p_2} + \dfrac{C}{s - p_3}$

$A = F(s)(s - p_1)]_{s=p_1} = K\,\dfrac{(p_1 - z_1)(p_1 - z_2)}{(p_1 - p_2)(p_1 - p_3)}$

$\qquad = K\,\dfrac{\Pi \text{ all zero lines to } p_1}{\Pi \text{ all pole lines to } p_1}$

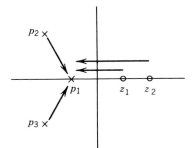

9-8.  c)

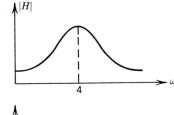

Sharper than (c)

Higher $Q$ factor

e)

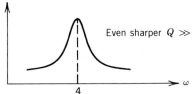

Even sharper $Q \gg$

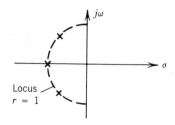

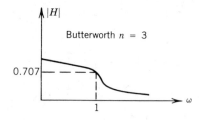

g)

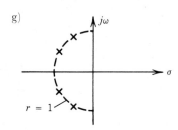

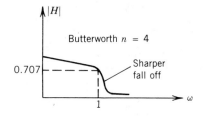

9-9. Review Bode corner frequencies

1) $(s + a) \to |\omega_0| = |a|$

2) $(s^2 + 2bs + b^2 + c^2) \to |\omega_0| = \sqrt{b^2 + c^2}$

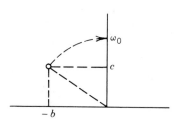

9-12. $|j\omega - a| = \sqrt{\omega^2 + a^2} = |j\omega + a|$

∴ Gain plot is same. But

$\tan^{-1} \dfrac{\omega}{-a}$ is in II quadrant, $90° < \theta < 180°$

$\tan^{-1} \dfrac{\omega}{a}$ is in I quadrant, $0° < \theta < 90°$

∴ different phase plots.

9-15. $H(s) = K \dfrac{s(s^2 + 2.5s + 50)}{(s + 5)(s + 200)}$  $\boxed{\phantom{x}}\!\!\downarrow\ \zeta = 0.05$

## CHAPTER 10

10-1. 5, 3, 4, 5, 3

10-2. a) $s^5 + a_1 s^4 + a_2 s^3 + a_3 s^2 + a_4 s + a_5 = 0$

10-3. $\dfrac{di_L}{dt} = \dfrac{R}{L} i_{in} - \dfrac{R}{L(1-\alpha)} i_L$    1$^{st}$ order, $\alpha \neq 1$.

10-5.  $\mathbf{i}_e = \begin{bmatrix} \mathbf{i}_{link} \\ \mathbf{i}_{branch} \end{bmatrix}$    $\mathbf{v}_e = \begin{bmatrix} \mathbf{v}_{link} \\ \mathbf{v}_{branch} \end{bmatrix}$

$\mathbf{Q}_f \mathbf{i}_e = [\mathbf{Q}_{f_{11}} \quad \mathbf{U}]\begin{bmatrix} \mathbf{i}_{link} \\ \mathbf{i}_{branch} \end{bmatrix} = 0$

$\therefore \mathbf{i}_{branch} = -\mathbf{Q}_{f_{11}}\mathbf{i}_{link} = \mathbf{B}_{f_{12}}^T \mathbf{i}_{link}$

$\mathbf{B}_f \mathbf{v}_e = [\mathbf{U} \quad \mathbf{B}_{f_{12}}]\begin{bmatrix} \mathbf{v}_{link} \\ \mathbf{v}_{branch} \end{bmatrix} = 0$

$\therefore \mathbf{v}_{link} = -\mathbf{B}_{f_{12}}\mathbf{v}_{branch}$

$\therefore \begin{bmatrix} \mathbf{i}_{branch} \\ \mathbf{v}_{link} \end{bmatrix} = \begin{bmatrix} 0 & \mathbf{B}_{f_{12}}^T \\ -\mathbf{B}_{f_{12}} & 0 \end{bmatrix}\begin{bmatrix} \mathbf{v}_{branch} \\ \mathbf{i}_{link} \end{bmatrix}$

$= \mathbf{x}$ includes these variables as state variables

Then use $\mathbf{i}_c = \mathbf{C}\dot{\mathbf{v}}_c$
$\mathbf{v}_L = \mathbf{L}\dot{\mathbf{i}}_L$
where $\mathbf{v}_c, \mathbf{i}_L \, \varepsilon \, \mathbf{x}$

10-7.

$i = \dfrac{v_{in} - v_c}{R}$

But $i = i_c = C\dfrac{dv_c}{dt}$

$\therefore C\dfrac{dv_c}{dt} = \dfrac{v_{in} - v_c}{R} = $ etc.

10-13.  $\begin{bmatrix} v_1 \\ v_2 \end{bmatrix} = \begin{bmatrix} r_{11} & r_{12} \\ r_{21} & r_{22} \end{bmatrix}\begin{bmatrix} i_1 \\ i_2 \end{bmatrix}$    $\mathbf{x} = \begin{bmatrix} v_2 \\ i_1 \end{bmatrix}$

$L\dfrac{di_1}{dt} = v_1$    $C\dfrac{dv_2}{dt} = -i_2$

$\therefore Li_1' = r_{11}i_1 + r_{12}(-Cv_2')$
$Cv_2' = C[r_{21}i_1' + r_{22}(-Cv_2')]$
Solve these two eqns. *algebraically* to get

$\left.\begin{array}{l} \dfrac{di_1}{dt} = \cdots \\[2em] \dfrac{dv_2}{dt} = \cdots \end{array}\right\}$ state equations

## CHAPTER 11

11-2. Decoupled equation for $v_c$!

$$\frac{dv_c}{dt} = -2v_c \qquad \therefore v_c(t) = v_c(0_-)e^{-2t}$$

11-5. $e^{\mathbf{A}} = \mathbf{U} + \mathbf{A} + \dfrac{\mathbf{A}^2}{2!} + \cdots$

$e^{\mathbf{B}} = \mathbf{U} + \mathbf{B} + \dfrac{\mathbf{B}^2}{2!} + \cdots$

Multiply: $e^{\mathbf{A}} \cdot e^{\mathbf{B}} = \mathbf{U} + \mathbf{A} + \mathbf{B} + \dfrac{\mathbf{A}^2}{2!} + \mathbf{AB} + \dfrac{\mathbf{B}^2}{2!} + \cdots$

But $e^{(\mathbf{A}+\mathbf{B})} = \mathbf{U} + (\mathbf{A} + \mathbf{B}) + \dfrac{(\mathbf{A} + \mathbf{B})^2}{2!} + \cdots$

$$= \mathbf{U} + \mathbf{A} + \mathbf{B} + \frac{\mathbf{A}^2}{2!} + \frac{\mathbf{AB}}{2!} + \frac{\mathbf{BA}}{2!} + \frac{\mathbf{B}^2}{2!} + \cdots$$

Compare:  $\therefore$ If $\mathbf{AB} = \mathbf{BA}$ both expansions are equal.

11-6. Use series for $e^{\mathbf{A}t} = \mathbf{U} + \mathbf{A}t + \dfrac{\mathbf{A}^2 t^2}{2!} + \cdots$

a) let $t = 0$

b) let $t = t_2 - t_0$, then $t = t_2 - t_1, t = t_1 - t_0$.

11-7. a) $\lambda_1 = -1 \qquad \lambda_2 = -2 \qquad \lambda_3 = -3$

b) $\lambda_1 = 1 \qquad \lambda_2 = \lambda_3 = 2$

11-9. c) $\boldsymbol{\Phi}(s) = \begin{bmatrix} s+1 & 2 \\ -2 & s+5 \end{bmatrix}^{-1} = \dfrac{1}{s^2 + 6s + 9} \begin{bmatrix} s+5 & 2 \\ -2 & s+1 \end{bmatrix}$

$$= \frac{\mathbf{K}_1}{s+3} + \frac{\mathbf{K}_{1,2}}{(s+3)^2}$$

Since $\sum \mathbf{K}_k = \mathbf{U} \qquad \therefore \mathbf{K}_1 = \mathbf{U} = \begin{bmatrix} 1 & 0 \\ 0 & 1 \end{bmatrix}$.

To find $\mathbf{K}_{1,2}$:

$$\mathbf{K}_{1,2} = \left\{ \begin{bmatrix} \dfrac{s+5}{(s+3)^2} & \dfrac{2}{(s+3)^2} \\ \dfrac{-2}{(s+3)^2} & \dfrac{s+1}{(s+3)^2} \end{bmatrix} \cdot (s+3)^2 \right\}_{s=-3} = \begin{bmatrix} 2 & 2 \\ -2 & -2 \end{bmatrix}$$

Check:

$$\frac{1}{s+3}\mathbf{U} + \frac{1}{(s+3)^2}\begin{bmatrix} 2 & 2 \\ -2 & -2 \end{bmatrix} = \boldsymbol{\Phi}(s)$$

$$\therefore \phi(t) = \begin{bmatrix} 1 & 0 \\ 0 & 1 \end{bmatrix} e^{-3t} + \begin{bmatrix} 2 & 2 \\ -2 & -2 \end{bmatrix} te^{-3t}$$

11-19.  a) $\dot{x} = -x + e$     $\mathbf{A} = -1$     $\mathbf{B} = 1$

   $r = x$     $\mathbf{C} = 1$     $\mathbf{D} = 0$

$$H(s) = \frac{1/s}{1 - \left(-\dfrac{1}{s}\right)} = \frac{1}{s+1}$$

11-24.  $(s^n + \alpha_{n-1}s^{n-1} + \cdots + \alpha_1 s + \alpha_0)R = (\beta_n s^n + \beta_{n-1}s^{n-1} + \cdots + \beta_1 s + \beta_0)E$

i.e.,
$$\begin{aligned}\beta_n &= a_n & \alpha_0 &= b_n \\ \beta_{n-1} &= a_{n-1} & \alpha_1 &= b_{n-1} \\ &\;\;\vdots & &\;\;\vdots \\ \beta_0 &= a_0 & \alpha_{n-1} &= b_1\end{aligned}$$

Advantage: diagram drawn *directly* from differential equation, without need of solving unknown amplifier gains.

11-27.  a) $\mathbf{x} = \mathbf{M}\hat{\mathbf{x}}$     substitute into $\dot{\mathbf{x}} = \mathbf{Ax} + \mathbf{Be}$
   $r = \mathbf{Cx} + \mathbf{De}$

$\therefore \mathbf{M}\dot{\hat{\mathbf{x}}} = \mathbf{AM}\hat{\mathbf{x}} + \mathbf{Be}$

$\therefore r = \mathbf{CM}\hat{\mathbf{x}} + \mathbf{De}$

$\therefore \dot{\hat{\mathbf{x}}} = \mathbf{M}^{-1}\mathbf{AM}\hat{\mathbf{x}} + \mathbf{M}^{-1}\mathbf{Be}$

$r = \mathbf{CM}\hat{\mathbf{x}} + \mathbf{De}$

$\therefore \hat{\mathbf{A}} = \mathbf{M}^{-1}\mathbf{AM}$     $\hat{\mathbf{C}} = \mathbf{CM}$

$\hat{\mathbf{B}} = \mathbf{M}^{-1}\mathbf{B}$     $\hat{\mathbf{D}} = \mathbf{D}$

   b) If $\hat{\mathbf{A}}$ is diagonal then

$\lambda_1 = a_{11}$     $\lambda_2 = a_{22}$     $\cdots$

# CHAPTER 12

12-2.  $R(s) = \dfrac{s}{s^2 + \omega_0^2} \dfrac{K\omega_0}{s^2 + \omega_0^2} = \dfrac{As}{(s^2 + \omega_0^2)^2}$

$\therefore r(t) = A_1 t \cos(\omega_0 t + \theta_1)$ unbounded!

12-3.  e)
$$\begin{array}{c|cccc} 4 & 1 & 1 & 2 \\ 3 & 1 & 2 \\ 2 & -1 & 2 & & \text{2 roots in rhp} \\ 1 & 4 \\ 0 & 2 \end{array}$$

   g) stable for $a > 0$

12-4.

henries
farads

12-5.

| 5 | 6 | $K$ | 2 | $\rightarrow \therefore K > 0$ |
|---|---|---|---|---|
| 4 | 2 | 3 | 10 | |
| 3 | $K - 9$ | $-56$ | | $\rightarrow \therefore K - 9 > 0$ |
| 2 | $3K + 85$ | $10K - 90$ | | $\rightarrow \therefore 3K + 85 > 0$ |
| 1 | $-56(3K + 85) - 10(K - 9)^2$ | | | $\therefore K > 9$ will satisfy these |
| 0 | $10(K - 9)$ | | | |

$$10K^2 - 12K + 5570 < 0$$
$$\therefore 10[(K - 0.6)^2 + 556.64] < 0 \therefore \text{Impossible!}$$

$\therefore$ No $K$ will do.

12-6.

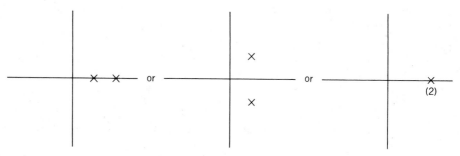

12-8.  b) third row starts with $-14$. Unstable

12-9.  $K > 2$

12-14.  a) Not d.p. (zero in rhp)

b) Not any $H(s)$ (imaginary coefficient)

c) Not any $H(s)$ (multiple pole on $j\omega$ axis)

d) O.K. both

g) O.K. both

## CHAPTER 13

13-1.  $\sin^2 t = \left(\dfrac{e^{jt} - e^{-jt}}{2j}\right)^2 = \dfrac{1}{2} - \dfrac{1}{2}\dfrac{e^{j2t} + e^{-j2t}}{2} = \dfrac{1}{2} - \dfrac{1}{2}\cos 2t$

Finite Fourier series $\longleftarrow$
$$T = \pi$$

13-4.  $f(t) = 5 + 4.502 \cos \dfrac{\pi}{4}t + 3.183 \cos \dfrac{\pi}{2}t + 1.501 \cos \dfrac{3\pi}{4}t + \cdots$

13-5.  $f(t) = \dfrac{A}{\pi} + \dfrac{A}{2}\sin \omega_0 t - \dfrac{2A}{\pi}\left(\dfrac{\cos 2\omega_0 t}{(1)(3)} + \dfrac{\cos 4\omega_0 t}{(3)(5)} + \cdots\right)$

13-6.  $a_n = -\dfrac{4A}{(4n^2 - 1)\pi}$    $b_n = 0$

13-11.

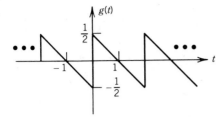

$\omega_0 = 2\pi$

$g(t)$ is odd $\qquad \therefore a_k = 0$

$$\therefore b_k = \frac{1}{k\pi}$$

$$\therefore f(t) = g(t) + \frac{1}{2} = \frac{1}{2} + \sum_1^\infty \frac{1}{k\pi} \sin k2\pi t$$

13-12. $\quad [f(t)]^2 = \dfrac{a_0}{2} f(t) + \sum_1^\infty a_k f(t) \cos k\omega_0 t + \sum_1^\infty b_k f(t) \sin k\omega_0 t$

$$\therefore \int_{-T/2}^{T/2} [f(t)]^2 \, dt = \frac{a_0}{2} \cdot \frac{a_0}{2} T + \frac{T}{2} \left( \sum a_k^2 \right) + \frac{T}{2} \left( \sum b_k^2 \right)$$

$$\therefore \frac{1}{T} \int_{-T/2}^{T/2} [f(t)]^2 \, dt = \frac{a_0^2}{4} + \frac{1}{2} \sum_1^\infty (a_k^2 + b_k^2)$$

If $i(t) = f(t)$ then

$$\frac{1}{T} \int_{-T/2}^{T/2} [i(t)]^2 \, R \, dt = \text{average power dissipated in 1 } \Omega$$

13-14. $\quad -\dfrac{2}{\pi} \cos 2\pi t + \dfrac{6}{\pi} \sin 2\pi t = \dfrac{\sqrt{40}}{\pi} \cos (2\pi t - 108.4°)$ etc.

13-18. $\quad P_{\text{avg}} = \dfrac{1}{T} \int^T \left\{ \sum_{-\infty}^\infty c_{v_k} e^{jk\omega_0 t} \right\} i(t) \, dt$

$$= \sum_{-\infty}^\infty c_{v_k} \left\{ \frac{1}{T} \int^T i(t) e^{jk\omega_0 t} \, dt \right\}$$

$$= \sum_{-\infty}^\infty c_{v_k} \left\{ \frac{1}{T} \int^T i(t) e^{-j(-k)\omega_0 t} \, dt \right\}$$

$$= \sum_{-\infty}^\infty c_{v_k} c_{i(-k)} = \sum_{-\infty}^\infty c_{v_k} c_{i_k}^*$$

13-21. $\quad c_k = \dfrac{Aa}{T} \left| \dfrac{\sin \dfrac{k\omega_0 a}{2}}{\dfrac{k\omega_0 a}{2}} \right|^2$

13-22.   $f(t - t_0) = \sum\limits_{-\infty}^{\infty} c_k e^{jk\omega_0(t - t_0)} = \sum\limits_{-\infty}^{\infty} \underbrace{c_k e^{-jk\omega_0 t_0}}_{\hat{c}_k \text{ for the}} \cdot e^{jk\omega_0 t}$

$\therefore |\hat{c}_k| = |c_k|$        shifted time function

# CHAPTER 14

14-1.   $|F(\omega)|$ same. Only $\phi(\omega)$ changes linearly.

14-2.   $F(\omega) = \displaystyle\int_{-\infty}^{0} e^{at} e^{-j\omega t}\, dt + \int_{0}^{\infty} e^{-at} e^{-j\omega t}\, dt = \cdots$

Area $= \pi$ as $a \to 0$    $f(t) = 1$
$$F(\omega) = \pi\delta(\omega)$$
$$\therefore \mathscr{F}1 = \pi\delta(\omega)$$

14-3.   $f(t) = \dfrac{1}{2\pi} \displaystyle\int_{-\infty}^{\infty} 2\pi\delta(\omega - \omega_0) e^{j\omega t}\, d\omega = e^{j\omega_0 t}$

14-4.   $\cos\omega_0 t = \frac{1}{2}(e^{j\omega_0 t} + e^{-j\omega_0 t})$ etc.

14-5.   $F(\omega - \omega_0) = \mathscr{F}f(t)e^{j\omega_0 t}$

14-6.   $\dfrac{dg(t)}{dt} = f(t) - \dfrac{1}{2}\delta(t - 2)$     $\therefore G(\omega) = \dfrac{1}{j\omega} F(\omega) - \cdots$ etc.

14-11.   Output $G(\omega) = H(\omega)F(\omega)$
$$= F(\omega)e^{-j\omega t_0}$$
$$\therefore g(t) = f(t - t_0) \text{ ideal delay}$$

14-12.   $e^{jk\omega_0 t} \to Ke^{jk\omega_0(t - t_0)}$    $k\omega_0 < \omega_c$
$\qquad\qquad \to 0$    $k\omega_0 > \omega_c$

$\therefore g(t) = K \displaystyle\sum_{k=-N}^{+N} c_k e^{jk\omega_0(t - t_0)}$

where $N$ is the largest harmonic contained in the passband (bandwidth)

$$N\omega_0 < \omega_c < (N + 1)\omega_0$$

$\therefore g(t) = Kf_N(t - t_0)$, a *partial* sum of the Fourier series of $f(t)$, delayed by $t_0$ and multiplied by $K$.

14-13.   $H(\omega) = [K - Kp_{\omega_c}(\omega)]e^{-j\omega t_0}$
where $p_{\omega_c}$ is shown $\to$

$\therefore h(t) = \mathscr{F}^{-1}H(\omega) = K\delta(t - t_0) - K\dfrac{\sin\omega_c(t - t_0)}{\pi(t - t_0)}$

14-17.   $F(\omega) = \dfrac{100}{10 + j\omega}$      $|F(\omega)|^2 = \dfrac{10^4}{100 + \omega^2}$

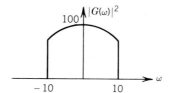

Total energy at input $= \dfrac{1}{2\pi} \displaystyle\int_{-\infty}^{\infty} \dfrac{10^4}{100 + \omega^2}\, d\omega = 500$ joules

Total energy at output $= \dfrac{1}{2\pi} \displaystyle\int_{-10}^{10} \dfrac{10^4}{100 + 10^2}\, d\omega = 250$ joules

$\therefore$ 50% energy output available

14-19.   b)  $F_L(s) = \dfrac{1}{\pi} R(\omega) * \dfrac{\sigma}{\sigma^2 + \omega^2}$

here $R(\omega) = \dfrac{\pi}{2} [\delta(\omega - \beta) + \delta(\omega + \beta)]$

$\therefore F_L(s) = \dfrac{2s}{\pi} \displaystyle\int_{0}^{\infty} \dfrac{\pi\delta(\lambda - \beta)}{2(s^2 + \lambda^2)}\, d\lambda = \dfrac{s}{s^2 + \beta^2}$

# CHAPTER 15

15-1.   $-\dfrac{d}{ds}(s^2 Y - s) + sY - 1 + \dfrac{d}{ds}[sY - 1] + 2Y = 0$

$\therefore \dfrac{dY}{Y} + \left(\dfrac{3}{s} - \dfrac{2}{s - 1}\right) ds = 0$

$\therefore \dfrac{Y s^3}{(s - 1)^2} = K_2$

$\therefore Y = K_2 \left(\dfrac{1}{s} - \dfrac{2}{s^2} + \dfrac{1}{s^3}\right)$

$\therefore y(t) = K_2 \left(1 - 2t + \dfrac{t^2}{2}\right)$

$y(0) = 1$      $\therefore K_2 = 1$

15-4.   $s^2 F - \dfrac{d^2 F}{ds^2} + kF = 0$

$\therefore$ same differential equation for $f(t)$ as for $F(s)$!

$\therefore \mathscr{L}f(t) = f(s)$ is its own transform.

15-6.   $r = \dfrac{1}{t^2 + 1}\left(\dfrac{t^3}{3} + t - 2\right)$

15-9.   $M_1(t; p)\{M_2(t; p)\} \neq M_2(t; p)\{M_1(t; p)\}$
But $M_1(p)\{M_2(p)\} = M_2(p)\{M_1(p)\}$

15-13.   $\mathbf{A}(t_1)\mathbf{A}(t_2) = \mathbf{A}f(t_1)\mathbf{A}f(t_2) = \mathbf{A}^2 f(t_1) f(t_2)$
$\mathbf{A}(t_2)\mathbf{A}(t_1) = \mathbf{A}f(t_2)\mathbf{A}f(t_1) = \mathbf{A}^2 f(t_2) f(t_1)$
But $f(t_1) f(t_2) = f(t_2) f(t_1)$ since they are scalars
Therefore O.K. and

$$\phi(t, t_0) = \exp \int_{t_0}^{t} \mathbf{A}f(\tau)\, d\tau = \exp\left[\mathbf{A} \int_{t_0}^{t} f(\tau)\, d\tau\right]$$

$$\begin{bmatrix} -2t & -t \\ 2t & -5t \end{bmatrix} = \begin{bmatrix} -2 & -1 \\ 2 & -5 \end{bmatrix} t = \mathbf{A}f(t)$$

$$\therefore \int_{t_0}^{t} f(\tau)\, d\tau = \int_{t_0}^{t} \tau\, d\tau = \frac{t^2 - t_0^2}{2}$$

$$\therefore \phi(t, t_0) = e^{\mathbf{A}} \cdot e^{\frac{t^2 - t_0^2}{2}}$$

where $e^{\mathbf{A}}$ was found in Chapter 11

# INDEX

**463**